THE RACE TO THE FUTURE

Also by Kassia St Clair:

The Secret Lives of Colour
The Golden Thread

THE RACE TO THE FUTURE

The Adventure that Accelerated
the Twentieth Century

KASSIA ST CLAIR

JOHN MURRAY

First published in Great Britain in 2023 by John Murray (Publishers)

2

Map and illustrations drawn by Nathan Burton.

A CIP catalogue record for this title is available from the British Library

Hardback ISBN 978 1 529 38605 9
Trade Paperback ISBN 978 1 529 38606 6
ebook ISBN 978 1 529 38608 0

Typeset in Plantin by Hewer Text UK Ltd, Edinburgh

Printed and bound in Great Britain by Clays Ltd, Elcograf S.p.A.

John Murray policy is to use papers that are natural, renewable and recyclable products and made from wood grown in sustainable forests. The logging and manufacturing processes are expected to conform to the environmental regulations of the country of origin.

Carmelite House
50 Victoria Embankment
London EC4Y 0DZ

www.johnmurraypress.co.uk

John Murray Press, part of Hodder & Stoughton Limited
An Hachette UK company

To O & P.

CONTENTS

CONTENTS

CARS AND COMPETITORS

CONTAL MOTOTRI
(6 hp)

Auguste Pons .... Driver

Octave Foucault .... Navigator

DE DION-BOUTON
(10 hp)

Georges Cormier .... Driver and
correspondent (*L'Auto*)

Jean Bizac .... Mechanic

DE DION-BOUTON
(10 hp)

Victor Collignon .... Driver

Edgardo Longoni★ .... Journalist (*Il Secolo*)

SPYKER
(15 hp)

Charles Godard .... Driver

Jean du Taillis★ .... Journalist (*Le Matin*)

Bruno Stephan★ .... Mechanic

ITALA
(35/45 hp)†

Prince Scipione
Borghese .... Co-driver

Ettore Guizzardi .... Co-driver/mechanic

Luigi Barzini .... Journalist (*Corriere della
Sera* and *Daily Telegraph*)

★ *This table represents the pairings when the drivers set off from Peking. During the journey, Jean du Taillis would replace Edgardo Longoni in the De Dion-Bouton. Bruno Stephan, a Spyker mechanic, would join Godard at Omsk.*

† *The prince's model was technically a 35/45hp, but contemporaries referred to it throughout as 40hp, so I have done likewise to avoid confusion.*

PEKING TO PARIS

ROUTE MAP

11 June–10 August

1907

(This map runs as the race did: east to west.)

KANSK

IRKUTSK

KRASNO

NIZHNEUDINSK

CHEREMKHOVO

LAKE BAIKAL

ULAN-UDE

4

TANKHOY

URGA

GOBI DESERT

3

2

BON VOYAGE

1

KALGAN

PEKING

CHINESE EMPIRE

❶
CONTAL MOTOTRI
Technically disqualified only a few miles north of Peking but takes train over mountains and continues.

❷
CONTAL MOTOTRI
Stranded in Gobi between camp and Pong-Hong. Out of race.

❸
SPYKER
Stranded in Gobi 150 miles from Ude.

R U S S I A N
E M P I R E

SIBERIA

MSK
KOLYVAN
OMSK
❺
TYUMEN
YEKATERINBURG
U R A L
PERM
MELET'
❻
PETROPAVL
CHELYABINSK
M O U N T A I N S
YELABUGA
KAZAN

(Continued overleaf)

0 km 200 km 400 km 600 km 800 km

❹
ITALA
Falls through bridge just before Tankhoy.

❺
ITALA
Takes the route recommended by the Russian organising committee.

❻
DE DION–BOUTONS
Take the official southerly, steeper but more direct route over Urals.

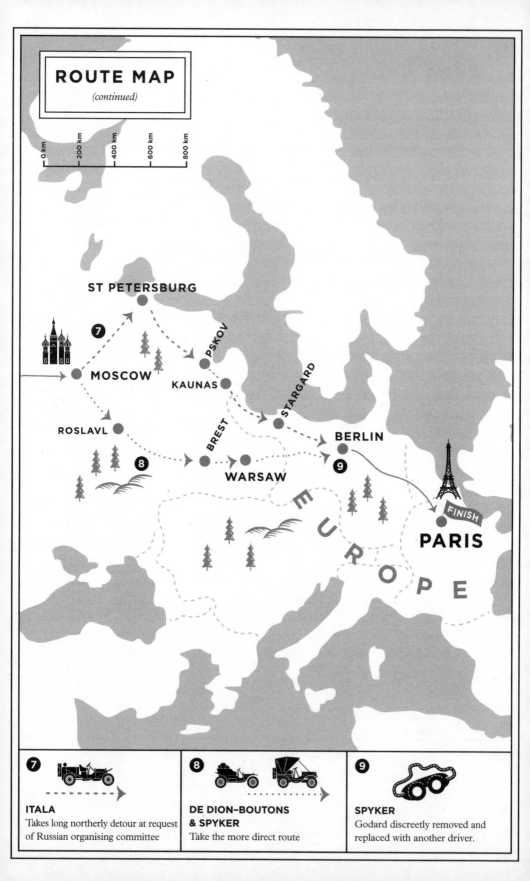

We invent the future out of fragments from the past.

Erwin Panofsky, 1892–1968

In the late nineteenth century, French illustrator Albert Robida imagined what a visit to the opera might look like in the year 2000.

PREFACE

He would be a rash man who would deny that within a dozen years or so, the entire balance of the world's population may not be shifted by means of the all-conquering motor.

Comyns Beaumont, 1907[1]

In January 1907 *Le Matin*, a Parisian newspaper, threw down a gauntlet to the world. What they proposed was an automobile race across two continents, a grand experiment that would push the boundaries of a fledgling technology. The trailblazing motorists would set off from Peking and drive westwards to Paris, passing through five capital cities and covering some 8,000 miles. What would unfold over the following months was both a rip-roaring adventure and a parable about the birth of a new world: ours.

While such a journey might be daunting even today, in 1907 the idea of driving so far and through such remote country was considered laughable. But, just as *Le Matin* had intended, it drew the eyes of the world. A week after the race was proposed, the *Sydney Morning Herald* called it 'such a bold and gigantic scheme that it almost takes one's breath away' and 'sufficient to frighten the pluckiest and most enterprising of promoters'. For the *Evening Bulletin* of Honolulu, it was 'the most stupendous automobile race ever undertaken'. According to one estimate, between 500,000 and 600,000 people thronged the rain-soaked streets of Paris to see the winner enter the city. Participants – several of whom nearly died in the attempt – were compared to Heroic Age explorers like Richard Francis Burton, Roald Amundsen and Ernest Shackleton.

The winner was 'a pioneer of civilisation whose name will be quoted by future generations'.[2]

The Peking–Paris was a decisive moment. It put the utility and practical significance of the automobile beyond doubt. Afterwards, even the most hardened sceptics could not ignore its potential as a rival to the railway or its applications in both civilian and military life. If the race was not quite the reason the automobile triumphed, it certainly smoothed its path. In practical terms, driving across two continents was the sternest test the technology had faced. Although there had been races and rallies before, they had been far shorter. One contemporary called it 'the greatest event in the automotive world, not just for the past year, but even for the entire existence of self-moving carriages'. The endeavour was adjudged to have 'worldwide economic and social significance'. The *Sunday Times* dared prophesy after the completion of the Peking–Paris that 'the automobile may be looked upon as one of the greatest mediums of civilisation in the future, as it will penetrate into the furthest corners of the earth long before such places are reached by regular trade routes.'[3]

The race also had potent symbolic resonance as a harbinger of a new era of globalisation. The distance between East and West had never before seemed so small and traversable. 'The Peking to Paris motor race may have something to do with the poetry of motion,' argued a writer in the *South China Morning Post*, 'but it helps to destroy the poetry of distance ... Locomotion and communication are making the world smaller.' Portions of the Silk Roads, which had begun to wither with the completion of the Trans-Siberian Railway, could suddenly be reimagined (as they are being once again today, thanks to China's Belt and Road Initiative). Automobiles could be agents for progress, civilisation and peace. One journalist opined that similar rallies – 'universal attempts to extend the civilisation of the world' – would be their own form of diplomacy. '[W]ars are likely to disappear since the common interests will become so strong in the future that they will easily absorb the separate interests of the different nations.'[4]

The Peking–Paris encapsulates the West's ideas about herself and her role in the world a few short years before these would be overturned by the First World War and the breakdown of empires. It is emblematic of the struggles of the Old World to retain technological and industrial supremacy. And it offers a window into Russia and China at pivotal moments in their history, moments that seem increasingly relevant to their own self-image and foreign policy today.

It is difficult to comprehend the scale of the challenge the Peking–Paris competitors faced. For a start, automobiles were in their infancy. Karl Benz's Motorwagen – usually celebrated as the very first automobile – was patented in 1886. It was a spindly contraption consisting of an engine, a seat and three wheels. It fielded around 0.6 hp* – like an underpowered hotel hairdryer on wheels – and managed a top speed of just under ten miles per hour. By 1907, the technology had progressed, but automobiles were still rare and considered unreliable. An article in *The Economist* castigated those who had 'rushed with such luckless enthusiasm to invest in motor-bus companies', arguing instead that 'the horse is coming triumphantly through the ordeal'.[5]

Although there had been races and rallies before *Le Matin* proposed the Peking–Paris, they had been far shorter and notoriously dangerous – usually involving fatalities. Automobiles were temperamental, prone to frequent breakdowns and punctures. (Most owners considered it essential to employ a chauffeur who could both drive and maintain the machine in working order.) Motoring manuals contained long lists of the tools, lubricants, spare tyres and parts that it was advisable to carry, even on short journeys.[†] They were also so expensive to buy, run and repair that they were out of reach of all but wealthy elites or manufacturers.

---

* Horsepower (hp) is a unit used to measure power, specifically of engines. The base calculation uses the power of a strong dray horse turning a twelve-foot mill wheel, so one horsepower is equivalent to 33,000 foot-pounds of work per minute. It was invented by James Watt, a Scottish engineer, in the late eighteenth century. The power of the vehicles that partook in the Peking–Paris ranged from 35/45 hp to just 6 hp. By way of comparison, today's Toyota Corolla makes do with 120 hp, the latest Ford F series starts at 290 hp and a Vespa scooter fields around 11 hp.

† This was a particular worry for the Peking–Paris racers, who would be physically unable to take the necessities for so long a journey with them and, for the first 5,000 miles or so, would be unable to procure them.

The year of the race, 1907, was one of the last when the automobile would be considered an experimental novelty rather than a practical form of transport. Radical sporting events like the Peking–Paris, which emphasised the exploratory and long-distance capabilities of the vehicle, rather than the more usual, humdrum ones, would also help nudge carmakers to go all in on combustion engines. It was also the final year when Europe – and France in particular – could claim to be the centre of the automobile industry. While their own manufacturers had been rapidly catching up, many Americans were still importing European cars. By October 1908, however, the very first Model T was completed at Henry Ford's plant in Detroit; already the automobile was settling into a new spiritual home.

What made the Peking–Paris so compelling was that the motor car was only half the challenge. Even the most bullish enthusiasts were intimidated by the route. Throughout western Europe, roads were generally poor outside major cities; and within them, many were cobbled or covered with tram rails, which hardly made for an enjoyable ride. Elsewhere, roads often existed in name only. From Peking to Moscow, the initial four-fifths of the route, the motorists would drive on tracks formed by the passage of donkeys, horses, carts and people on foot. These would be ankle-deep in dust or knee-deep in mud, depending on the weather, and littered with trees, roots, boulders and other debris. Where there were major obstacles – for example, in the mountainous passes north of Peking – vehicles would need to be lifted bodily over obstructions. Where even this proved impossible, drivers would be forced to make their own road, using pickaxes, axes and shovels.

Most of the land through which the motorists would drive was remote and sparsely populated: accidents or miscalculations about equipment, food or fuel supplies could prove fatal. Participants would have to scale mountain ranges and cross a desert. The vehicles in the race were phaeton-style – the most common at the time – which meant that apart from flimsy canvas canopies (soon discarded to keep the weight down), they would be open to the elements. Since there were no petrol stations, fuel would need to be sent out in advance and left in strategic locations. For a portion of the route, this meant transporting fuel by camel. If motorists failed to locate a cache, or if one was pilfered before they reached it, they would be stranded.

Despite the doom-mongering, participating in so historic an event proved a strong lure: the starting line would feature a compelling and varied cast of characters. Among their number were an Italian prince and his chauffeur, a former professional French racing driver and a conman, as well as journalists whose job was to write and transmit regular reports on their progress.

The Peking–Paris was a news age phenomenon, garnering headlines from Melbourne to New York, Manchester to Shanghai. Newspapers revelled in the glamour and drama of the race, spinning it as a triumph of modernity over antiquation, of speed over sloth, of Western civilisation over Eastern barbarism. There was also a sense that it was a triumph of journalism. Proposed by a canny French newspaper, the story held the news-reading world captive for months, elevating the newspapers and journalists involved as surely as it did the automobile itself. And it was the telegraph that would prove crucial to the whole endeavour. Its physical structures – offices, wires and poles – helped shape the route while also transmitting updates to readers around the globe.

Although focused on the immediate difficulties and dangers of the race, the Peking–Paris motorists were witnesses to seismic historical events. China – although vast and the source of many luxury goods coveted by other nations – was largely viewed with contempt by the West. During the previous half century she had endured a number of humiliating and expensive military defeats, leaving her with enormous debts and a tattered reputation. Relations with many Western powers were strained. A vivid image by French political cartoonist Henri Meyer in 1898 neatly sums up the state of China's foreign relations. In it, Queen Victoria, Kaiser Wilhelm II of Germany, Tsar Nicholas II of Russia and a Japanese samurai are seated around an enormous pie labelled 'China' (see page 44). While they wield knives and squabble over who will take the best slices, a Manchu gentleman stands behind them with his hands raised in impotent horror. To make matters worse, the Qing dynasty, which had ruled China for over two hundred and sixty years, was faltering from one crisis to the next. It would fall four years after the Peking–Paris, ending more than two millennia of imperial rule, ushering in a period of turmoil and paving the way for the rise of Mao Zedong and the Communist Party of China.

The Russian Empire too was in tumult. In the late 1800s, it was the largest empire in the world, stretching from the Black Sea in the west to the Bering Straits in the east, with a population of around 135 million. Its sheer size – not to mention its geographic, cultural and linguistic diversity – made it challenging to govern. Like China, it had been ruled by the same dynasty since the seventeenth century, and the system they were using to control their territory was beginning to show signs of strain. The empire was reeling after the revolution of 1905 and a bruising defeat in the Russo-Japanese War (1904–05). Although, in theory, the unrest had been quelled and concessions made, including the formation of an elected parliament, the situation was anything but peaceful. Arson, murder and armed robbery were daily occurrences; soldiers were stationed at every bridge and patrolled city streets. Aristocratic estates were regularly overrun, pillaged and burned by the peasantry. A few days before the first motorists crossed the border, Tsar Nicholas II and Prime Minister Stolypin seized back political power from the Duma in what is often called the June Coup.

Things were scarcely less troubled in the West. In terms of size and scope, this time is often seen as the high-water mark for colonial expansion. From the late nineteenth century, both older colonial powers – including France, the Netherlands and the United Kingdom – and newcomers – Japan, Italy and the United States – had scrambled to acquire influence and territory in Asia and Africa. Technological developments such as the telegraph and railways had allowed them to acquire and exert greater control over conquered land and people. However, holding on to these empires was beginning to create more and more problems. Colonial expansion and retention was costly, both financially and in terms of lives lost to uprisings and disease. Squabbles between colonial powers were also escalating as they tried to secure raw materials, trade routes and fertile land in which to grow cotton, tea, indigo and other valuable crops. The dawn of the new century would see imperial rulers engaged in tense diplomatic wrangling, crystallising the alliances and enmities that would play a pivotal role in the outbreak of the First World War seven years later.

My fascination with the Peking–Paris story began with a mistake. Researching the origins of *rosso corsa*, Italy's racing red, for *The Secret Lives of Colour*, I came across numerous references to Prince Borghese and his pioneering run from Peking to Paris. The story went that Italy selected the colour of the prince's car as its national racing colour in tribute to his success. Captivated, I wrote the story in my book, dutifully emailing the museum that houses the car today to confirm that the Itala, now a dull grey, had once been bright red. (Legend had it that the car had been hastily repainted to prevent rusting after being accidentally dropped off a boat.) To my horror, it would turn out to be untrue. The museum tested the paint all the way down to the metal and not a scrap of scarlet was found. Worse, I received this information after *The Secret Lives of Colour* had gone to press.

This was far from the first red herring I would encounter. Those participants who later published books about the Peking–Paris race had a habit of obfuscating the truth, prompted by rivalry, loyalty, censorship, the desire for a good story or patriotic fervour – one simply turned out to be an inveterate liar. A resurgence of interest in early automobiles and races during the 1950s and 1960s, particularly in Britain, produced several more books about the race. While dashing reads, these helped enfold the Peking–Paris in further layers of myth.

Researching this book has been an exercise in pruning back legends and unearthing previously overlooked primary sources. By going back to and cross-checking against contemporaneous telegrams and news reports, I hope to put a few hardy Peking–Paris fables to rest, including the one about *rosso corsa* that I unwittingly helped fuel.

The desire to dig deeper into the race and explore its historical context took me to a host of unlikely locations: a family home on an island in Lake Garda, archives at Hong Kong University and Queen's University Belfast, automotive museums in Turin and at

Brooklands, and an otherwise unremarkable hill in Camborne, Cornwall. I spoke to family members of the original racers, who generously shared photographs, documents and troves of family lore. I interviewed enthusiasts who have organised or participated in recent Peking–Paris rallies.* Some of the research and most of the writing of this book took place during the Covid-19 pandemic and the early months of Russia's invasion of Ukraine. This made my plans to retrace the original route with my husband in a vintage Toyota Land Cruiser impossible. It is my sincerest hope that we will one day fulfil this dream, although by that time we'll be taking our daughter with us. (She was born a week after the 116th anniversary of the start of the race.)

Doing justice to the Peking–Paris story has meant learning about the development of the automobile, which led, in turn, to finding out more about horse-powered transport, road accidents, telegraphy and the politics of oil. So that I could get a better understanding of the people and places the Peking–Paris participants encountered, I unearthed accounts by others who had travelled the same way in the years before and after 1907. Within the book, these themes and others interleave the primary narrative of the race in the form of standalone essays, which you can read in turn, skip to or read later, as you wish.

What writing this book has taught me, and what I will share with you in the following pages, is that the Peking–Paris race is a parable about a world teetering on the very edge of the most consequential century in human history. We are the inheritors of the world the Peking–Paris race helped create.

Kassia St Clair
March 2023
Vauxhall, London

---

* Many Peking–Paris enthusiasts will wonder why this book uses the term 'race' rather than 'rally', which is preferred today. My reason is simple: it was predominantly referred to and experienced as a race by contemporaries – including participants.

Men travel faster now, but I do not know if they go to better things.

Willa Cather, 1927

Charles Godard in the driver's seat of the red, blue and white-striped Dutch Spyker, shaking hands with admirers before the start of the race.

PROLOGUE

Peking: 10 June

It is three minutes to nine on the morning of 10 June 1907. Gloved hands elegantly raise a white flag high into the air and the crowd below hushes. Or hushes as much as a crowd of that size ever could. Within the parade ground of the French Legation in Peking, the panting of several stray dogs and the yips of a pet can still be heard, along with a chuckling coloratura of caged larks held by a few Chinese dignitaries. Horses' harnesses, spurs and dress swords clink. Ladies' skirts rustle. The click of cameras. A breeze jostles colourful bunting and a large banner over the gate reading 'BON VOYAGE!' The wind brings with it the quickening patter of a June rainstorm that might, with luck, fall long enough to slake Peking's dust but cease before it renders the streets an oozing, mustard-hued morass. Discernible above, below and overwhelming all other sounds is one never heard on the city's streets before: the throbbing, thrillingly modern basso profundo of five automobile engines.

The automobiles had gathered in the misty parade ground just before seven thirty, expecting an eight o'clock departure, but at ten minutes past the hour they remain stationary. Tedious formalities gobble up the minutes. The crowd, buzzing like an overheated hive, is polyglot and overwhelmingly European. The band of the French colonial infantry hustle through their repertoire. Diplomats and businessmen from France, Holland, America, Russia and Britain talk shop in a confusion of tongues. Some had

1

arrived by train earlier that morning or the day before from the business centres of Tientsin and Shanghai to witness the historic start of China's first automobile race. Most drolly predict it will end before the cars reach the Great Wall, a day or so's ride north. They regale one another with a crescendo of travellers' tales: the rocky pinch of mountain passes, infested roadside inns, ruts like moats, the parched vastness of the Gobi.

Heads turn to watch the slow progress through the crowd of a dignified, white-bearded figure. Sir Robert Hart, Inspector General of China's customs service and the most revered foreigner in China, nods and exchanges greetings with acquaintances, leaving a respectful hush in his wake. He stops to talk to a statuesque couple wearing well-tailored, ankle-length travelling coats. It is Prince Borghese, driver of the Itala, and his wife, Anna Maria. She is vivacious, putting both hands out in greeting and smiling. Borghese inclines his head.

Nearby, Jean du Taillis, a correspondent for *Le Matin*, due to travel with the Spyker, perches on a packing case like an ungainly heron. An assortment of clothes, long limbs and wire-framed spectacles, he is completely absorbed in scribbling a last-minute dispatch. Luigi Barzini, his slight, beaky colleague from the Italian newspaper the *Corriere della Sera* and a passenger in the Itala, stands a few feet away, one ankle crossed over the other, surveying the scene around him.

Of the five automobiles in the parade ground – all heaped with luggage and supplies – two attract the lion's share of comment. The first is the tiny, three-wheeled Contal Mototri. Despite being thought of as an automobile, it is clearly more akin to a motorcycle, with two wheels at the front, one at the rear and two seats, one in front of the other. From the moment its participation was announced months before, experts and casual observers had agreed that its lightness will be an asset on poor roads. Besides, it is French, and therefore hails from the heart of the motoring world. Even so, it now seems implausibly frail and overburdened beside its burlier fellows, scarcely robust enough to leave the city, let alone travel 8,000 miles westwards. The Dutch Spyker draws a larger group. Whether this is because of its outlandish paint scheme – the blue, white and red stripes of the French flag – the imposing height of its green canvas cover, its smart Louis Vuitton travel trunks or the volubility of its barrel-chested, bandy-legged French driver, Charles Godard, is uncertain. A small contingent from the Dutch Legation stand beside it, determinedly ignoring the snub implied by its paintwork and loudly talking up the Netherlands' motoring industry to anyone who will listen.

Beside them are the favourites. A twin pair of vehicles from the French manufacturer De Dion-Bouton. Their pedigree is impeccable: not only are they from the largest and most successful marque, but also their predecessors triumphed in the world's first motor race from Paris to Rouen on 22 July 1894.

Nearby is the stripped-down carcass of the prince's enormous vehicle, the Italian Itala. Its bodywork has been removed, and so have the seats and practically everything else. In place of mudguards are long wooden planks, on which are stowed a few lengths of rope, a pickaxe and blankets. It is easily five times the size of the smallest competitor, the Contal, and roars loudest, its engine packing nearly three times the horsepower of the next most powerful vehicle, the Spyker. Notwithstanding the hauteur of its driver, it seems a hulking, skeletal carthorse next to its sleeker, leaner rivals.

Something Godard says – one hand tented on his chest, the other raised in declamation – elicits laughs and cheers of 'Bravo!' A soldier, wandering over to du Taillis, smirks and waves a cigarette in Godard's direction, wondering how the journalist will fit into the Spyker, overflowing as it will be with the ego of his travelling companion. Eyeing Godard, du Taillis shrugs. *'C'est juste "Le Fanfaron du Gobi".'** If he is privately less sanguine, he consoles himself with the thought that Godard will at least make for compelling copy. Speaking of which, might the soldier be able to take this report to the telegraph office for him? Today, if possible: he wants to make *Le Matin*'s front page. Du Taillis gets up to stretch his legs and say his goodbyes. Catching his eye, Barzini smiles, shakes his head and turns towards the imposing figure of the Italian prince, in whose vehicle he will travel.

* 'That's just "The Braggart of the Gobi".'

3

Finally, the preliminaries are complete. Each driver has been solemnly handed a note for 2,000 francs by a representative of the Russo-Chinese Bank, the return of a deposit they paid months before to secure their place in the most daring automotive challenge the world has ever known.* The time is 8.50 a.m. Passengers extricate themselves from conversations and stride to their vehicles, fighting to squeeze in amidst luggage and equipment. Cavalry officers mount their horses: they will escort the cars out of the fortified Legation Quarter and into the city beyond. The band form up by the gate. 8.54 a.m. Drivers move to the front of their vehicles, insert crank handles and give them a sharp, clockwise turn. Engines snort into life. Plumes of smoke erupt from five exhausts and fill the air with the acrid smell of combusting fuel. 8.56 a.m. The starter is Madame Boissonnes, wife of the first secretary of the French Legation. She takes up a position in full view, holding her white flag in one hand and a bottle of champagne in the other and looks round at a carpet of expectant faces. The scene settles. Voices mute. Faces crane towards her. Hands tighten on wheels. Down flies the flag. Smash goes the bottle. Shards of emerald glass and champagne spume catch the light. The race from Peking to Paris has begun.[1]

---

* This was a huge sum, which explains why there were only five people at the starting line when forty or so had showed interest. At the time, 2,000 francs would be around double the annual salary of a well-paid university professor and three times the annual salary of a skilled telegraph operator.

1

HITTING THE ROAD

Birth of the Automobile

Of all inventions, the alphabet and the printing press alone except-
ed, those inventions which abridge distance have done most for
the civilisation of our species. Every improvement of the means
of locomotion benefits mankind morally and intellectually as well
as materially.

Thomas Babington Macaulay, *The History of England*, 1848

By the time everything was ready, it was dusk, Christmas Eve,
1801. A burly group of men emerged from John Tyack's smithy
at a crossroads in Camborne, Cornwall. Into the glowing red
square of light cast by the forges they dragged an enormous, puffing engine mounted on a wooden chassis with thick, creaking
carriage wheels. The air smelt of woodsmoke, hay and hot metal.
Captain Richard 'Dick' Trevithick, engineer at the Ding Dong
mine near Penzance and proud inventor of the contraption
about to be unleashed on its maiden voyage, supervised proceedings. A stoker shovelled wood and coal into the machine's fiery
drum. Trevithick examined the gauges. Satisfied, he mounted
and pulled his cousin, Andrew Vivian, up beside him. Seven or
eight men scrambled aboard too. Vivian pulled a lever and the
machine moved steadily off, its pistons pumping the crosshead
up and down like a guillotine. Whooping with delight, the party
chugged along the steep, curving lane up Camborne Hill towards
Beacon, puffs of steam emerging from the stack into the frigid

air.* This journey – neither particularly long nor picturesque, and accomplished at the same pace as a trotting horse – was nevertheless the very first undertaken in a passenger vehicle powered by an engine.[1]

Ever since our species got up on two feet and stayed there, being tied to one place has been anomalous. Charles Pasternak, the British biochemist, argued in his book *Quest* that humans' urge to travel is both innate and essential to our well-being. Certainly for most of our existence, until around 12,000 years ago, we were largely nomadic, drawn onward by sources of forage or an abundance of game. Thereafter, trade with other communities near and far allowed people access to highly prized textiles, perfumes, pigments, spices and other goods.

Means of swift travel are a staple of myths, legends and sagas worldwide. Khonsu – the name means 'traveller' – was the ancient Egyptian god of the moon, described at one point as 'the greatest god of the great gods'. The Greek god Hermes, patron of roads, travellers and luck, moved between the mortal and immortal realms borne by winged boots. Vahanas, the divine mounts of Hindu deities, were worshipped in their own right. Shiva has a bull; Vishnu flies on Garuda, a kite; Ganesh rides a fleet-footed mouse. Heimdall, a Norse deity, was charged with protecting Bifröst, the rainbow bridge between the worlds of gods and men. Skidbladnir, the finest vessel in Norse mythology, could sail in air as well as in water and was roomy enough to accommodate the pantheon but could be folded up like a handkerchief.

Away from the mythical world, overland locomotion was slow to evolve and dependent – until the invention of the railway – on animals. Given the expense and constraints this reliance entailed, the idea of horseless carriages piqued the interest of inventors long before Trevithick. In around AD 550 a Chinese

---

* 'Goin' Up Camborne Hill', a folk song commemorating Trevithick's moment of triumph – 'The 'orses stood still; the wheels went aroun'' – is still popular in Cornwall. Popular enough that it popped into the head of Rick Rescorla, a Cornishman and security manager at the World Trade Center in New York during the 9/11 attacks. He sang it repeatedly through a megaphone while urging some 2,700 people to safety from the South Tower. He was last seen going back into the South Tower to help the evacuation efforts.

scholar described wind-driven carriages capable of transporting thirty people. Leonardo da Vinci conceived of a vehicle propelled by clockwork around 1478. Sir Isaac Newton predicted steam-propulsion carriages in 1680. The following century, a three-wheeled, steam-propelled cart was constructed by Nicolas-Joseph Cugnot for the French army to transport cannons, but it could only run in twenty-minute bursts. An 'improved' version, built in 1771, overturned rounding a corner near the Place de la Madeleine in Paris and was promptly sequestered in the Arsenal in the interests of public safety.

Trevithick's 'Puffing Devil' did not fare much better. A few days after its successful trip around the Camborne countryside, it broke down and was hauled off to a wooden shed. According to an eyewitness, the driver and passengers 'adjourned to the hotel, and comforted their hearts with a roast goose and proper drinks, when, forgetful of the engine, its water boiled away, the

The very first automobiles, like this 1886 Benz Patent-Motorwagen, clearly showed their 'horseless carriage' origins.

iron became red hot, and nothing that was combustible remained, either of the engine or the house'.[2]

Perhaps thanks to this inauspicious beginning, the automobile was still in its infancy more than a century later, when the Peking–Paris took place. During the nineteenth century, focus had shifted to the development of first railways and then the bicycle, but eccentric, automotive-minded inventors in Europe – France and Germany, in particular – continued to make scattered breakthroughs. Etienne Lenoir and Gottlieb Daimler, for example, worked separately on developing fuel-powered engines in the 1860s and 1870s. However, it wasn't until the late 1880s that the automobile gained real traction. On 29 January 1886, Karl Friedrich Benz, the son of a Mannheim engine driver, was granted a patent (number: DRP 37435) for a three-wheeler with a four-stroke internal combustion engine and integrated chassis, the world's first production automobile. Two years later, Bertha, Benz's wife and investor, would undertake an historic, sixty-five-mile journey from Mannheim to Pforzheim in an updated 'Model III' in what would prove a resounding publicity coup. Afterwards, she let it be known that as well as securing fuel from apothecaries along the way, who sold it as cleaning fluid, she had cleared the clogged fuel line with a hat pin, insulated the faulty ignition with her garter and devised the world's first pair of brake pads – made from leather obtained from a cobbler – when the original solid wooden ones wore down.

Benz began selling his automobiles in 1888. By 1895, around 500 from various makers had been sold in Europe, and American inventors, including Charles and Frank Duryea, were starting to exhibit prototypes. Early models were still explicitly 'horseless carriages', constructed exactly like horse-drawn vehicles with an engine in place of an animal. It was not until 1901, when Benz moved the engine under the bonnet of one of his models, that the automobile developed its own identity.[3]

These pioneering years laid the groundwork for the way the industry would develop and how people would ultimately view the automobile. From the beginning, this would contain a slew of contradictions. The automobile was simultaneously the ultimate luxury consumer good and a practical invention, capable of hauling great loads and transporting people in comfort, cleanliness and safety from A to B. It was also redolent of risk, adventure, romance, masculinity and a rugged spirit of independence. To drive one was to harness a machine capable of catastrophe at any moment. From the beginning, brands adroitly harnessed this contradiction, parlaying it into outrageous publicity stunts and lucrative sales.

At the turn of the century only a small, elite minority owned an automobile. In 1906 Woodrow Wilson, president of the United States, accused motorists of spreading 'Socialistic feeling', claiming that for ordinary Americans 'they are a picture of arrogance of wealth with all its independence and carelessness'.* Since hygiene and presentation were important to this class, advertisements often emphasised the relative cleanliness of a marque. Graceful maidens attired in spotless gowns and veiled hats were shown as passengers or drivers. While speed and dependability were trumpeted, aesthetics mattered too. The Mors was 'The Silent Car'; the Spyker, 'dustless'. Similar attributes were also major selling points for steam and electric cars. 'NO SHOCKS IN STARTING OR STOPPING. NO NOISE. NO SMELL' read one advertisement for a Rexer Steam Car.[4]

Yet, while some manufacturers and users saw automobiles as essentially 'domesticated' – clean, safe and soothing extensions of affluent private households – to many others danger and disreputability were part of the allure.

Michelin, the likely suppliers of the Spyker's tyres during the Peking–Paris, were masters of exploiting such marketing opportunities.† Founded in 1830, the firm was a global industrial

---

* Unsurprisingly, the president of the American Automobile Association rejected the charge, saying, 'Scores of well-to-do farmers are now automobile owners ... Just as soon as any class of people realize that the automobile is useful for the ordinary work and duties of life apart from its pleasurable attributes as a touring car, it will be regarded with favor.'

† Several sources mention that the *spare* tyres bought by Godard in Amsterdam (on the Spyker account) and later sold in Paris to purchase his steamer fare were Michelins. Identifying the *original* tyres is more difficult, but the likelihood is that Godard bought spares that matched the originals. However, du Taillis, in his book, wrote that the Spyker's tyres were Hutchinsons. This firm, based in France and best known for its rubber shoes and bicycle tyres, was founded by an American, Hiram Hutchinson, in 1853.

success story by the turn of the century. In 1906, their total sales were 37 million francs. Michelin was aspirational, taking care to associate their products with modernity, speed, progress and rakish male sociability. Daring race victories on Michelin tyres were highlighted in newspapers and magazines to demonstrate their reliability and endurance. A brand mascot – Bibendum, also known as the Michelin Man – was conceived as a conspicuous *bon vivant*, depicted puffing away on cigars and enjoying good food and the company of beautiful women. In France, the firm appealed to buyers' patriotism by positioning themselves as a uniquely successful national brand, and stoked their potential consumer base by encouraging car ownership and tourism using their Michelin area guides, founding a tourist office and advocating politically on behalf of car users.[5]

In addition to direct advertising, Michelin were adept at using print media in subtler ways too. In 1900, the firm helped finance *L'Auto-Vélo*, a new sports magazine, aimed at expanding and glamorising the motoring and cycling market, even if Michelin was not always directly mentioned. Three years later, the newly abbreviated *L'Auto* sponsored the first Tour de France amidst a shower of publicity. In 1907, the publication engaged Georges Cormier, driver of one of the De Dion-Boutons, as their Peking–Paris correspondent.

When motoring sports and events appeared in early specialist magazines and newspaper motoring pages, they generated reams of surprisingly unflattering coverage, focusing on the difficulties and mishaps. Narratives of the Peking–Paris – even those sponsored by automotive brands – were typical. Attention was lavished on the hardships the motorists endured and the obstacles they overcame: the slow pace, getting soaked through when it rained, encounters with recalcitrant inn owners and hostile cart drivers, and above all, the mud that would encrust and begrime the long-distance traveller.

The dirt, odours, elbow grease and even brute force involved were as much part of the automobile's glamour as its contradictory promise of clean, seamless transport. Struggles were a point of pride for early motorists, especially in hindsight. Hiram Percy Maxim, an

American engineer in the 1890s and early 1900s, fondly reminisced in 1937: 'One wore old clothes and carried a full assortment of tools and spare parts when one ventured forth in horseless-carriage days.' One former Model T owner proudly described the cussedness of the car's transmission as 'half metaphysics, half sheer friction'.

In some instances, inventions that might have made automobiles more user-friendly were resisted, most famously when it came to methods of starting them. All the Peking–Paris vehicles were equipped with crank handles. These were inserted into a small hole at the front of the car and sharply turned to start the engine, at which point the driver could remove the handle, get in the car and set off. Although this sounds simple, they were temperamental, inconvenient and even dangerous. Often requiring two or three turns and a certain knack to catch, the handles could kick back unpredictably as the engine fired up. Stories about broken wrists, arms and jaws abounded. A columnist for *Motor* magazine vented his feeling about cranking in 1911, calling it 'the most humiliating requirement that still remains for the driver of a machine' and, in a shower of heart-felt plosives, a 'needless producer of perspiration, profanity and pernicious contusions'.[6]

Engines in electric automobiles had been switched on from the driver's seat for years, and electric ignitions for combustion engines were patented in 1903 and 1911. However, it would take nearly a decade for them to become standard. (Some vehicles, especially utility ones, were still routinely being sold with a crank-handle option into the 1970s.) Strangely, their ease of use seems to have *contributed* to their slow adoption, as along with other easier-to-use starters, they were associated by manufacturers and the general public alike with electric vehicles and female drivers, which undermined motoring machismo.

Many automotive firms, including Opel, Peugeot, Rover, Pirelli and Michelin, drew much of their initial experience from the world of bicycles, where they had faced similar challenges. After 1885, when the 'safety bicycle' was invented (as an alternative to earlier, more dangerous designs including the penny-farthing, or high-wheel model), the popularity of cycling exploded. Manufacturers presented them to potential buyers as a form of transportation and as a symbol of emancipation, electrifying modernity and the new ethos of speed. They placed advertisements, founded specialist publications, sponsored clubs and lobby groups, and organised rallies, exhibitions and other events.* Bicycle races became popular and profitable spectator sports. The first city-to-city race, organised by *Le Vélocipède Illustré* and held in November 1869 between Paris and Rouen, was something of a sensation, with over 300 entrants (although only a hundred turned up and just thirty-four completed the course). The same route would be used, twenty-five years later, for the first motor trial.[7]

Even allowing for the fact that it was the first of its kind, *Le Petit Journal*'s 1894 eighty-mile Paris–Rouen motor trial was unconventional. A motley assemblage of 102 vehicles entered, fielding between them twenty different methods of propulsion. Steam squared up to compressed air; clockwork to gravity; petrol to a mechanism powered by pendulums. In the end, only twenty-five machines were deemed safe enough to compete. An early and inelegant De Dion-Bouton steam-powered entrant was among them, looking like a dog cart hitched to a tractor. It remained in the lead for the duration, crossing the finishing line in around six hours and forty-eight minutes.† De Dion was denied first prize, however, on the grounds that his vehicle required, in addition to the driver, a stoker to shovel fuel into the engine. This was ruled impractical.[8]

The petrol engine was the true victor of the race. While most of the vehicles broke down at some point, all thirteen of the petrol-powered ones completed the course. Soon after, the De Dion-Bouton firm abandoned steam in favour of petrol and went on to be among the world's foremost car manufacturers. The marquis helped found the Automobile Club de France and became a tireless motoring advocate, frequently in the pages of *Le Matin*. It

---

* One concrete way in which the bicycle paved the way for the automobile was in creating a need for smooth thoroughfares better suited to tyres than hooves. In America, for example, the League of American Wheelmen was a tireless advocate through the 1880s and 1890s, founding *Good Roads* magazine, lobbying cities and states and, in 1893, pressuring Congress to spend $10,000 studying and effecting improvements.
† The competitors had stopped for a leisurely lunch in Mentes, making precise timekeeping difficult.

was clear that races, rallies and similar exhibitions were powerful tools for the emerging industry. They were excellent at generating news coverage, not least because they were often sponsored or arranged by newspapers or magazines. America's inaugural motor race, for example, which was modelled on the Paris–Rouen, was organised in 1896 by the *Chicago Times-Herald*. Motoring pioneers – including Émile Levassor, Marcel Renault, Frank Duryea and Henry Ford – raced their own automobiles as a way of attracting investors and generating name recognition. The events proved so popular and lucrative that they endured even as automobiles became commonplace.*⁹

From very early on, motor sports fell into two broad camps. The first were held on public roads, often between towns or cities, and were usually organised with great fanfare in the hopes of stoking international interest and ginning up sponsorship and advertising. The years following the Paris–Rouen saw dozens of such events. The Paris–Bordeaux–Paris (1895), the Chicago Times-Herald race (1895), Paris–Trouville (1899), the Gordon Bennett Cup (1900–1905) and the Paris–Vienna (1902). They were pell-mell affairs, with wildly varying criteria for entrants, idiosyncratic judging and often the expectation that competitors would stop for long, liquid luncheons together. In 1899, for example, a handicap race held in France pitted walkers, horsemen, cyclists, motorcyclists and cars against each other. Horses, incidentally, finished first and second.

During this time there was also an appetite for solo endeavours. Georges Cormier, a De Dion-Bouton driver during the Peking–Paris, undertook several journeys across Europe and into north Africa in the early 1900s, which he wrote about for French sporting publications. Far more notorious was an around-the-world trip undertaken in the spring of 1902 by Dr Edward Ernest Lehwess and Max Cudell. Their vehicle, an enormous, yellow 25 hp Panhard-Levassor bus modified into a motorhome, was christened

---

* Similar events promoting horses, bicycles and motorcycles were still popular in 1907. Midway through the Peking-Paris, *Le Matin* would organise a week-long Festival of the Horse that included a Paris–Deauville–Paris race involving sixty-six contestants, pitching horses and riders against sundry horse-drawn vehicles, including 'phaetons, tilbury gigs . . . spider carriages, and even a three-wheeler'.

the Passe Partout after the manservant in Jules Verne's *Around the World in Eighty Days*. The pair departed from Hyde Park Corner in London on 29 April to great fanfare, but made embarrassingly stately progress: it took them nine weeks to get past Berlin. That autumn, the Passe Partout would run into a snowdrift just outside Nizhny Novgorod and the trip was quietly abandoned.[10]

Although popular with the public, large automobile races proved less so with authorities, especially after the 1903 Paris–Madrid. Dubbed 'The Bloody Race' by *Le Matin*, it was a catastrophe, and is credited with putting a temporary stop to the tradition of large, inter-city races on public roads in western Europe. The 220-odd participants set off from Paris on 24 May 1903, watched by enormous crowds estimated to be around 2 million strong. Dry weather meant the roads were dusty and visibility was poor, while enthusiastic spectators crowded onto the roads and drivers seemed determined to set new speed records no matter how bad the conditions. Afterwards, even the most ardent elements of the motoring press would admit it was shambolic. In all, there were eight fatalities during the first day of racing and innumerable accidents, mechanical and technical faults. One car swerved to avoid a dog and instead hit a tree; another overturned on a rail crossing and caught fire, killing the chauffeur and three spectators. Marcel Renault, co-founder of the eponymous automobile company, missed a turn and was hurled from his open vehicle; he died two days later. The race was called off in disarray after the first stage by the French and Spanish governments and the surviving vehicles were towed in disgrace to the nearest railway stations by horses, forbidden from even having their engines started.[11]

The Peking–Paris was a kind of response to this debacle: since it began in China and the majority of the journey would run on quiet roads far away from populated areas, organisers were banking on there being fewer contestants and fewer embarrassing accidents. (This, as the entrants would learn for themselves, would prove a forlorn hope.)

The second kind of motor sport was, in some respects at least, more controlled. Held on privately owned land and, later, on dedicated tracks, they would evolve into circuit races such as Formula

One. Brooklands, completed in July 1907 in Surrey, England, was the world's first purpose-built oval racing circuit.* Steeply banked, it provided a thrillingly modern arena for the new sport. Days after it opened, the pioneering motorist Selwyn Francis Edge completed a twenty-four-hour race there. Brooklands would also play host to the first person to cover a hundred miles in an hour and several land-speed records. Other historic tracks include the Knoxville Raceway in Iowa and the Milwaukee Mile – both originally built in the 1870s for horse racing – and the Indianapolis Motor Speedway, the second purpose-built circuit (and the first to be called a speedway).

From Bertha Benz to today's Monte Carlo Rally, publicity and plenty of money have fuelled motors and motor sports. But the Peking–Paris came at an auspicious moment. From rackety beginnings and following a long gestation, automobiles were beginning to win over the sceptics. A race of this kind was a perfect vehicle for showcasing and drumming up public support for the technology after previous disappointments and disasters. The scope of the challenge and the ambitious, romantic route appealed to a wider and deeper audience. The industry that has been built up in the years since is a testament to this pioneering era. A 2021 study commissioned by the Fédération Internationale de l'Automobile (FIA) estimated motor sports' annual gross output at 160 billion euros, responsible for generating 1.5 million paid jobs and involving well over 2.5 million participants.[12]

---

* A slightly earlier one, the mile-long circuit at Aspendale in Victoria, Australia, was added inside an existing horse racing track in 1906 and abandoned soon after.

2

Paris – Peking

30 January–9 June

Enthusiasm grows stronger by action, but weaker through words.

Luigi Barzini, Itala team

The first inkling the world had of the race were three words in dense, inch-high newsprint in the French newspaper *Le Matin* on 31 January 1907: 'PARIS–PÉKIN AUTOMOBILE'. On a front page dominated by news items about record numbers of warships being built and a fatal mining accident at Saarbrücken, Germany, this headline, with its alliteration and tang of adventure, demanded attention. Curious readers who dived into the column beneath while sipping their morning coffee, however, might have been bemused by the diatribe they found there.

The automotive industry, the editors argued, pride of the French nation, was at risk of being stifled by the very events in which automobiles should be proving their mettle: races and rallies. Not the 'entertaining spectacles' they ought to be, these events, held on 'carefully selected small circuits', confined cars to infantile, circular routes. This was an insult to the technology's potential. Unlike horses, trains or bicycles, automobiles offered seductive visions of untrammelled, easy movement over great distances. The ability to set out spontaneously for far horizons. 'Progress,' the article continued, 'cannot make concessions to mediocrity.' What was needed, if motor cars were ever to prove themselves to a dubious public, was a demonstration that 'with an automobile one can do

17

anything and go anywhere. Anywhere – yes, truly anywhere.' To that end, the editors of *Le Matin* wished to throw down a gauntlet before the motoring industry of France and, indeed, the globe: 'Is there any person, or perhaps several people, who will accept the challenge of travelling from Paris to Peking this summer by automobile?'[1]

The itinerary too would have been familiar to *Le Matin*'s readers. In addition to the alliteration, Paris and Peking formed a satisfying pair: both capital cities, they were seen as the respective hearts of West and East. Worlds – as well as continents – apart. Overland journeys between the cities using combinations of horses, dog sleds, riverboats or, where they existed, railways, had become something of a late nineteenth-century travel trope. Prominent examples included Victor Meignan's 1885 *From Paris to Pekin over Siberian Snows*, a chauvinistic account that leaves the reader wondering why the author ever left France, and, published four years later, Harry de Windt's swashbuckling *From Pekin to Calais by Land*.

The idea of such a motor trial was evidently alluring. Before the day was out, *Le Matin*'s editors had received telegrams from the Marquis de Dion, of the De Dion-Bouton automotive firm, and Mr Camille Contal, of Contal motors. Mr Contal, accepting the challenge, wrote that 'my valiant little Mototris . . . will not be daunted by this race, however terrifying it may be.' De Dion, with equal bravado and an eye firmly on the marketing opportunity, proclaimed it to be 'an idea worthy of Jules Verne', adding that the 'roads are abominable and often exist only on the map. Yet I nonetheless consider that if any automobile can make this journey, the De Dion-Bouton will make it.'[2]

It was inevitable that the Marquis de Dion would be among the first to sign up and would influence proceedings thereafter. A boisterous, roly-poly of a man with deep pockets and a taste for well-groomed moustaches and fast vehicles, de Dion developed an interest in automobiles after discovering a miniature steam engine in a toy shop in 1882. Inspired, he hired the toymaker Georges Bouton to make a scaled-up version capable of powering a two-seater tricycle. By 1889, he was able to exhibit an early De Dion-Bouton model at the Exposition Universelle in Paris and in 1894 he drove one of his vehicles in the world's first automobile race.[3]

Contal and de Dion's involvement in the race had the desired effect. By 15 February 1907, eighteen cars* were pledged to take part, with more signing up each day. One crucial change was the direction of the race. Originally advertised as the 'Paris–Pékin', it would now begin in China and head west, ensuring a triumphal finish in Paris outside the offices of *Le Matin*.† Would-be participants were informed they would need to be ready in time to ship their cars to China by 14 April, giving them just two months to prepare.

Of four possible routes under discussion, the organisers selected the most northerly, striking out north-north-west from Peking, traversing Mongolia, entering the Russian Empire before turning westwards at Lake Baikal, a remote region in eastern Russia.‡ This was neither the most direct nor the most glamorous itinerary. It would take them ten degrees further north than necessary, avoiding western China and the fascinating but troubled Balkan and Caucasus regions. It would also mean coaxing notoriously delicate and puncture-prone automobiles up steep, rugged passes north of Peking and across the inhospitable Gobi Desert, before condemning them to thousands of miles along muddy Siberian tracks. The advantage was that it hugged the telegraph and Trans-Siberian

---

* These included several more French vehicles, including a Porthos, an 18 hp Panhard and a Passe Partout voiturette. There was also a 'Belgian car of unspecified make' and a lone British entrant, a 'C.V.R.' from the St James Motor Company, based at 199 Piccadilly, next door to the famous church designed by Christopher Wren and around 300 metres from where I write these words.

† The reversal was made between 1 and 10 February but remained somewhat ambiguous. An article on the 15th was headlined 'Peking–Paris' but, in the first line, referred to the 'prodigious Paris–Peking challenge'.

‡ The paper promptly dispatched a man called Eugène Lelouvier, an employee of De Dion-Bouton who had some experience exploring Russia, to scout the route. Lelouvier accomplished much of his eastwards journey by cart and on foot, and is said to have taken extensive notes, although these have never been published. The conditions, it still being the middle of winter, were extreme. By Ulaanbaatar, the tip of his nose had succumbed to frostbite and he was barely able to walk. A litter was fashioned for him by the resourceful Mr de Stepanoff, who ran the local branch of the Russo-Chinese bank. The following year he would enter the New York–Paris race, quitting after two days following arguments with his teammates.

Railway lines. From a news-making and publicity perspective, this would enable the journalists in the cars to send back regular reports. For the drivers, it meant that, if disaster struck, they could more easily send off for spare parts or simply return home by train.[4]

Auguste Pons and Octave Foucault pose with the heavily laden Contal Mototri, the smallest and least powerful vehicle in the Peking–Paris.

From the beginning, *Le Matin*'s enthusiastic bulletins were met with near universal howls of scepticism, cynicism and derision. A writer for the British magazine *Autocar* sniped that the event 'is one of those hardy annuals that crop up . . . when there is nothing else to talk about'. Russia's *Avtomobil'* (Автомобиль) was even more damning, claiming that *Le Matin* was a newspaper 'renowned for its tirelessly inventive self-promotion', and that the French, 'with their usual frivolity', were like Don Quixote, refusing to put their lovingly crafted idea to a proper test, such as 'a preliminary, trial expedition as far as Moscow'. If they did, they would surely abandon the whole idea in utter dismay at the terrible roads, filthy hotels and lack of basic amenities. The writer allowed that it might be possible for 'us, the Russians, or the Americans, but not the French. They are far too spoiled by their wondrous roads and consider it some sort of "Tour de France" as a grand voyage.' The *South China Morning Post* called the Peking–Paris 'a colossal joke', akin to a 'motor-boat race under water from Berlin to Rangoon', or a 'trip to the Moon by wireless telegraphy – competitors not to exceed 2,000,000,000 miles an hour'.[5]

Whether as a result of this deluge of scorn or, perhaps, the imposition of a 2,000-franc deposit, the Peking–Paris committee and the entrants began to lose heart. A statement published on 22 February concluded that the 'extraordinary challenge' was really 'an unachievable mission'.

A month later, things were bleaker still. Costs were ballooning, and they received telegrams from alarmed diplomats and hastily arranged local organising committees in China and Russia warning that the enterprise was preposterous. In large sections of the specified route there was nowhere to obtain fuel; there were, indeed, few roads. Enthusiasm at *Le Matin* began to curdle. Of the two dozen or so individuals who had shown serious interest, only a handful were willing to ship their vehicles to China. De Dion would send two cars but declined to drive one himself. The venture seemed destined for ignominy, which was more than the editors could bear. Telegrams were dispatched to the entrants informing them of a momentous decision: as abruptly as it had been proposed, the Peking–Paris was off.[6]

Luckily fate, and the logistics of early twentieth-century travel, intervened. By the time *Le Matin* had sent their message, one participant – Prince Scipione Borghese, an Italian aristocrat – had already committed. His vehicle, a patriotically all-Italian Itala fitted with Pirelli tyres, was steaming towards China aboard a fortnightly Norddeutscher Lloyd boat under the care of Ettore Guizzardi, the prince's mechanic and chauffeur. Borghese himself was also packed and ready to go. His response to the Peking–Paris committee and the editors of *Le Matin* was curt: 'I sail tomorrow from Naples.'

The die was cast. Politesse and national pride would not allow the organising committee to be humiliated by the Italian nobleman. The remaining entrants were peremptorily directed to depart for China on 14 April, as previously arranged.

The day after they set off, *Le Matin* published a defiant article written by Georges Bourcier Saint-Chaffray, general commissioner of the organising committee and a close associate of the Marquis de Dion.* The piece passed over the last-minute change of plans and took pains to patriotically contrast Prince Borghese – 'confident in his good fortune' – with the French: 'The intrepid explorers of the Mongolian desert, the valiant champions of mechanical locomotion.'[7]

For the participants of the race, the month ahead would be spent navigating the waterways between Marseilles and Tientsin aboard the *Océanien*. A twenty-three-year-old French mail steamer, she was long and low-slung, with three large masts and two stubby black chimneys that belched forth smoke from her coal-fired boilers. Comfortable rather than luxurious, she was also ponderous, with a top speed of sixteen knots. *Océanien*'s most notable passenger thus far had been Paul Gauguin, who in 1891 had spent 'sixty-three days of feverish expectancy' aboard, bound for Tahiti. While their own journey was half that of Gauguin's, it was still sufficient to give the motorists ample time – while they walked the decks or relaxed in the salons – to reflect on the folly of attempting the rugged return journey overland in vehicles built for the smoothest of European roads.[8]

Spyker

For Jean du Taillis,[9] the weeks on the *Océanien* were largely spent observing his future travelling companions. Born Fernand Joseph Octave Boulet in a small town in Normandy in 1873, he had begun styling himself Jean du Taillis – adapted from his mother's maiden

---

* Bourcier Saint-Chaffray's involvement boded ill. True, he had experience lobbying for the French motor industry, but he had also been responsible for organising the Course Alger-Toulon, a motorboat race from France to Morocco in May 1905, a disaster of truly farcical proportions. Launched during a storm, the vessels foundered one by one and each needed to be rescued in turn. Not one vessel completed the course.

name, Havas Dutaillis – in his twenties. Now thirty-four, he was blonde-bearded, gangly and generally affable, although prone to squalls of irascibility when he felt cheated. In a culture notable for its sartorial conformity, his dress sense was *outré*: often incorporating distinctive headwear from countries he had previously visited. His eyesight was poor and he always wore gold-framed glasses. His skin was tanned and weathered for his age, the result of having fallen in love with North Africa, where he had spent several years working and travelling. Although he liked grandly styling himself an *homme de lettres* on official forms, he was not particularly successful or well known. He sustained himself by writing and editing books about France's colonies, taking photographs of Morocco and Algiers for news agencies, and writing for national and local publications. The high-profile role of Peking–Paris correspondent for *Le Matin* was a career coup, one he was determined to grasp with both hands.

Charles Godard, with whom du Taillis would be spending the journey back to Paris in a 15 hp Dutch Spyker automobile, was his primary focus. Godard was born in 1877 in Pesmes, a small village a day's ride east of Dijon, in Burgundy. His was not a prestigious or wealthy family – his father was a gardener – nor does it appear to have been a close one. By the time he was twenty, they had scattered: he was working in Reims, ninety miles north-east of Paris, while his parents remained in Burgundy.

Godard should have been unremarkable in appearance. He was five foot three (a shade below average for the time), with blue eyes, dark hair and an oval face. By 1907 – his thirtieth year – he had filled out. Suit buttons strained; his hands were plump, with short, blunt fingers. His hair was receding, and he wore it combed back. Yet there was a childish elasticity to his features. Moustaches quivered, nostrils flared, eyes crinkled, rolled or widened beneath eyebrows that bobbed up and down like storm-tossed boats. This restlessness found an echo in his character: Godard's energy and ebullience were so potent that he tended to be remembered, even after the most casual of acquaintances. The *Océanien*, it seemed to du Taillis, might have been powered by Godard's *joie de vivre* alone. 'The purser and the excellent captain were at his command, pianos travelled from the second-class lounge to the promenade deck, accordions were commandeered, all at the say-so of Mr Godard. Everyone laughed, danced and whirled around thanks to him.'

Aboard, less endearing traits began to emerge. Godard snored, which did not augur well for the weeks ahead when they would be sharing tents and hostel bedrooms. More ominously, he was suspiciously close-lipped about his life before entering the Peking–Paris. He had revealed his birthplace to du Taillis, but on everything else relating to his youth, previous employment or even his experience with automobiles, he remained silent. At odds with this reticence, and making it all the more suspect, it soon became clear that he was boastful, vain and incautious, earning him the soubriquet 'The Braggart of the Gobi' long before they reached Peking. Worse still, he also showed himself to be unprincipled, lacking in foresight and, du Taillis increasingly feared, a penniless scoundrel.

Godard offhandedly mentioned to du Taillis that he had convinced the Dutch car manufacturer Spyker to loan him an automobile in which to participate. (They would later allege that he had assured them that *Le Matin* had promised to cover costs and that there was a large cash prize, both of which were untrue.) And it had been his idea to deck out the Spyker in the eye-catching stripes of red, white and blue of the French flag: an insult to his Dutch benefactors. Godard also told the journalist that he had managed to get hold of the great number of spare parts and tyres they would need for the journey. He was vague on details, but it later transpired these had been illicitly acquired by assuring a shop that the Spyker firm would settle the bill. This was moot anyway: Godard had promptly sold the lot to fund his steamer fare to China. First class, naturally.[10]

Perhaps Godard's most intriguing characteristic was his knack for landing on his feet at the very moment it seemed his chickens had come home to roost. On debarking at the port of Tientsin on 1 June with du Taillis and overseeing the removal of the Spyker from the hold, he was accosted by a representative of the shipping firm. The paperwork was all in order; there just remained the small matter of the 3,000-franc bill before the automobile could be handed over. Mr Godard surely remembered he had promised payment for the transport of the vehicle from Marseilles on delivery? Du Taillis, mortified by association, looked on, his ears reddening, as the heads of their fellow passengers began to turn. Godard, however, seemed almost to relish the confrontation. Chest thrown out, gesticulating wildly and insensible to reason, he demanded the Spyker's immediate release.

Godard, as du Taillis knew, would not pay because he could not pay: he was flat broke. At the very instant that du Taillis was resigning himself to the idea that he would not be driving back to Paris after all, a man materialised from the crowd, like a fairy godmother, and placed a soothing hand on Godard's elbow. This, it transpired, was the Dutch consul, anxious to forestall a diplomatic incident involving the first vehicle from his home country to arrive in China. 'Without hesitation,' du Taillis marvelled, 'the Consul offered his bond for the payment and with that freed both Godard from a crushing nightmare and the Spyker from the hold of the steamer.'*

That problem dealt with, at least until the consul demanded they settle the debt, all Godard and du Taillis had to worry about was how on earth they were going to get the car on a train to Peking and buy the fuel, food and everything else required to take them the 8,000 miles back to Paris. 'Trifles,' Godard crowed to du Taillis, rubbing his hands together in a pantomime of glee. They would work something out.[11]

Itala

Also on 1 June, nine days before the start of the race, Luigi Barzini reached Peking. Travel-stained and hollow-eyed, he was deposited by a guard from the Italian embassy at the formidable Grand Hotel des Wagons Lits in the Legation Quarter of Peking. His journey had been especially gruelling as a result of him having been sent westwards from Italy by the *Corriere della Sera* to report

---

* News reports claimed that Godard induced the consul to cover the costs by explaining that his letter of credit had not yet arrived. Since no bank would offer him one, it never would.

on the rebuilding of San Francisco after the 1906 earthquake and the finale of the Harry K. Thaw trial, a celebrity-strewn cause célèbre that had been engrossing the news-reading public for months. He checked in and swayed up to his room behind the bellhop, bearing a letter from his employer and a handwritten note from the man with whom he would shortly be sharing the adventure of a lifetime.

Although in his early thirties, Barzini was slender, almost boyish. Observers often noted how frail he looked, an impression bolstered by slight shyness and a tendency to look anxious. Appearances notwithstanding, he was easily the most renowned of the Peking–Paris journalists, and was reporting for both the *Corriere della Sera* and the *Daily Telegraph*. His career had taken off in 1899 after he persuaded a reclusive Italian opera star to give an interview. This had impressed the editor of the *Corriere* so much that Barzini was hired as London correspondent on the spot.[*] Armed only with a new suit and a dictionary, he headed to Britain and proceeded to write, to great acclaim, exposés of London's Chinatown and its opium dens. From there, he was sent to China in 1900, to cover the Boxer Rebellion; Russia in 1903, to report on political unrest; and Japan in 1904, as war correspondent during the Russo-Japanese War.[12]

Although better known than his peers, like them he was well travelled and well read, but solidly middle class.[†] The note that he had found waiting for him at the front desk of the Grand Hotel des Wagons Lits, which he clutched as he sat on his bed looking out at the city beyond, was from a man of a very different order. Written in a stiff, aristocratic hand, it formally welcomed

---

[*] His son, who would later follow in his footsteps as a journalist and author, liked to recall that his father had fibbed in the interview about his grasp of English, claiming to be fluent when he barely spoke a word.

[†] A third correspondent, Edgardo Longoni, was also at the starting line. A sportswriter for the Italian *Il Secolo* newspaper, he had been due to ride in a Fiat belonging to Count Groppello, which never arrived in Peking, possibly because Groppello had taken *Le Matin*'s attempt to call off the race seriously. Longoni began the Peking–Paris riding in the De Dions but ultimately returned home early. He would go on to buy *Il Secolo* during the First World War with Gian Luca Zanetti, founder of the Unitas publishing group.

Barzini to Peking and invited him to meet its author, Prince Scipione Borghese, on 6 June, by which time the prince and his wife would have returned from a horseback recce of the route northwards.

Born in 1871, the thirty-five-year-old Prince Luigi Marcantonio Francesco Rodolfo Scipione Borghese came from one of Italy's most illustrious families. One ancestor had been pope. Another – a seventeenth-century cardinal and Scipione's namesake – had filled the Villa Borghese, the family palazzo in Rome, with paintings by Caravaggio and Raphael and sculptures by Bernini. Borghese, however, had grown up at a time when his family's power, influence and funds were waning. Masterpieces such as a famous painting of Cesare Borgia – inspiration for Machiavelli's prince – had still been on the walls when he was a boy, but many had since been sold. So too, just four years previously, had the enormous grounds – tastefully landscaped in the eighteenth century in the English style – of his family's villa itself. This protracted humiliation had a profound impact on Borghese's character, inculcating self-reliance, aloofness and the occasional flash of ruthlessness.

Privately, however, Borghese was also sensitive and cultured; he had an eye for art and enjoyed sketching places he saw on his travels. (Among the possessions he kept with him until his death was a collection of watercolour vistas he had painted while travelling in the Middle East.) Educated in the Classics, he spoke excellent French, English, German and a little Russian. And although each passenger of the Itala was restricted to only fifteen kilograms of luggage, to keep weight down, he allowed himself a set of miniature, leather-bound books, most of them in Italian or Latin, but one in English: a Lilliputian copy of *Gulliver's Travels*.[13]

The prince did not wear his family name or history lightly, which had a powerful effect on those around him. His family tree was a veritable who's who of European nobility: de Nagy-Appony, de La Rochefoucauld, von Benckendorff. Barzini – not one to be easily intimidated but nevertheless a man who could appreciate the importance of a well-placed patron – treated Borghese with deference. He took care never to criticise or contradict the prince in print, submitting to his will without demur and faithfully casting him in the role of hero.

It helped that the prince looked the part. He was physically fit – he had attended the Military Academy in Turin as a youth and would later enjoy mountaineering – and was easily the tallest of the group. His nose was rather large and his hair, which had long since receded, was combed straight back from a looping widow's peak. In the few group photographs taken of drivers and passengers in Peking before their departure, Borghese stands slightly aside, resisting photographers' attempts to bracket the motorists together.

Even for his contemporaries, he was a contradictory, difficult man to understand. Unlike the majority of his family, whose lives centred around Italy and Rome in particular, he was adventurous. In 1900, he had embarked on a journey through the Middle East from the Persian Gulf to the Pacific. He was interested in politics, particularly those on the far left. He had been elected to parliament in 1904 after spending a number of years helping to set up agricultural workers' cooperatives near Lake Garda. His politics, however, did not make him approachable or particularly sympathetic. Italian observers often noted that Borghese behaved and dressed like an Englishman, by which they meant that he was reserved and self-contained and had a fastidious taste for tailored suits – he favoured a creamy linen one with a mandarin collar while in Peking.

On the other hand, he prized loyalty and was most comfortable amidst an intimate circle of family and friends. Ettore Guizzardi, the mechanic-cum-chauffeur Prince Borghese chose to travel with, had been with the family since he was fifteen. Around the turn of the century, a train had jumped the rails at Albano, on the border of one of the prince's properties. The train's driver was killed outright; his son, Ettore, seriously injured. Borghese ensured the boy was cared for at his villa and, when he recovered, took him on as a member of his household.

A few years after the race, Guizzardi would marry in the grounds of the palace on Lake Garda that belonged to the prince's wife,

Anna Maria de Ferrari.* Commemorative photos of the reception, with Guizzardi smiling next to his bride, were carefully pasted into leather-bound family albums beside images of the prince, Anna Maria and their two daughters, Santa and Livia, on camping holidays and visits to the Venetian Lido. In one set of photographs, taken in June 1904, the prince wears a pale suit and white bow tie and mugs for the camera, eyes scrunched up, lower teeth protruding, lips pulled towards his chin like a grouper.[14]

De Dion-Boutons

At the other end of the social scale was Jean Bizac, the De Dion-Bouton mechanic. Nicknamed the Groundhog by his team-mates in recognition of his capacity for daytime naps, he looked slightly sinister, habitually holding his chin tucked down towards his chest, his eyes shadowed by asymmetrically scribbled eyebrows. He dressed, whatever the weather, in the French workers' uniform of rough woollens and blue flannel jacket. He sweated profusely, never complained – although he had no time for those he considered dilettantes – and made it his business to wake the others in his group each morning. While he had spent years making automobiles, his experience of being in one was limited. 'Bizac had never sat in an automobile!' one driver noted with delighted astonishment. 'Taking a seat in one, feeling the wind against his face at high speed, was something he had never experienced.'

The others lay between these two social poles. Georges Cormier, Victor Collignon (the De Dion-Bouton drivers), Charles Godard (team Spyker), Auguste Pons and Octave Foucault (Contal) were

---

* Anna Maria de Ferrari Borghese was a keen photographer and traveller in her own right. Daughter of the Duke Gaetano de Ferrari and Maria Sergeyevna Annenkova, an adopted daughter of Tsar Nicholas I, Anna Maria married Prince Borghese in 1895, bringing with her as a dowry an elaborate Venetian Gothic palace on the tiny Isola del Garda in northern Italy, where the couple spent their summers and largely raised their two daughters. She would later disappear under mysterious circumstances. She set out from the palace on 24 November 1924 to plant some oak trees in the grounds nearby and never returned. The waters around the island were searched by divers, but a body was not recovered. Incidentally, the de Ferrari family were not related to Enzo Ferrari, the racing driver who would found the Italian motoring marque in the late 1930s.

lower middle or skilled working class, and most were involved in the automotive industry, either as sportsmen, mechanics or dealers. Introducing them to his middle-class Parisian readers, du Taillis poked fun at their use of sporting jargon. '"No need to *lay it on thick.*" Which might be said to mean: "Why are you exaggerating?"'[15]

As 10 June approached, differences in class and character were becoming ever more evident. While Godard fancied the race already won and the prince was displaying his habitual sangfroid, it was clear to du Taillis and Barzini that others were less sanguine.

Georges Cormier – who had chosen to travel to Peking by train, because he despised steamer travel – was the most openly daunted by the whole affair. A sleek otter of a man, with dainty ears, a drooping moustache and eyes that sloped up towards the bridge of his nose, Cormier was deeply unimaginative in the way that only dedicated sportsmen can be. At thirty-eight he was, like his benefactor the Marquis de Dion, a leading light in the French motoring world, with a reputation to uphold.

Motorised vehicles were the fixed point around which his world revolved. He was the proud proprietor of Agence Cormier, an automobile dealership in Paris's fashionable eighth arrondissement. Du Taillis – who did not like him much, finding him morose, fussy and a little sly – joked that, after years of successfully competing in motorbike and automobile races, latterly Cormier had 'developed the unfortunate habit of travelling alone'. In 1902, he spent thirty-nine days touring western Europe in a new 6 hp De Dion-Bouton on loan from the Marquis. Cormier's account of this trip for the magazine *La Vie au Grand Air* was for the consumption of purists only. He professes to have found Vienna 'rather disagreeable, involving twelve kilometres of uneven cobblestones', while in Utrecht the 'roads were paved with tiny little bricks laid on the fields'. Geneva was dismissed with the brief note that 'torrential rain forced me to stop [there] for twenty-four hours'.*[16]

---

* He would bring much the same vim and stylistic flair to the book he published. In his foreword to the 1954 edition of Cormier's *Le Raid Pékin–Paris en 1907*, Marcel Reichel generously calls it 'lively' and 'direct' – flattery we can excuse because Reichel was a sports journalist and by this time Cormier was a founding member of the influential Salon de l'Automobile and former president of the first trade union of automotive

Given his extreme preference for driving down the finest French tarmac on his own, the 8,000-mile journey from Peking in the close company of others was giving him pause. So too, from the very instant the automobiles were unloaded from large wooden packing cases on the platform of Chien Men Station, did the logistical challenges that began sprouting like summer weeds.

In the fortnight before the race began, Cormier scurried around Peking procuring all the tools and supplies he could think of. Soon a hillock appeared in the French Legation composed of cans of preserved meat, condensed milk, 110 litres of petrol, 20 litres of oil,* ships' biscuits, camping stoves and mattresses, tents, sleeping bags, hoists and hammocks. Before long, the pile of equipment dwarfed the little three-wheeled Contal Mototri parked beside it.

Cormier also dispatched Victor Collignon, driver of the second De Dion-Bouton, to reconnoitre the route out of Peking. Collignon was a deferential, unassuming man, which suited Cormier perfectly. With dark hair, ample cheeks and ears that stuck out slightly from beneath his favourite flat cap, Collignon looked rather like an overgrown schoolboy. Perennially cheerful – one acquaintance described him as 'a good mood in motion' – he was a veteran sportsman. Before driving automobiles he had raced motorcycles, even coming second in a race during the 1900 Olympics in Paris. Like Cormier, he was an old De Dion-Bouton hand: he had driven an early model in the infamous 1903 Paris–Madrid race. Since then he had become better known as a mechanic than as a driver, which suited his easy-going nature and his knack for tinkering. 'The engine is his passion,' du Taillis wrote. 'He is sworn to the engine. He knows all its secrets.'

His loyalty, once given, was unwavering. Of the Peking–Paris competitors, Cormier was the only person to earn it. Collignon

professionals – but even he admits the author 'dispenses with any rhetorical or literary embellishments'.

* Cars in those days usually operated on a 'total-loss system' of lubrication, which meant they needed an extraordinary amount of oil to run. In 1907 Spyker were proudly advertising in the motoring pages of the *Daily Telegraph* that their cars were 'Dustless, smokeless, odourless, guaranteed using only ½ pint of oil per day!'

would later open his account of the race with a few touching lines about his colleague. 'We were also dear friends. I trusted him. I know him to be the most capable driver and the best companion for the long journey that it was possible to have . . . I was happy to believe that we would remain together all the way from Peking to Paris.' When Cormier told him to inspect the route on horseback as far as the Nankou Pass, Collignon did so without question.

Prince Borghese had recently returned from a similar reconnaissance expedition with his wife and one of her friends. They had carried bamboo rods, cut to the width of the Itala – at 40 hp and more than two tons, it was by far the largest and most powerful of the automobiles taking part – to check whether it could fit along the narrow, rocky tracks that had previously accommodated only mules, donkeys, horses and travellers on foot. It could not. Pickaxes and shovels were added to the list of tools the Italians would need to bring.[17]

Contal

Léonard Louis Auguste Antoine Pons and Octave Foucault, driver and crew respectively of the Mototri, were both small men – essential for riding the minute Contal – and shared the same tough, wiry physique and predilection for flat caps.* Foucault, who had a mop of dark curly hair, liked to wear his pushed right back with a crest of locks escaping over his forehead. Pons, at thirty-two, was succumbing to male-pattern baldness and preferred to wear his cap low, clamped tightly to his head with a pair of driving goggles. Still, he was handsome, with large, wide-set eyes and a moustache that swooped over his top lip.

Of the two, Pons was the real presence. (His mechanic was barely mentioned by other participants. News reports variously gave his first name as either Octave or Oscar, and his surname as Foucaud or Foucault.)† Pons had a sing-song southern French accent and a keen sense of both honour and grievance. He was

---

* To the contemporary eye, they are oddly reminiscent of Nintendo's Mario and Luigi.
† I have used the spelling Pons preferred, since he knew him best.

stubborn, competitive and a true believer in the power of light, nimble vehicles over long distances and rough country. This made him a perfect fit for the Contal Mototri, a unique vehicle that took a bit of getting used to. It was small and three-wheeled, more akin to a long motorcycle than an automobile, with a single, rear driving wheel and two wide-set front wheels used to steer, with a long pair of handlebars curling back and round. The driver sat at the back, over the rear wheel, the engine between his legs. The passenger perched, fully exposed, in an open bucket seat between the front wheels and directly ahead of the driver, rather compromising the latter's view of the road.[18]

After they entered the Peking–Paris, Pons took the Contal up Mont Ventoux, a Provençal mountain north-west of his home town of Draguignan, for practice runs. An extraordinary photograph from this trip shows him and his passenger crouched low over the speeding machine, both leaning into a sharp bend with a solid wall of rock to their right, a spuming trail of road dust in their wake. Notably absent are the spares and equipment they would need and the modifications they made for the race, including the addition of a fold-out bridge – composed of four wooden planks – that doubled as mudguards.

For Pons, the Peking–Paris offered both an opportunity to make his mark and a glamorous escape from humdrum family life with his Italian wife, Maria, and their nine-year-old daughter, Alice. He was determined to make the most of the opportunity. Long before their departure he assiduously courted the press. Several publications used photos of the modified Contal, with Pons at the wheel, showing off its adaptations and proclaiming that it was this automobile – the lightest by far – that would triumph over the following months.

At the starting line, the contrast with the other vehicles could not have been more marked. While all the passengers had forgone luxuries and non-essential supplies in the battle against superfluous weight, in the Contal this had been especially extreme. 'There is no debating the heroism of Pons and Foucaud [sic],' wrote du Taillis. '[They] had sacrificed everything: their personal luggage (only one small suitcase was given to Godard), their sleeping equipment (camp beds had been abandoned), their very food supplies (hard tack and corned beef).'*[19]

Back in Peking, over the first days of June, du Taillis and Barzini were delighted to be reunited. They had met the previous year while attending the Algeciras Conference.[†] They spent time together in the sumptuous lounge of the Grand Hotel or exploring the city, talking shop and gossiping about their fellow travellers. While du Taillis was full of colourful stories about his French compatriots, Barzini did not have much to report. The prince, enjoying the company of his wife and friends, had little time for a jobbing writer, no matter how renowned. When not with du Taillis, Barzini spent the days enjoying the comforts of his hotel and reacquainting himself with Peking, which he had first seen while reporting for the *Corriere* during the 1900 Boxer Rebellion.

As the country's capital and centre of dynastic power, the city had an indisputable allure. 'The approach,' according to one visitor, 'is tremendously impressive. Lying in an arid plain, the great, grey walls, with their magnificent towers, rise dignified and majestic. Over the tops of the walls nothing is to be seen. There are no skyscrapers within; no house is higher than the surrounding defending ramparts.' Inside, the fortifications multiplied. Sarah Pike Conger, wife of an American diplomat, noted in 1898:

Peking is composed of four walled cities ... The Tatar City has massive walls, bastions, heavy gates, and immense gate-towers that can be seen miles away ... In the south wall of the Tatar City are three gates, opening into as many broad thoroughfares. The middle gate, or the Ch'ienmen, is the largest and in every way the finest. It is protected by a walled court with four heavy gates.

Within, Peking was full of scuttling paths and enclosed spaces:

---

* 'Hard tack' is a tough biscuit or cracker, commonly used by explorers and seafarers, that is very long-lasting.

† This was convened to discuss the relationship between France and Morocco, and to try to ease tensions arising from competing French and German interests in North Africa.

cities within cities. The Legation Quarter, from where the race would start on 10 June, was set within the Tatar City. Just to the south, the boundary between the Tatar and Chinese cities was flanked by one of the country's earliest railway lines. Chien Men, Peking's first public station, named after the large gate in the city's wall, had opened in 1901. It was located just outside the Tatar City's fortified southern gate, a stone's throw from the American Legation. Leading off Chien Men and occupying a large area north-west of the Legation Quarter was the Imperial City. Nestled inside that was the Forbidden City: the vast seat of imperial power and home to Empress Dowager Cixi, the woman who had helped govern China for forty-six years.[20]

Within the walls, Peking's streets teemed. 'A few are wide,' an American visitor to the city wrote, 'but the majority are narrow, winding alleys, and all alike are packed and crowded with people and animals and vehicles of all kinds. Walking is a matter of shoving oneself through the throng, dodging under camels' noses, avoiding wheelbarrows, bumping against donkeys, standing aside to let officials' carriages go by.' A peculiar traffic composed of blue-canvas-covered carts, camel trains and water-bearers snarled at busy intersections. Half-wild dogs and birds picked at rubbish thrown into the streets. After dark, some dogs grew bold enough to attack lone travellers: pedestrians were advised to carry sticks.

During the summer, a miasma rose from the heaps of refuse and the city's age-old sewage system to mingle with the thicket of scents peculiar to Chinese cities. (Paul Claudel, a French consul, noted in Shanghai the mingling odours of incense, earth, burning chestnuts and 'a heavy perfume, powerful, stagnant, strong as the beat of a gong' from an opium den.) Shops selling everything from jade snuffboxes to stuffed ducklings and women's hair ornaments were tucked behind ornately carved wooden shopfronts.* A good deal less picturesque was the city's infamous ankle-deep yellow dust: 'dense, impalpable, penetrating, invasive, foul-

---

*These carvings entranced du Taillis. 'The most peculiar divinities parade or fight with many-shaped dragons, horned or clawed, and all in the midst of the strangest flowers and most fantastical birds! Art is very much present in these astonishing compositions, full of imagination and fantasy.'

smelling,' according to du Taillis. '[It] grips you by the throat, gets in your eyes, and in your nose.'[21]

Although well travelled, Barzini shared the prevailing Western view of China in general and this city in particular. 'A proud stronghold of changelessness,' was his estimation. A dynasty mired in its past that was being dragged by more civilised nations – much against its will, but for its own good – into the bracing current of modernity. Many of the ideas and qualities that Europeans like Barzini prided themselves on seemed to find a dark echo in China. To them, Europe was civilised; China was not. They were refined; China uncouth. They were powerful; she weak. And where they were fast and efficient, she was slow and corrupt. 'The very presence of a motor car in that ancient town,' Barzini wrote, 'seemed more absurd than would the sight of a palanquin going over London Bridge. To feel Peking around you is like feeling yourself launched backwards through time to some remote, immutable form of life fixed long ages ago.'[22]

The Peking–Paris participants flattered themselves – and certainly later led their readers to believe – that theirs were the very first cars on Chinese soil. This myth has endured. In fact, the early progress of the motor car in China can be traced in the pages of Shanghai's *North-China Herald*, which took a special interest in the topic. On 8 January 1897, it noted the scarcity of these faddish new contraptions on London's streets. Precisely four years later, an editorial opened with the 'remarkable' fact that 'there is not, as far as we know, a single motor-car in China, though there was at one time a French motor-tricycle on our streets, which suddenly disappeared under a cloud of its own evil-smelling smoke ... [I]t will be strange if before long our streets are not full of motor-cars, which are faster, take up much less room, are cleaner, and cheaper to run than horsed vehicles.'

By June 1902 there were indeed several in Shanghai. The editor of the *North-China Herald* received a letter complaining about 'a motor car ... rushing along the Bubbling Well Road, sounding the horn vigorously', which had caused a rickshaw pony to bolt onto the pavement and collide with a lamp post. By the summer of 1907, an Automobile Club had been established in the city and the shine had rather worn off this supposed harbinger of European superiority. When a motoring reader wrote in to promote the liberal use of the horn to prevent accidents, a curt rejoinder was issued: 'Our correspondent makes the elementary mistake of supposing that roads exist solely for the use of motor-cars and that accidents in crowded thoroughfares can be avoided by horns or sirens instead of by driving at a legitimate speed.'[23]

Such gripes notwithstanding, both the newspaper editors and the Peking–Paris teams would have been put out to discover that they had probably all been beaten to it by Empress Dowager Cixi herself. A smart 1901 American Duryea touring car, coated with imperial yellow lacquer and decorated with a snarling dragon and throne-like seat, was given to her by the consummate courtier General Yuan. Admittedly, this splendid vehicle was rarely used, and only ever within the palace grounds in Peking.[*] The Peking–Paris cars could claim the limited laurels of being the first known to christen the roads of Peking *without* the Forbidden City.[†24]

During the last week of May and the first week of June, a range of topics filled the mouths and ears of the residents of Peking's Legation Quarter: opium prices and proposed anti-opium regulations; railway building; threats of cholera and famine to the south; and, of course, the Peking–Paris race. Grist to the mill on the latter was an unlooked-for impediment: the Chinese government was refusing to issue passports to the participants.

Although the motorists seemed wrong-footed by the diplomatic

---

[*] Anecdotally, this was because imperial etiquette demanded no one sit higher than the empress, meaning that the chauffeur would have had to drive the car kneeling on the base boards. There are stories that, to get around such difficulties, the car had silken ropes attached to it so that it could be pulled along by eunuchs. This is a wonderful story but unlikely to be true. The Duryea, still on display at the Summer Palace in Beijing, was designed with the luxurious rear passenger seat rather higher than the driver's seat. Yet another fable connected with Cixi's motor car is that it was the cause of the country's first motor accident when the court chauffeur ran down one of the palace eunuchs after drinking too much rice wine.

[†] The precise date the Duryea arrived in Peking is unknown. The order the Peking–Paris cars drove in the city was: 1) Itala, 2) Spyker, 3) the De Dion-Boutons, and 4) the Contal.

wrangle over their passports, they should not have been. Allowing foreigners freedom of travel around China's interior had long been resisted, and the route north from Peking was especially sensitive. Sir John Walsham, Britain's ambassador to China in 1887, had been doubtful he would be able to secure Harry de Windt permission to travel this route by horse and camel. The automobile added a new dimension. China was wary of Russia's acquisitiveness, and the route through Mongolia seemed a likely line of attack if it could be made faster and less arduous. Why, the Chinese officials wanted to know, if the motorists were going to Paris, were they taking such an indirect course? (This, incidentally, closely followed a Chinese railway line, then under construction, from which individuals within the government might profit. If a rival automobile route opened, it would be financially ruinous and a great embarrassment.) How many passengers could one car take? Why were diplomats from so many European countries applying pressure to ensure the race went ahead? Why did so many of the participants – Cormier, Collignon, Borghese and Bizac – have military experience?*

For those involved in the race, such questions seemed absurd. Barzini believed the government was, as ever, being 'watchful against the profanations of the west'. Du Taillis, more sympathetic, still scorned their 'puerile scaffolding of imaginary dangers', their 'childish' concerns. Tensions seethed. French, Italian, Dutch and Russian ministers besieged their Chinese counterparts with messages, visits, entreaties and threats. The more pressure they applied, the more suspicious the Chinese ministers became.[25]

On 9 June, the day before the race was due to start, passports had still not been delivered. Cormier, whose feet were already getting rather cold, called a meeting and suggested selling the vehicles and calling the whole thing off; Collignon, following his lead, agreed. Once more the fate of the Peking–Paris race hung in the balance. Again, one participant remained unmoved. 'Gentlemen,'

---

* Both du Taillis and Barzini mention the same specific concerns the Chinese raised with Western diplomats. Unfortunately, I was not able to find sources that illuminated this row from the Chinese point of view.

Prince Borghese said, 'whatever you decide, my resolution is made: no one can change that.' Godard, ignoring his lack of funds, fuel, spare parts and tyres, agreed, perhaps hoping that making himself agreeable to the wealthy prince would pay dividends if he got into financial trouble or needed supplies.

That evening, the passports were delivered. The participants went to bed battling a sense of unreality. It was happening: five motor cars would set off the following day heading west, towards Paris, the self-proclaimed capital of modernity.[26]

3

HART AND HUMILIATION

China's Troubled Foreign Relations

A good traveller leaves no track or trace.

Lao Tzu (*c*.604–*c*.531 BC)

At seventy-two, balding, stooped and troubled by chilblains on his toes, it was all the once keen walker Sir Robert Hart could do to hobble around his garden. When visitors asked him what to do while in Peking, however, his answer was immediate: take a leisurely stroll around the city walls at dusk. From up there, the city would spread out at their feet like a picnic blanket. The watercolourist Thomas Hodgson Liddell described it as a woodland city. 'Seen from this height it has the effect of great masses of green, with the roofs of houses peeking up.' Most were grey, but here and there other colours gleamed: green, indicating that the buildings belonged to princes, and yellow, reserved for royal palaces, catching the last of the sun's rays.[1]

In 1907, night was drawing in on both Hart's career and that of the Qing dynasty he served. The following year he returned to the United Kingdom for a brief retirement. He died on 20 September 1911, twenty days before the dynasty collapsed after 267 years in power, ending over two millennia of imperial rule in China.

Hart had arrived in China in 1854 to work for the British Consular Service and quickly distinguished himself. By 1863 he was Inspector General of the Chinese Maritime Customs Service (CMCS), a mammoth institution organised by Western nations

41

to collect tariffs on foreign imports for the Chinese government.* Under his leadership, the CMCS would grow to employ 3,500 people, of whom 700 were Westerners, and collect 27 million taels annually. This would become the principal financial support of the Qing regime. It also amassed a vast portfolio in addition to tax collection: managing government ports, lighting waterways, organising a postal service and imperial navy and advising the dynasty itself. But his career, glittering as it was personally, spanned a period of ignominy for his host country. The years 1839–1949 would come to be known as China's 'Century of Humiliation' and have a profound impact on its attitudes to foreign intervention and its own integrity.[2]

A portion of the issues facing the country during the late nineteenth and early twentieth centuries were structural. The Qing rulers were Manchu, a minority ethnic group from Manchuria in north-eastern China, with customs distinct from the majority of their subjects, who were Han.† True, their rule had lasted over two and a half centuries, but cultural differences were still cited by those unhappy with the regime. This latter group had ballooned as the population surged from around 300 million in 1790 to 400 million by 1850, straining the country's economy and leading to shortages of land and civil unrest.

The Taiping Rebellion, which lasted from 1850 until 1864, claimed between 20 million and 70 million lives – making it the bloodiest civil war in history. It rattled Chinese society to its foundations. Hong Xiuquan, who led the rebellion, came from a small village near Canton (now Guangzhou), in southern China.‡ He

---

* Under Hart, the CMCS was theoretically subordinate and loyal to the Chinese government. The reality, given the heavy foreign influence, was rather more complicated; both Western and Chinese scholars have since argued that it served as a tool of imperialism.

† Traditionally Manchus spoke their own language and had a distinct form of education, and the women did not generally practise foot binding as the Han did. Some Manchu customs had, however, been forced on the entire population, causing long-standing resentment. For example, since the seventeenth century Han men had been forced by law to adopt the Manchu-style 'queue' hairstyle, which involved regularly shaving the top and front of the head and braiding the remaining hair into a long plait.

‡ After failing the exams needed to enter the imperial civil service multiple times, Hong Xiuquan had a breakdown. He had visions of himself ascending into heaven, meeting an old man with a golden beard, and came to believe that he was the younger brother of Jesus Christ, tasked with ridding China of demons and the Manchus.

was a charismatic preacher, and cultivated a large, loyal following among the poor, who were numerous and primed against the Qing by a series of natural disasters, which led to famine, and resentment against the emperor after he raised taxes. Hong and his adherents began conquering territory, over which Hong proclaimed himself Heavenly King; victories were followed by brutal counter-attacks.[3]

The country was facing external difficulties too. China produced luxurious goods such as silk, tea and fine porcelains, which were in great demand internationally; trading access was prized by foreign powers. Import duties on tea alone brought Britain's Exchequer £3 million annually, roughly half the annual expenses of the entire Royal Navy in 1800. China, however, was a wary trading partner. Approached by an envoy from King George III in 1793 desiring greater access, the emperor coolly replied that his 'Celestial Dynasty possesses all things in prolific abundance and lacks no product within its own borders.' This was not quite true. China also profited greatly from trade, but rival powers had historically proved troublesome, unwilling or unable to see events from her point of view.

Most infamously, British merchants had been smuggling Indian-grown opium into China for years. The resultant economic and social problems, including a rise in addiction, led to a prohibition by the emperor in 1800. Britain went to war – the Opium Wars of 1839–42 and 1856–60 – to defend the trade and her merchants' profits. Both conflicts resulted in resounding defeats for China. Britain demanded indemnities (around £5 million in 1842), the right to more trading ports in which Western law would be applied, the right of missionaries to travel and proselytise, and other legal and territorial concessions.[*] They also ramped up opium imports. In 1860, almost 48,000 chests were being shipped from Bombay and Calcutta. A few decades later the explorer Isabella Bird reported that, in some provinces, up to 80 per cent of men and 50 per cent of women smoked opium. In Szechuan, where it was grown, the opium shops were 'as thick as the gin shops in the lower parts of London'.

Opium had a long history of being used medicinally in China and some elites took moderate amounts of opium socially. Although many

---

[*] The 1842 treaty also ceded Hong Kong to the British.

of the more lurid stories about hordes of emaciated opium fiends subsisting in squalid dens were overblown, it did become a serious issue. The flood of cheaper imports during the nineteenth century, coupled with domestic woes including natural disasters, farming crises and rebellions, created a wider, persistent and more troublesome group of opium users. These were often shunned by their families and peers – especially after opium growing, selling and smoking were made illegal. They were also more at risk, since they were poorer, of being unable to afford their opium habit, leading to serious withdrawal symptoms, the inability to retain a job or a turn to crime.

Over the latter half of the nineteenth century, Western nations and Japan continued to apply military pressure and their demands multiplied. By 1900, China had effectively lost control of her most lucrative seaports and many regions were under the partial control of foreign powers. Two recurring points of friction were the

Henri Meyer, who illustrated the works of Jules Verne, created this political cartoon depicting foreign powers hungrily slicing China up like a cake.

building of railway and telegraph lines. 'They deface our land-scape, invade our fields and villages, spoil our feng-shui, and ruin the livelihood of our people,' wrote one anguished minister. Once again, China was forced to capitulate. By 1889 two lines of rail-way and a hated telegraph system were up and running. Reforms that might have helped ordinary people, such as the building of new schools and colleges, were proposed and then abandoned. Humiliation heaped upon humiliation.[4]

Hart himself may have been personally popular in Qing circles – some fondly called him 'our Hart' – but the CMCS was inextricably linked with hostile foreign machinations in China. Mutual mistrust, resentment and even loathing simmered. To the emperor at the time of the first Opium War, international relations became 'unspeakable bullying'. 'So much anger and hate bottle[d] up inside me,' he wrote. Empress Dowager Cixi, who came to power in 1861, harboured her own anti-foreign sentiments, along with many of her subjects. She, in turn, was disliked and doubt-ed by foreign powers, who believed her rule was illegitimate and accused her of stoking up hatred against them and obstructing reforms.* By the turn of the century, China was a country of kin-dling waiting for a spark.[5]

For residents of the Shandong province in northern China, the years 1898 and 1899 were memorable for all the wrong reasons. After years of neglect, the dykes around the Yellow River failed. A flood fol-lowed by severe drought decimated harvests, leaving many homeless and nearly the entire population on the brink of famine. 'For the first time since the great famine in 1878,' wrote one missionary, 'no winter wheat to speak of had been planted . . . The ground was baked so hard that no crops could be put in.' Food prices doubled. Gangs of armed young men and women – known as 'Boxers' in English because many

---

* Westerners found it peculiarly difficult to understand the loathing in which they were often held in China. 'This hatred of the foreigner is a very curious characteristic of the country,' Isabella Bird noted in 1900. 'No one can tell how it has arisen, for though one can understand that the attempt of Western nations to force open the ports of the country, and the seizure of territory by certain of them, and perhaps the advent of the missionaries, are causes enough to provoke opposition and hatred, they do not account for its ferocity.'

practised martial arts – began attacking churches and missionaries, clashing with conservative government troops and marching on ports and cities. Banners calling for the killing of all foreigners were hung in the city streets. The Legation Quarter was under siege for fifty-five days during the summer of 1900 while other detested symbols of foreign meddling – especially Western infrastructure – were targeted. 'Little by little connection with the world beyond Peking has been cut off,' the American Sarah Pike Conger wrote to her sister in early June while holed up in the city. 'Both telegraph lines are gone; the railroad is gone. No Legation mail pouch; there is but little mail, and that is brought by couriers.'[6]

The Boxer Protocol, signed after the rebellion had been quelled by an alliance of foreign armies, was rapacious.* A total of 450 million silver taels (four times the Chinese government's annual revenue) was demanded. Hart was appalled and, like many in China, thought it unfair this did not take into account the substantial plunder seized by allied armies while looting Peking and Tientsin after the uprising. He sought to mitigate the burden by offsetting it against increases in taxation on imports and exports in China's favour. Nevertheless, he wrote in 1902: 'The future looks very dark indeed and I fear nothing but bad will result.'[7]

From the Qing dynasty's point of view, Hart was correct. The Boxer Rebellion and concessions proved the final straw for China's last imperial dynasty, weakened economically and politically by decades of bruising interactions with foreign powers. It would be swept from power by a series of revolts in 1911 and 1912, leaving a vacuum filled first by the Kuomintang nationalists and then, in 1949, by the Communists under Mao Zedong.

Mao was a supreme product of this era. Born in Hunan province in 1893 to a relatively prosperous peasant family, his formative years were set against a backdrop of Chinese weakness. The imperial regime was tottering on feet of clay, plagued by civil unrest, famines, a series of natural disasters, salted by military attacks and financial demands by foreign nations. He left home

---

* This coalition was called the Eight-Nation Alliance and included troops from Germany, Britain, Japan, France, Russia, Italy, America and Austria-Hungary.

young, rebelling against his strict father, and moved first to the provincial capital and then to Beijing.

By his late teens, the Qing dynasty had toppled, and his country was in turmoil. In 1919, at the Paris Peace Conference, the Allies decided to award former German concessions in China to Japan, rather than return them to China. This provided the spark that launched the student protests known as the May Fourth Movement, in which Mao played a leading role and which would result in the formation of the Chinese Communist Party (CCP). In a revolutionary magazine at the time, Mao wrote:

> As a result of the world war and the bitterness of their lives, the popular masses in many countries have suddenly undertaken all sorts of action. In Russia, they have overthrown the aristocrats and driven out the rich . . . The army of the red flag swarms over the East and the West, sweeping away numerous enemies . . . The whole world has been shaken by it . . . The world is ours, the nation is ours, society is ours. If we do not speak, who will speak? If we do not act, who will act? If we do not rise up and fight, who will rise up and fight?

Three decades later, on 1 October 1949, Mao would stand on top of Tiananmen Gate – just a few hundred yards from the Legation Quarter where the Peking–Paris began – overlooking the square. In his speech, he announced the birth of the People's Republic of China and the beginning of Communist rule.[8]

Over the next quarter-century, until Mao's death in 1976, China was transformed through continual upheaval. Groups who had previously held little power – first the peasantry, then the young – were held up as examples for the rest of society. During Mao's Great Leap Forward initiative (1958–61), land was collectivised and speedy industrialisation pushed, resulting in a famine that killed tens of millions of people.* The Cultural Revolution

---

* Estimates of the death toll of the Great Leap Forward vary wildly, from 3 or 4 million to 45 million; a plurality of historians and researchers argue a figure in the tens of millions.

(1966–76), a call by Mao to destroy the 'four olds' of Chinese society – customs, habits, culture and thinking – rent the country's social fabric still further. Radical young Red Guards, encouraged by Mao's rhetoric, destroyed schools and temples, and denounced, attacked, tortured and even killed teachers, parents and officials.

Today's China is unrecognisable. Since the death of Mao in 1976 it has turned its face to the world and experienced an economic miracle. In 1981, 88 per cent of its population were living in extreme poverty; now the figure is less than 1 per cent. In 2010, it overtook Japan as the world's second-largest economy. More billionaires live in Beijing than New York. Beyond its borders, China flexes increasing political clout: it became a member of the World Trade Organization in 2001, and leverages trade deals and development partnerships all over the world.

Despite this, the memory of China's Century of Humiliation remains potent. It helps explain the country's continued diligent pursuit of influence and power abroad. President Xi Jinping (much like Russia's President Putin) has made little secret of the fact that he would like to see the world carved up into large spheres of influence, with China freely able to exercise power beyond her current borders.

Although it was Russia who formed the Eurasian Economic Union (EAEU) in 2015, to strengthen the relationship between itself and post-Soviet states including Belarus, Kazakhstan and Kyrgyzstan, it is China that has worked to reap the rewards. Central Asian countries receive staggering largesse from Beijing. In Tajikistan, for example, which became independent in 1991 after the collapse of the Soviet Union, China has paid for new schools, a power plant, traffic cameras and police cars and lent the Tajik government $1.3 billion. As one official put it: 'China is doing what the Soviet Union used to do.'[9]

Public framing of these manoeuvres shares intriguing similarities with the grandiose and optimistic sentiments evoked by the Peking–Paris motorists in 1907. In 2013 President Xi launched what he called the 'project of the century'. China's Belt and Road Initiative is a sprawling set of infrastructure investment and regional development drives the aim of which is improving and strengthening connections between China and 160 other countries. While spending has been diffuse – recipients include Myanmar, Pakistan and Greece – the overland route from Beijing, through Mongolia and into Russia has been designated as one of six key 'economic corridors', with plans for transport infrastructure, ports, industrial investment and 'people-to-people exchanges'.

What is clear is that, just as in 1907, progress, travel, trade and the fostering of fraternal links between countries remain powerful rhetorical motivators. This time, however, China is determined that she does not need the help of outsiders. There will be no Hart pulling the strings; the Party will be in the driver's seat.[10]

4

Peking – Kalgan

10–14 June

The lighter you are, the greater the chance of getting through.

Jean du Taillis, Spyker team

Seen from above, the French Legation parade ground – starting line of the Peking–Paris – was large, perhaps the size of two football pitches. It was fenced around by formal, three-storey buildings encircled in turn by a high stone wall with few apertures. Men in uniform grouped into irregular fields of colour: black, blue, khaki, white, red, green. Hats were all but obligatory. Pith helmets, kepis and peaked caps bobbed and weaved. Dotted among them were a few conical bamboo hats known as *dǒulì*, civilians' jaunty boaters and wide-brimmed ladies' confections overlaid with thick dust veils.

After Madame Boissonnes dropped the flag, the vehicles passed out of the French Legation barracks through gates bristling with bunches of tricolour flags. The Spyker moved off first, followed by the Itala, the two De Dion-Boutons and the Contal. The French and Italian vehicles had acquired little national flags, which flapped jaunty semaphores as they drove. The pace was slow. A nattily dressed man, resplendent in white cotton trousers, boater, bow tie and cane, was able to jog up the line of vehicles, past Collignon's De Dion-Bouton and up towards the Spyker. The 18th Colonial Corps, marching ahead of the vehicles, played the oddly frenetic 'Sambre et Meuse'. Mist softened the edges of the city. The cavalcade turned right onto Legation Street and then right again, heading north and skirting the eastern wall of the Imperial Palace. After a few hundred yards, the band and observers peeled away. As kepis were thrown into the air with a hoarse volley of farewells and cries of '*Vive la France!*', the heavily laden automobiles were swallowed by mist and mizzle.[1]

Straw-hatted Peking policemen held back pedestrians and pointed out directions to the drivers with their batons. Earlier, city workers armed with wicker watering cans had moved through the streets, laying the dust.* The rain rendered their work unnecessary: already the yellow dust was thickening into mire, mixing with the seep of open cesspits. Five sets of new tyres left deep, skidding imprints in the malodorous ooze.

The territory between Peking and Kalgan (now Zhangjiakou) had long given the race planners pause. A manager from the Urga (now Ulaanbaatar) branch of the Russo-Chinese Bank had

The proposed route, which included few roads for the first 6,000 miles, was enough to daunt even the most daring explorers.

---

* Lightly wetting the road so that wheeled vehicles would not kick up clouds of dust as they drove.

written to *Le Matin* to state – incorrectly – that there was not 'a single river' between Urga and Peking. Drivers were told that they would need to carry their own water and set up depots, which they had done; a convoy of fourteen mules had been dispatched from Peking to Kalgan. All the vehicles were expected to fend for themselves and each other until then.

The route through the vertiginous mountain passes would be treacherous. It was littered with boulders and there were no roads, just pack animal tracks. Collignon confided to du Taillis that the way over the Nankou Pass was 'made up of steps, each as high as two ordinary steps and around three metres deep'. Hardly an ideal surface on which to drive.[2]

The cars differed enormously in size and power. Received wisdom had it that the smallest, lightest vehicles would be best suited to the tough terrain. Although they were less powerful and slower on good roads, they would have several advantages on the bad ones: they would sink less in sand or marshes, and they would be easier to lift. Auguste Pons, driver of the Contal Mototri, was predictably the most committed disciple of the small automobile. Yet Pons's Contal had only 6 hp to rely on, while the De Dion-Boutons had 10, the Spyker 15 and the Itala 35/45.

Despite the disparities in their relative power, all were heavily laden with water, fuel and luggage* and carried at least two passengers. The Itala, for the initial stage, had five. The Prince, who drove the Itala out of the city, was joined by his wife, Anna Maria, and his brother, Don Livio, attaché at the Italian Legation.† The additional passengers scrambled onto the humped back of the Itala beside Ettore Guizzardi and Barzini, like children mounting an elephant.[3]

At the moment of departure, rivalries were subsumed beneath the veneer of shared purpose and identity, and the giddiness of having begun. They called out to each other, joking that the

---

* Barzini reported their starting weights as: Contal 700 kg; De Dion-Boutons 1,400 kg; Spyker 1,400 kg; Itala 2,000 kg. But when power was factored in, the Itala regained the advantage: it weighed 50 kg per 1 hp while the Spyker, De Dion-Boutons and Contal weighed 98 kg, 140 kg and 116 kg per 1 hp respectively.

† These extra passengers disembarked after the first day and returned to Peking.

last-minute arrival of their passports was an omen. They were invincible. They were harbingers of change. As cars, drivers and passengers swayed through Peking's streets, du Taillis fancied that 'ancient dust fell from the curved roofs'. Perhaps, he thought, this was 'a little bit of the decay of the celestial dynasties that was falling, while the new China shook off its torpor and awoke at the sound of the progress that was passing by: our cars on the move'. His first race report was triumphal: 'Our departure itself attests to the truth of the axiom that nothing is impossible – not for a Frenchman or for *Le Matin*.'[4]

There are many points on which the first-hand accounts of the Peking-Paris rally disagree. One point of perfect harmony, however, is that the scale and scope of the challenge on which they had embarked forcefully struck the motorists the moment they left Peking. Latent tensions between the motorists burst into open irritation; plans and promises that had seemed firm collapsed; and it became obvious that China's terrain was as ill-suited to automobiles as the race's naysayers had warned from the start. 'The perfect swimmer,' du Taillis later philosophised, 'at the edge of the sea or a lake, knows that the best way not to shiver in the water is to jump into it with his head down and in a single bound. And it was the same for us: to forget the ease of a smooth ride on tarmac, we put our heads down and got straight into the potholes.'[5]

Kalgan, gateway to the Gobi Desert, was around 180 miles from Peking as the mule trudged. Reaching it meant heading north through a densely populated plain towards a spur of the Great Wall. Rivers would have to be crossed or forded, and rocky, sandy and marshy terrain navigated. As the drivers pressed on, the earth would boil up around them, first into undulating hills and then into mountains. Just as the way became impossibly steep and rocky, they would be met by the teams of men and pack animals they had engaged at Peking. Together, they would have to tackle three successive mountain passes. The most southerly, the six-mile Nankou gorge, cut through the landscape like a rocky *V*, offering views of the crenellated hilltops beyond. (It was a favourite haunt of sightseers and photographers, who could reach it in a day's ride from Peking.) Automobiles, drivers, passengers and their local porters would spend their first night in the small town of Nankou. In all, it was thought it would take them a week to traverse the 110 miles from Peking to the northern edge of the mountain range. On horseback, the journey typically took four days.[6]

Just three or four hundred yards outside Peking's northern gate, the vehicles halted in a choking yellow cloud that swirled and eddied around them before subsiding in the worsening rain. As it cleared, the motorists looked around in chagrin. Of the five vehicles that had set off from the French Legation less than an hour before, two – the De Dion-Bouton piloted by Collignon and Auguste Pons's Contal – were missing.

Prince Borghese raised his driving goggles, revealing a cartoonish oval of clean skin. His brows were drawn together, mouth tight. He turned in his seat to take in the grimy faces of his wife and brother, then levelled his gaze at the two other cars. 'I am resolved to drive on to Nankou,' he called out to the Spyker and Cormier's De Dion-Bouton in lightly accented French. Not waiting for comment, he put the Itala back into gear and moved off. Barzini – half embarrassed by the prince's peremptory manner, half exasperated that Collignon and Pons had managed to get lost in a city they'd had weeks to familiarise themselves with – aimed an apologetic salute in du Taillis's direction from his perch on the Itala's running board. As the car jounced, he swiftly resumed his grasp and turned his face in the direction of travel.

Du Taillis watched as the Italians turned the first corner and disappeared from view, leaving only deep tyre tracks in their wake. Five minutes passed, then ten. Drops of water rolled down the canvas sheets tied over the luggage of the stationary vehicles, off hat brims and reddened nose tips. Godard had dressed in summer attire to match the jolly stripes of the Spyker: both now looked as if they'd gotten lost on an ill-fated jaunt to the seaside. The supplies in the back, in contrast to those in the De Dion-Boutons, were jumbled in haphazard disarray.

As neither Godard nor Cormier were keen on the faintly humiliating job of returning to the city they had just left to find out what had happened to the others, they decided to draw straws. When

Godard lost, du Taillis dismounted and wandered over to a roadside stand to buy a cup of tea while the Spyker headed back through the gate into the city. Sipping morosely, he listened to Cormier vent his frustrations to Bizac: Collignon's foolishness, Godard's hogging of the spotlight, the prince's lack of team spirit. The litany of complaints worsened du Taillis's mood. That morning had already seen one humiliating spectacle. Godard had argued with their hotelier over the bill; once more, the Dutch minister had interceded. Du Taillis felt heavy, deflated. 'Only just out of Peking, whose walls could still be seen, and we had already lost each other.'[7]

It was clear to all that the collective spirit and enthusiasm had shattered as soon as they encountered the realities of the road. Later, Pons, Cormier and du Taillis confirmed that the plan had been for the group to stick together through the initial stages of the race, at least as far as the Gobi Desert; Collignon never acknowledged this pact. Barzini skated over the issue entirely in his book, saying that the prince was invited to go on ahead, but his contemporaneous articles imply otherwise: 'To-day the participators . . . met together and drew up the general regulations. All the motors will proceed together as far as Irkutsk,' he wrote in a piece published on the day of their departure. He also mentioned that provisions were agreed upon in case either a car or a person were incapacitated, although what exactly had been agreed was not specified.

In time, all the participants would fail to adhere to any such standards, but Borghese was the first to break them. He had spent months preparing for the journey. He believed wholeheartedly that it was possible and was determined to complete it. He had little patience with what he clearly took to be the ineptitude of some of the others in the party.

Le Matin, for its part, had muddled the rules during the rushed planning phase. At one meeting it was suggested that participants should stick together until Berlin; something clearly never agreed upon by the drivers. An editorial on 17 February simply stated that 'There are no formalities to complete, no rules to trouble yourself with. You must simply set off from Peking in an automobile and arrive in Paris. That is the challenge.'[8]

After three hours – during which Godard corralled the errant vehicles – and a rather strained lunch stop, the French motorists were back on the move, largely travelling in silence. The pace was glacial: slower than walking. The problem was the three-wheeled Contal, which skidded, bucked and sank. Its rear wheel was forced onto higher ground between the double line that cartwheels had cut into the earth. A few miles passed, with the De Dions and Spyker having to wait at each bend in the road for Pons and Foucault to push, carry and cajole the Mototri, every yard hard-fought.

After an hour, enough was enough. Cormier, Collignon and Godard rounded on Pons: continuing together was intolerable; he would have to abandon the race. Pons was first shocked, then furious. He flatly refused to give up, but did eventually agree to return to Peking. From there he, Foucault and their Contal would take the train to Nankou. This would technically disqualify them, meaning that the Peking–Paris had claimed its first victim after less than a full day's driving. Pons, however, remained optimistic about the Mototri's prospects beyond the first pass. Fuelled by bravery, stubbornness and pride, he declared he would drive, if not from Peking, then from *Nankou* to Paris.[9]

Had those in the remaining four vehicles been in the mood to appreciate it, the landscape through which they drove might have enchanted them. 'I have seldom seen lovelier scenery,' wrote Harry de Windt, a British explorer who had travelled that way two decades previously. 'The dark wooded mountains in the distance standing out in striking contrast to the green plains of maize and barley, the clear sparkling streams, spanned by picturesque bridges glittering with enamel and porcelain.' The automobilists, however, could only see tortuous obstacles to be overcome in the name of progress.[10]

To the north, Barzini bounded and swayed aboard the Itala. 'The roughness of the road and the undulations of the ground made it lurch and skid. Over the sand it had the elastic bounding of a feline animal.' The Itala's bodywork had been stripped back for the first leg of the journey and it was fantastically uncomfortable. There was no longer a floor: it had been removed to reduce the vehicle's weight. Beneath the passengers' feet, the transmission shaft whirred and the dusty ground spooled away. The driver and one passenger could enjoy the luxury of sitting on an upturned wooden packing case, but everyone else had to hold on as best they could to the ropes and straps securing the luggage. Guizzardi made space on one of the two fuel tanks, sitting with knees to ears in the middle of a spare tyre. Barzini thought he resembled 'a wrecked mariner suspended on a lifebelt'.[11]

The motorists crossed two picturesque marble bridges, arched so high over the water that their reflections formed perfect circles. Cormier confidently dated the first one, at Tsing-ho, to 'the days before Christ'. To Barzini's eyes, it had an elegance that was 'almost European'. Perhaps this was why he airily informed his readers that its construction had been attributed to Marco Polo. Du Taillis, yet to recover his humour, was less sentimental. 'The bridges were like everything else in China: superb debris, magnificent witnesses of ancient splendours . . . a tangle of huge slabs, disjointed, collapsed here, overlapping there, in a gigantic pile.'

The Frenchmen deployed their pulleys, hoists and ropes, hammering iron stakes between marble slabs and heaving their cars over first one bridge and then the other. The Italians, after trying and failing to use the wooden mudguards of their vehicle as a ramp, attempted a full-frontal dash. The wheels spun uselessly, 'shooting off sparkling bits of stone with the nails of the tyre-covers', while the engine emitted 'panting clouds of dense, white, acrid gas'. At last, the tyres bit and the car mounted the stonework, 'slowly, awkwardly . . . like an enormous tortoise, whose low shell almost grazes the ground, and whose four paws are spread out, strong and wary.'[12]

If the plain had proved challenging, it was as nothing compared to the passes. The steep-sided Nankou gorge, which the Italians reached on the first day, had been cut through layers of granite by the force of spring snowmelts.* In summer, people and animals

---

* Only the Itala – and the Contal, which arrived by train – made it to the night's first agreed stop, in Nankou village itself. The French lost the light and camped in a spot they realised the next morning was just a few hundred yards in front of the entrance to the gorge.

picked their way along the dried watercourse; squalls could lead to flash floods. Du Taillis thought it 'grandiose in its harshness, made even more wild by the expiring remains of ancient fortifications: the rubble of the towers and ruined walls was piled up over the chaos of the blocks torn from the sides of the mountain.' Yet, for all their picturesque allure, there was a reason the Chinese considered these passes vital strategic defences.* For Harry de Windt, who had travelled on horseback, the going had been so rough his party had dismounted and led their ponies, while 'the mule-litters plunged and rolled about among boulders like ships in a storm'.†

Over the following days, the automobiles were not driven but dragged, pushed and even carried. The French party had hired 150 men and several mules. The Italians, who at the first night's stop removed yet more of the Itala's bodywork‡ – as well as tools, spares and luggage – to be carried on mule carts, had been faithfully promised twenty-five men and four mules by the Peking Transport Company, but received instead the men, only one mule, a horse and a small white donkey.

At one point the path was 'set between walls of rock so near to one another that we could touch both by spreading out our hands'. The Italians had no choice but to widen it with pickaxes and shovels, inch by ringing inch, to ease the car through. Even so, the rear wheels 'were sometimes forced together so strongly that they inclined inwards in the shape of a V. The chassis creaked and groaned under the strain. Following the course of another river,

---

* Remnants of the Nankou gorge's defensive use and importance as a trading route were both hewn into the landscape itself. Near Nankou, a gate 'adorned with sculptures of figures borrowed from Hindu mythology' towered over the pass. Farther on were the remains of several abandoned defensive encampments.

† De Windt kindly took the trouble on 14 June 1907 to write to the *Daily Telegraph*, with the air of a man glad his own ordeal is over, to say that, as difficult as things currently were for the automobilists in China, they would really 'have a bad time of it in Siberia'. He reported having been delayed by floods for days and having been pelted with stones and rotten eggs.

‡ This greatly irritated Cormier, who wrote that Borghese rendered his car 'merely something to sit in', while the other vehicles were 'fully loaded'. He neglected to mention that these fully loaded vehicles required ten more men each and several more mules to assist them.

du Taillis watched the Spyker cleaving through brown water, sending up waves 'as if at the prow of a ship'.

A rainstorm turned the ground into dark, greasy mud deep enough to reach halfway up the wheels. Passing by a 'gigantic willow . . . joyously decked with green' in a village, the wheels of the Itala became lodged in the roots. They put down their iron levers and took up their hatchets. 'The severed roots were tied up, pulled apart, torn out, broken and twisted until the wheels were completely free.'[13]

The men the motorists had hired chattered and called out a distinctive chant as they worked. *Laè, laè, la! Laè, laè, la!* On the first day, the toot of the Itala's horn had added to the cacophony; by the second, it had grown hoarse and lapsed finally into silence. Where the road narrowed or large boulders blocked the way, they would all have to 'become road workers, breaking the rocks using our pickaxes and lifting them with a large wrench'.

The going was slow. Overhead, the sun crept by, its path abbreviated by the high walls of the passes. The De Dions, Contal and Spyker travelled just nine miles over nine hours on 11 June; the next day, it took Borghese six and a half hours to cover thirty-seven miles. The contrast between the seamless, smooth and speedy ideal of the automobile and the tedious, grinding reality could scarcely have been more marked. Perhaps, at such moments, the Europeans had occasion to remember their emotions on leaving the capital – the triumph, their confidence in their automobiles' speed and power. They were kept too busy to dwell on it. 'We are not,' du Taillis admitted, 'strictly speaking, motorists.'[14]

For a few hours at a time, the ground would level out and become sandy, giving way under the weight of the vehicles. At such moments, 'Nothing was heard but the shuffling and trampling of feet, the panting of the men, the creaking of the sand under the heavy wheels of the car.' They would begin to long for the hectic, breath-quickening immediacy of the vertiginous sections and the hard, rocky surfaces. Soon enough, they would be back on a path coiled narrowly through a defile a hundred feet high and nearly a dozen miles long. The tools would come out again and the ropes would tighten. *Laè, laè, la! Laè, laè, la!*

Teams formed around each vehicle as drivers, passengers, a handful of soldiers from the French and Italian Legations and labourers spent hours each day engaged in a battle against gravity and the surrounding terrain. Barzini quickly came to admire Guizzardi's quiet competency and mechanical knack. Enveloped in the folds of a large mackintosh cloak to keep off the rain, the chauffeur would stand inside the Itala, grasping the steering wheel with one hand and gesticulating to the porters with the other. When not chatting to the soldiers or helping heave the Itala over or around particularly tricky corners, Barzini and the prince took it in turns to ride a pair of donkeys so tiny that their 'feet touched the ground on either side . . . [and] our raincloaks hid them with the grand lines of classical drapery'.

Du Taillis, travelling with the other Frenchmen a little behind the Itala, made copious notes. He admired 'the dark line of the Wall that wound along the tops of the mountains', peered into the kitchens of the inns they ate at, and praised the dedication of the soldiers, assigned from the 18th Colonial regiment. He scribbled long dispatches for *Le Matin*, entrusting them to the soldiers. He largely avoided Godard, who was increasingly volatile due to the slowness of their pace.

On 12 June, the French automobilists woke to find their porters had overindulged in opium. 'They looked like they had come from beyond the grave and seemed unable to make any effort at all,' du Taillis wrote. Godard flew into a rage. 'It was all very well me scolding [him], recommending that he display the impassive calm that is the only thing the people of the yellow race take as a sign of a high-class person. Godard fulminated. Godard rebuked. Godard dared to snatch a hashish pipe from the mouth of a coolie.'[15]

On 14 June, the day the motorists would have arrived in Kalgan had they been riding horses, only the Italians had reached their destination. It would take the Frenchmen a further two long, slow days, each silently champing at the bit. Godard preoccupied by the necessity of obtaining supplies and spares without payment. All of them trying not to 'think too much – already – about Borghese's lead.'[16]

5

AGE OF INVENTION

A New Era of Communication

Wireless is all very well, but I would rather send a message by a boy on a pony.

Lord Kelvin (1824–1907)

Residents of Bath Place, an unassuming row of cottages in Slough, England, heard pitiable screams issuing from one of the houses between six and seven o'clock on the evening of 1 January 1845. Rushing in, past a shaken-looking Quaker hurrying towards the station, they were stopped short. Their neighbour, Sarah Hart, was writhing on the floor. She was moaning loudly, a bloody foam crusting her lips, 'her clothes nearly up to her knees, and the stocking on her left leg nearly down and torn'. On the table nearby was a bottle of beer and a half-empty tumbler. Sarah died before a doctor could be summoned.

Surmising what had happened, someone rushed after the Quaker, just in time to spot him climbing onto the 7.42 p.m. train to London. That might have been that, had it not been for the new telegraph system installed at the station. It was rudimentary – an early William Cooke and Charles Wheatstone model – and not capable of transmitting lower-case letters, punctuation or the letters *J*, *Q* and *Z*. The hurried message sent by rail officials from Slough to Paddington read:

A MURDER HAS GUST BEEN COMMITTED AT SALT HILL AND THE SUSPECTED MURDERER WAS SEEN TO TAKE A FIRST CLASS TICKET TO

63

LONDON BY THE TRAIN WHICH LEFT SLOUGH AT 742 PM HE IS IN
THE GARB OF A KWAKER WITH A GREAT COAT ON WHICH REACHES
NEARLY DOWN TO HIS FEET HE IS IN THE LAST COMPARTMENT OF
THE SECOND CLASS COMPARTMENT.

Despite some difficulty deciphering 'kwaker', police were
called and were waiting for the suspect when he disembarked at
Paddington. His name was John Tawell, a convicted forger, Hart's
lover and the father of her two children, in debt to her for child
maintenance. He had poured a bottle of Scheele's Prussic Acid, a
treatment for varicose veins, into her beer.

*New forms of communication like the telegraph allowed information to
travel further and faster than ever before.*

The telegraph, like many inventions, had multiple ancestors
and had been gestating for several decades before finally being
'invented' in the mid-1830s. One problem that had dogged its
introduction was that contemporaries struggled to grasp precisely
why or when such speedy communication would be needed. Postal
services were reasonably efficient and in Great Britain they had
recently been reformed, making them both simpler and cheaper.

The Salt Hill murder, however, proved a perfect cause célèbre. Here was a clear-cut instance in which sending information faster than a letter borne by train, horse or foot was essential. Over the following decades, investment flooded in and the telegraph network spread over the globe like the filaments of a fungus through leaf litter. A transatlantic cable was laid in 1858, and then again in 1866 after the first broke. By 1875, telegraph cables linked Britain, America, India, Europe, the Far East and Australia. To Westerners, the speed at which messages could be sent across the world, and the idea of those messages speeding through wires under oceans and across continents, became, along with the electric light and the railway, a touchstone of progress and a crucial facet of their identity.[1]

The years around 1900 were marked by cultural anxieties about the qualities that Europeans and Americans believed defined them. An 1892 short story, 'Number Twenty', imagined the dying century as an old man at 11.30 p.m. on 31 December 1900: 'He grumbled at the magnitude of the wealth he had acquired by trade; he grumbled at the extent of territory he had opened up to commerce; he grumbled at the result of his inventions and the fruits of his scientific inquiries.' It seemed he was particularly irked by the latter, because they 'had revealed to him almost everything except what he most wanted to discover, and what alone he cared to know'.

Most people, however, believed in the progress, speed and invention they felt undergirded civilisation, transforming and propelling it into a glorious future. One engineer wrote that 'distance and time have been so changed to our imaginations, that the globe has been practically reduced in magnitude'. Others declaimed that the telegraph and railway had 'annihilated time and space'. Commentators likened the telegraph variously to society's nervous system, a civilising tool of empire and a kind of scientific, Western magic. In a speech to the Institution of Electrical Engineers in 1889, Lord Salisbury, the prime minister of Great Britain, announced that the telegraph was 'a discovery which operates . . . immediately upon the moral and intellectual nature of mankind', assembling every living soul 'upon one great plane, where they can see everything that is done, and hear everything that is said'.[2]

The Peking–Paris – and the automobile itself – seemed to many to be the next step in this sequence. An editorial in the *Daily Telegraph* called the race 'the triumph of speed' and marvelled that regular telegrams – a 'potent invention of the West' – were enabling 'the world at large to enter into the kaleidoscopic whirl of an unprecedented experience'. 'The forces of progress,' the writer concluded, 'are on the march.' The participants certainly thought of themselves as agents of progress, finding kinship with any emblems of modernity they came across – railway lines, electric lights, stone buildings and, of course, the telegraph. Du Taillis, sleeping on the floor of a remote telegraph office at Udde in Mongolia, was comforted by 'the lullaby of that soft divine music that electricity makes when harnessed for the minds of men, who can reach out to one another across the hemisphere'.[3]

At the time of the Peking–Paris, this Western sense of inexorable scientific development had been heightened by the successful wireless experiments of Guglielmo Marconi. Born near Bologna, Italy, in 1874, Marconi possessed a passion for practical science and a remarkable knack for intuiting the flaws in existing systems. Marconi's technical assistant from 1926 to 1936, G. A. Isted, evocatively described him as having 'wireless greenfingers'. By 1895, drawing on the work of others in the field, Marconi had created a transmitter capable of sending signals over one and a half miles using electromagnetic radiation, or radio waves. This device had obvious and practical uses that the cable system did not, including the ability to send messages to ships.* Marconi patented his wireless communication system in London in 1897 and, in December 1901, he stunned the world by announcing that he had succeeded in transmitting the Morse letter *S* (three short pips) 2,000 miles across the Atlantic, from Poldhu in Cornwall to

---

* It was the presence of a Marconi wireless system aboard the *Titanic* in 1912 that allowed the radio operator to send out distress signals to nearby vessels. As a result, some 700 passengers in lifeboats were rescued the following day.

St John's in Newfoundland.* 'In the history of modern scientific development four great epoch-marking events may be recorded,' a correspondent for *The Times* reported, 'the perfecting of the electric light, the laying of the Atlantic [telegraph] cable, the inventing of the telephone, and the discovery of the Röntgen [X-] rays. To these may be added a fifth – that represented by Mr. Marconi's exploit last week.'[4]

Cable telegraph operators, who saw themselves as an established technology vulnerable to technological advances, were openly hostile to Marconi. At a public demonstration in June 1903, Marconi was mortified to find that the wavelength on which he was supposed to be sending a message from Cornwall to London had been hijacked by a mischievous employee of the Eastern Telegraph Company. 'Rats rats rats rats,' ran the impostor message being received in front of a tittering audience. 'There was a young fellow of Italy, who diddled the public quite prettily . . .'

The unease felt by individual cable operator employees in the face of Marconi's successes can also be deduced from a series of poems sent by the operators in Liverpool to their counterparts in North Sydney at Christmas in 1901:

> Don't be alarmed, the Cable Co.'s
> Will not be dead as you suppose.
> Marconi may have been deceived,
> In what he firmly has believed.
> But be it so, or be it not,
> The cable routes won't be forgot.
> His speed will never equal ours,
> Where we take minutes, he'll want hours.
> Besides, his poor weak undulations

---

* Some believe that he either lied or was mistaken because the equipment he was working with was so basic – at the Newfoundland end he was using a wire held up by a kite as an aerial. Both he and his assistant, Kemp, however, remained convinced of what they had heard. Either way, he certainly was able to send two-way messages by 1907. In December 1906 one of his rivals, Reginald Aubrey Fessenden, presented the world's first radio broadcast using wireless.

Must be confirmed to their own stations.
This is for him to overcome,
Before we're sent to our long home.
Don't be alarmed, my worthy friend,
Full many a year precedes our end.

This elicited the reply:

Thanks old man, for the soothing balm
Which makes me resolute and calm.
I do not feel the least alarm,
The signal S can do no harm,
It might mean sell to anxious sellers,
It may mean sold to other fellers.
Whether it is sold or simply sell,
Marconi's S may go to – well![5]

Because it was conceived as a spectacle by a newspaper, the Peking–Paris was as much a creature of the fully fledged telegraph age as the hatchling motor one. Newspapers had exploded in the nineteenth century, as technological developments such as speedy rotary press printing and mechanical typesetting, faster transport and communications all collided with growing urban populations and literacy rates. The media industry, whose business depended on the timely delivery of powerful stories, was an early adopter of the telegraph. The Associated Press was formed in America in 1848 to help spread telegraph transmission costs. Paul Julius Reuter, a former bank clerk, formed his press service in Europe soon thereafter. The firm relied on the telegraph network to move commercial news between banks, brokerage houses and newspapers. In areas where the lines were not yet complete, Reuter used carrier pigeons to bridge the gaps, ensuring an uninterrupted flow of information.

Then, as now, certain stories did better than others. Murder, celebrity and scandal were popular. In the early months of 1907, for example, front pages from Ontario, Canada, to Tuapeka, New Zealand, plastered details of the Harry K. Thaw trial. Thaw, an unstable American playboy, had shot society architect Stanford White at Madison Square Gardens in front of hundreds of witnesses. The murder had been a belated act of revenge for White's rape of Evelyn Nesbit – a showgirl and later Thaw's wife – when she was sixteen. The story, which was one of the reasons Barzini arrived rather late in China, became an international sensation, part of the ongoing fascination with 'yellow journalism' – a taste for lurid stories that exploded during the 1880s and 1890s thanks to American newspaper barons, most notably Joseph Pulitzer and William Randolph Hearst.

Exploratory expeditions were another popular genre, presenting as they did a flattering view of ineluctable progress. The North-West Passage was finally vanquished by Roald Amundsen between 1903 and 1906, to much public acclaim. In 1907 his attention, along with that of many other explorers, was turning to new conquests, especially the North and South poles. (This new obsession is perhaps why, when *Le Matin* announced the Norwegian's return from the North-West Passage in February 1907, it titled the piece 'Return from the North Pole'.) Proposals for thrilling and improbable expeditions – 'Airship to the Pole: A Jules Verne Project' – frequently appeared in print. The Peking–Paris had a similar flavour: a novel type of adventure for the reading public. The automobile, *Le Matin* informed readers in its 31 January article announcing the event, was 'intended to give man mastery over distance' and 'allow us to travel hitherto unexplored routes'.[6]

News coverage thereafter used the Peking–Paris as a modern foil for the wilderness through which they travelled. 'It was an intoxicating journey over the thick grass,' Barzini wrote on 18 June, 'and along tracks beaten by the feet of thousands of camels, zigzagging through ground now covered with bush, now sandy, and with the infinite vistas of telegraph poles which point the way to civilisation.' Collignon described the Gobi as 'an enormous plateau, bare and undulating like the sea'. Where they passed settlements, horse riders often jumped into saddles to race alongside the cars. '[F]or some minutes the automobile had to be driven with a regular escort of cavalry amidst savage shouts of joy, which really resembled war-cries, and the undulation of silken clothing swept by the wind.' Readers of the *Corriere della Sera* were treated to an image of the Itala speeding through the desert at sixty miles per hour, across ground 'as level as a billiard table', occasionally start-ling antelope into headlong flight. Herds of wild horses, drawn by some instinct, would gallop towards and alongside the cars as one, before turning and scattering like an earth-bound murmuration.[7]

Such anecdotes appeared in regional newspapers as geographic-ally diverse as the *Lancashire Daily Post*, the *South China Morning Post*, the *Hawaiian Gazette* and the *Sydney Morning Herald*. On 30 June, the *New York Times* ran a story, under the headline 'Urga Lama has an Auto', about a vehicle belonging to 'the Lama of Urga, who is second only in importance to the Grand Lama of Tibet'.[8]

Yet, as thrillingly revolutionary as the telegraph was, it had its drawbacks. At one telegraph office Barzini visited in China – a three-mile trek from where the Italians had made camp – the clerks 'were smoking opium, lying flat upon their *kang* with pipes in their hands and wrapped in a cloud of the fragrant, thick, slow smoke of the narcotic'. They were too stoned to operate the equipment and his report went unsent. At the Pong-Kong* office, which had been open six or seven years, the motorists were the very first to send messages. They all proudly mentioned it in their next dispatch-es: evidence of their role in pushing the boundaries of progress eastwards.[9]

When stations were out of easy reach, communication slowed and stuttered. Telegrams had to be entrusted to acquaintances or strangers, along with the money needed to send or deliver them. 'A newspaper correspondent is always inclined to consider the loss of his communications as a serious misfortune,' Barzini wrote. 'Journeys, expenses, difficulties, all these can be rendered useless by some futile contingency by which a dispatch is left in the bot-tom of somebody's pocket or by the side of the path.'

His grousing is understandable when you consider what went

---

*The exact whereabouts of this telegraph station – variously referred to by the journal-ists as Pong-Kong, Pong-Kiong and Pong-Hong – and its name today are mysterious. Maps from the period show a telegraph line snaking north into Mongolia, but none that I could discover depict these northern relay stations.

into their creation. On smoother sections du Taillis and Barzini were able to write on the move, but often they wrote by candlelight while their companions slept. Barzini told a colleague that in China he wrote his dispatches 'vertically, in columns, in letters half a centimetre large, not joined up, because lines of joined-up writing looks very different to Chinese eyes'. And, to avoid confusion between the Italian verb è, 'is', and the conjunction e, 'and', he used the French *est* and the English *and*. Later, in Russia, the travellers were officially monitored. They probably suspected that their telegrams were read and censored: such fears would explain the oblique tone with which politics and unrest were discussed. At one office, a Russian official simply defied international convention, flatly refusing to send one of Barzini's reports because it was in a foreign language. 'We do not telegraph in Italian,' the clerk said, handing back the inky sheaf of papers. 'No one can understand Italian here.'[10]

The glossy facade of progress to which Westerners were wedded was also sometimes undermined by their physical unreliability. Electricity, telegraphy – wireless or not – railways, the automobile and every other symbol of progress could be foiled by the raw materials required to assemble them. Wood rotted and splintered; leather cracked; rubber perished; metals corroded; gutta-percha (a tree sap from Malaysia used to insulate cables) was expensive and finite. The progress of electrifying London, for example, was marked by countless humiliating and petty reversals because of the difficulty and expense of building and maintaining the networks. The first electric lighting station in London – opened at Holborn in 1882 – was abandoned four years later after financial losses. A second attempt in the City a decade later was shambolic and much derided by the public, who hated the inconvenience of having the roadways dug up to lay cables almost as much as they feared accidents caused by the collapse of overhead wiring. After storms in 1887, even the *Electrician* bemoaned 'broken and bent telegraph posts on the roofs, wires rolled up and put away in corners, with ends hanging or tied round railings'.[11]

Some of these problems were soon to be remedied. A Belgian-born inventor, Leo Hendrik Baekeland, who was himself an avid motoring enthusiast, was sequestered in his laboratory between

18 and 21 June 1907.* While the Peking–Paris drivers were struggling two continents and an ocean away, Baekeland was in Yonkers, New York, experimenting with mixtures of formaldehyde and phenol, obtained from coal tar or petroleum. 'An exceedingly active period,' he wrote in his diary. 'I consider this days [*sic*] very successful work which has put me on the knot of several new and interesting products which may have a wide application as plastics and varnishes. Have applied for a patent for a substance which I shall call Bakalite.' By 11 July, after more experiments, he was fizzing with excitement, agitating over the timing and whereabouts of his patent applications, fearful that someone else might have pipped him to the post.† 'Unless I am very much mistaken,' he wrote, 'this invention will prove important in the future.'[12]

He was right. Bakelite – an *e* was later substituted for the second *a* – was the first fully synthetic plastic and ushered in the 'Age of Plastics'.‡ If it was not as pliable as Baekeland might have wished, it was easier to work with and cheaper than celluloid. It also dried hard and was an excellent insulator. Bakelite would go on to be used to make consumer items like phones, radios and jewellery, but was also, because of its ability to withstand heat and electricity, immediately invaluable to both the automobile and the communications industries. Cars would soon be bedecked with Bakelite

---

* His first motor, acquired in 1899, 'a two-cylinder touch-spark affair' of American manufacture that 'made noise enough to awaken a whole cemetery', earned him the enmity of his neighbours and drove the man he hired to service it to drink. Undeterred, he bought several more. In 1906 he embarked on a sightseeing trip through Britain, France and down to Naples with his wife and children in a luxurious 'limousine type' vehicle upholstered in green leather.

† He was right to worry. James Swinburne, a Scottish chemist, had discovered a similar method of combining formaldehyde and phenol – using heat and caustic soda as a catalyst – a couple of years previously. However, he did not file a patent application until 1907 because he was hoping to discover a better method and did not want to draw attention to his research. When Swinburne finally did file, it was only to discover that he was 24 hours too late: Baekeland had filed first.

‡ To give some idea of the scale of the production of plastics today, 360 million tonnes of plastic were produced in 2018; 62 million tonnes in Europe alone. The proportion of oil being used to produce plastics is expected to rise during the twenty-first century, even as demand for oil for motor fuel decreases.

knobs and steering wheels; telegraph keys – used to transmit messages – would have Bakelite covers and buttons.

For the Peking–Paris automobiles – still built of wood, rubber, metal, canvas and leather – journeys through difficult terrain, with extreme temperature fluctuations and repeated exposure to water, were especially challenging. While experienced coachmakers tried to ensure parts could withstand the elements as best they could – by, for example, varnishing the wood to prevent water soaking in, and painting metal parts to stop them rusting – rough conditions and the tinkering, assembly and disassembly necessitated by the route meant the vehicles were especially vulnerable. The motorists were all too aware that at any moment parts could seize, tear, perish or shatter.

Repairs to a vehicle would mean a delay and, whatever the promises made before departure, the crew of the damaged vehicle would be left behind to fend for themselves. From the first, dread of some accident hung over the drivers. Except for Godard, who had sold his spares in order to afford the passage to China, crews carried essential tools and spare parts with them. When things broke or malfunctioned, as they frequently did, the crews had to rely on their own technological expertise for repairs. Where this failed, they would telegraph the manufacturers and beg them to send replacements on the railway or hope that they could manufacture a replacement in the nearest town.* Dreams of winning – of even completing – the race would be over.

---

* This had been another compelling reason for the route to cleave so close to the telegraph lines.

6

Kalgan – Urga

14–24 June

> At our current rate, travelling the 12,000 kilometres to Paris
> would mean driving continuously for three hundred days.
>
> Jean du Taillis, Spyker team

Set on a wooded river in a cleft between mountains to the south
and east and the Gobi to the north, Kalgan was a wealthy trading
town and, historically, China's northern gateway.* Whitewashed
buildings enveloped in vines and striped awnings tumbled down
to a stone bridge over the river – 'as if waiting to cross it', Barzini
thought. Nearby, the riverbanks foamed with piles of freshly
washed fleeces and ill-smelling effluent from the tanneries.

The city's principal business was tea. In the recent past some
350,000 chests of it passed through the city annually, although
the trade had been buffeted by the building of the Trans-Siberian
Railway. Kalgan also boasted markets for ponies, wood, fleeces
and furs. The continual flow of goods and traders made the city
lively: mingling with the river's reek were the scents and sounds of
a fairground. 'Street tumblers, jugglers, and fortune-tellers plied
a brisk trade ... heedless of the yelling, wild-looking Mongols
galloping madly about on ponies.' Canvas-draped booths selling
iced drinks, tobacco and scallion pancakes studded the streets.

* Today, Kalgan is called Zhangjiakou and covers some 14,000 square miles between
mountains and ski resorts to the east, Beijing to the south-east and Inner Mongolia and
the Gobi Desert to the north.

The city's Chinese inhabitants favoured blue or blue-and-white garments; crowds were enlivened by reds, pinks, buttercup yellows and greens worn by Mongolians and Russians.[1]

The Italians reached the city late on 14 June; the De Dion, Spyker and Contal teams two days later. With the rainy season approaching, time was not on their side, yet they couldn't afford to rush their preparations. Although much of the Gobi was grassy and there was plenty of wildlife in some regions, water was scarce. The wells, dug every twenty to thirty miles, were often muddy, salty and barely drinkable. Nevertheless, they were the desert's lifeblood: missing one might prove terminal.

Prince Borghese puts the Itala through its paces with the Chinese governor of Mongolia as his passenger while the governor's officers ride alongside.

In the 1880s it took the explorer Harry de Windt a month to travel between Kalgan and Urga (Ulaanbaatar) in a camel train. 'Fourteen days of undulating grass plain, monotonous and unbroken, save by an occasional "yourt" or encampment, four days of deep, sandy desert interspersed with two ridges of rock . . . Five more days of green plain with intervals of gravel, thickly covered with the brightly coloured transparent stones, for which the Gobi is famous.' During the day, the heat stifled, and the sun scorched skin like crackling.[2]

Previous travellers had spent a fortnight or more in Kalgan preparing for the ordeal, acquiring camels, translators and carts that had to be built from scratch. Even for automobiles, preparations took time. The region had few settlements, meaning that they would have to be self-sufficient. Carefully calculated loads of fuel and oil were loaded onto a convoy of nineteen camels to be distributed between three depots at Pong-Kong, Udde and Urga. The local representative of the Russo-Chinese Bank instructed the automobilists in the finer points of Mongolian currency. This usually involved bartering large bricks of cheap black tea, but since space aboard the automobiles was at a premium, silver bars would be used instead. Paying for things required chipping 'taels' (around 37 grams) from the bars using Swiss army knives and minute weighing scales designed for the purpose.[3]

Mongolians in caps and wrapped gowns, 'gathered at the waist with a cord, pinned on the shoulder like a Greek dalmatic', bartered for sheepskin rugs and furs. Desert temperatures could drop from 40 degrees Celsius in the day to near freezing at night. A pair of well-worn fur mittens of almost comically large proportions, possibly purchased in Kalgan, were retained by Prince Borghese until his death.[4]

Two tracks led from Kalgan to the relatively large Mongolian settlement at Urga, a journey of 600 miles through the desert. The first track, although longer, was more frequented: strung between settlements and relay stations and reliably beaded with carts and travellers. The second, mostly used by camel trains, arrowed northwest through the Gobi's desolate heart, a journey of around twenty-five days at a camel's shuffling pace. The latter route had been chosen. It hugged telegraph lines, allowing the journalists to send off reports from the heart of the Gobi or to call for help if needed. From Urga, it would be only a few hundred miles to Maimachen (now Altanbulag) and Kiakhta (now Kyakhta), twin cities that faced off over the border between the Chinese Empire and the Russian.[5]

The motorists planned to set off on the same day – 17 June – to travel in convoy, but the pattern of acting as two independent groups, defined by nationality, had calcified. Du Taillis referred to the 'three French cars' being 'a pack' or a 'little company', ignoring the fact the Spyker was Dutch and eliding the disqualified Contal Mototri entirely.

Prince Borghese – solitary by inclination and autocratic by upbringing – made little pretence of being willing to stay with and help prop up the efforts of the others. He was frustrated by the refusal of Pons and Foucault to bow to inevitable defeat, and by the fussing and ineptitude of Cormier and Godard respectively. The Itala's early lead, which had taken the patriotic French motorists by surprise, had not improved relations. Barzini's short 12 June bulletin, dashed off from Nankou, could not avoid a trace of smugness: 'No other automobile had reached here yesterday evening until six o'clock, when the Contal tricar [*sic*] arrived, but by train.' Du Taillis, writing several hours later, was less flattering: 'We are all doing well. Prince Borghese is still ahead of us, alone.'[6]

The sight of Prince Borghese and Guizzardi painstakingly reassembling the bodywork of the Itala onto the chassis on the morning of their departure, a job that took several hours, irked the French drivers further. Cormier groused about the time wasted, but insisted on waiting for them anyway. Describing the scene, du Taillis referred to 'the princely driver of the Itala' and contrasted him with the French, who 'considered it to be a more convincing demonstration of endurance . . . to load up each car at the start with everything required for the whole course of the journey'. As a patriot, he ignored the greater number of mules and men that had been needed by the French, as well as the fact that the Spyker carried Pons's luggage.

Resentment of the Italians, however, was insufficient to unify the French. The Itala once again sped off without a care in the world, while the two De Dions and the Spyker halted at every bend, waiting for Pons and Foucault to putter into view, red-faced and sweating as they manoeuvred the three-wheeled Contal over rocks and potholes. The stopping and starting in the parched landscape acted as an irritant, fraying the nerves. As they waited, the others inspected impressionistic maps, agonising over northern rivers that would soon be impassable, pointing at sparse dots, each theoretically indicating the presence of a life-saving well in the Gobi.

While most travellers hired guides, the motorists had decided not to. In Kalgan, they had been cheerfully assured of two infallible navigational aids: the telegraph wires and the camel bones that lined the route. Cormier, who, du Taillis drily noted, was constantly catastrophising, threatening to throw in the towel and comparing every dirt track they encountered with the paved roads of France, had been telling everyone since before they left Peking that 'the rain means mud, swamps, certain failure'. But the slowness of their progress as they left Kalgan made the journalist less sanguine. 'The drivers of the larger cars,' he wrote to *Le Matin* on 17 June, 'are troubled by the delays caused to the entire caravan by the Contal Mototri.'[7]

On the night of 17 June, all five vehicles camped in the grasslands at the edge of the Gobi, the only time they would spend a night together on the road. 'It was a deliciously soft, light evening,' du Taillis recalled. 'The silvery reflections of the moon were scattered over the tall grass that swayed in the breeze.' The men pottered around their cars amicably enough, lighting fires, putting up tents, examining and discussing incipient mechanical issues. The journalists were instructed to cook. Du Taillis rustled up a novel soup, made from 'cakes, like chocolate . . . contain[ing] all the principles of nutrition'. Barzini's efforts, by his own admission, were less successful: the Italians went to bed hungry.

As the light faded, and despite their proximity to one another, a feeling of melancholy settled over the camp. 'Darkness seems to isolate and separate one,' Barzini wrote. 'It ends by making one silent because it gives the sensation of being alone. And when in the darkness nothing else could be seen distinctly but the tiny red speck of our lighted cigarettes . . . we became entirely silent.'

The next morning, the Contal Mototri set off in the cool predawn. They had agreed to this staggered departure over dinner the previous evening, so that the others would spend less time waiting for Pons and Foucault to catch up. Even so, the De Dions and the Spyker passed their compatriots after just twelve miles. Soon after, they were passed in turn by the Itala. Godard, resentful of the prince's refusals to lend him money and fuel, turned and yelled across to the De Dions, 'Borghese has only one idea: scratching all of us.' Scratching, he explained to du Taillis, was motor-racing slang, meaning 'passing a competitor and then not bothering about him so you can arrive first'. Unhappily preoccupied by this idea, the De Dion and Spyker teams drifted onto a path that veered right, away from the telegraph poles. This did not trouble them at first. 'In countries like this,' Cormier yelled to the others, waving expansively, 'there is only one road with little variations on it, which always end up meeting each other.'[8]

Time and miles passed, but the tracks did not converge. The teams decided, rather than turning around, to cut due west across the virgin plateau. Around them stretched a monotonous vista, wide and softly undulating, tufted with dusty green vegetation and overtopped by a hot blue sky. The ground beneath their wheels looked like pale sand but on closer inspection was composed of thousands of minute, multicoloured stones. Conversation had to be conducted in shouts over the roar of the engines, though they had little to say to each other anyway.

After half an hour, they were once again beside the guiding thread of the telegraph wire and on a trail bearing the 'very visible imprint of the Itala's gripping tyres'. Here, they turned off their engines to conserve precious fuel and waited for Pons and Foucault, eyes trained back down the open track, passing around a pair of binoculars. An hour lumbered by. Cormier and Godard, who rarely saw eye to eye, set up a bass duet bemoaning the delay. No matter how hard they looked they saw 'nothing but the powdery sky and the endless green of the grass'. More time passed in resigned silence. Du Taillis and Godard, irritated that Cormier's airy arrogance had cost them time and valuable fuel, brooded. Eventually, worn past endurance by the tedium and heat, they convinced themselves that Pons could not be behind them at all: he must have passed this way during the time they were on the wrong road. They got back in their cars and headed on to Pong-Kong telegraph station, their first planned stop in the desert. Pons was not there.[9]

After the race, when it had become clear that blame was to be apportioned, the participants' accounts were peppered with self-justification. Predictably, the Frenchmen found the prince responsible. Cormier wrote that they 'decided to place our trust in Prince Borghese's judgement'. 'Well,' the Italian replied in this telling, 'I'm leaving tomorrow at three o'clock,' only grudgingly agreeing to leave money for supplies for the missing drivers. Du Taillis wrote that the prince refused to discuss Pons at all, leaving the decision entirely in their hands; he also implied that Cormier, Collignon and Godard were blinded to the possible fate of the Mototri crew by their own ambition:

> Not having any sporting interest, nor any commercial responsibility, I had to limit myself to recording the decisions of others, other than to slip a human note into these conversations where there was a lot of professional rivalry, it must be admitted . . . if everyone was to comply with the commitments they had made, they would have to stop completely, and to agree on a methodical search plan to find Pons before going any further.[10]

Barzini would tell a different story: 'Our companions expressed their settled conviction that Pons had turned back . . . We had no sort of anxiety as to the fate of Pons, and of his companion; they were still in the inhabited zone, and could, no doubt, easily find hospitality and help.' The telegrams sent by both du Taillis and Barzini on the evening of the 18th support this claim. In a message sent to *Le Matin* at 6.25 p.m., du Taillis said only that 'the Contal Mototri is lagging behind'. Two hours later, in a fuller and more descriptive missive, the Contal team was not mentioned at all. Barzini telegraphed the *Corriere* and *Daily Telegraph* to say, 'The Contal tricar has found it impossible to proceed any further. The competitors are therefore reduced to four . . . It is hoped that we shall now be able to accelerate our pace.' The motorists, despite later protestations, seem to have had no fears on Pons's behalf and instead felt rather relieved. They assumed he had resigned from the race and turned back, leaving them to carry on unencumbered.[11]

Setting off the next day, 19 June, Godard and du Taillis were preoccupied and irritable, barely even on speaking terms. Just before their departure, the journalist had surprised the driver and Borghese 'in the midst of a secretive, private conversation about petrol and supplies running out'. Du Taillis was mortified.[12]

Rather than carrying extra fuel from Kalgan, Godard had offered to take some of Pons's luggage and supplies, a decision that du Taillis bluntly called foolish in the extreme. Godard, betraying some uncharacteristic worry, tried a witticism at his own expense: 'What will I have to do now to justify the nickname you gave me, the Braggart of the Gobi?' He paused. 'Not that I am sure that I know what "a braggart" is.' Du Taillis, unsmiling, pointed to a pale smudge near the horizon, all that could be seen of the Itala. 'Follow them. We cannot afford to get lost today. It is bad enough that Pons and Foucault have been left behind.'

Just a mile and a half after leaving, du Taillis realised to his horror that something else had been forgotten in their haste. Pékin, a small Pekinese dog Godard had acquired on his arrival in China and which had become a great pet to the French crews, was not in her customary spot at Godard's feet: she had been abandoned at the telegraph station. 'Let's go back and get her. It's so close,' du Taillis implored. 'Never!' Godard replied, his usually elastic face screwed into an unaccustomed grimace. 'I would *never* sacrifice the success of this race for a dog; I would *never* expect my brave Spyker to expend unnecessary effort before we reach Paris. It is a great pity, but too bad.' Du Taillis suspected Godard's vehemence had more to do with his pride and lack of fuel than his feelings about Pékin. The pair relapsed into silence.[13]

Fifty miles into the 175-mile journey to the next telegraph station and fuel depot at Udde, the engine of the Spyker let out a series of choking coughs and then, anticlimactically, died. The sun was rising steadily, the air above the sandy expanse surrounding them beginning to shimmer. Du Taillis turned to Godard, who, studiously avoiding his eye, was trying to restart the automobile and muttering into his moustache. They were out of fuel. The two men got down without a word and turned in the direction of the column of dust behind them that marked the progress of the De Dions. Both cars, they knew, had full reserve tanks. Enough, if shared with the Spyker, to get all three cars to Udde. Godard, his face pressed into a scowl, would still not look at du Taillis, but stared resolutely towards the twin cars, hailing them as they approached. 'We have no more fuel!' he yelled.

He had misread his mark. The lead De Dion, with Cormier at the wheel, slowed but did not stop. The driver roared with laughter at the sight of the woebegone motorists. 'You don't need anything, do you?' he called out with a wink. 'I'll have your petrol sent to you as soon as we get to Udde.' With a cheery wave, he sped off, still chuckling. Collignon, in the other De Dion, tamely followed close behind.*[14]

Safe in the lead vehicle, which the meticulous prince and his chauffeur had ensured was amply provisioned, Barzini found the passage through the Gobi tortuous. 'Our skin was parched as with fever, and the sun beat so hotly on our face[s] and hands that it was as if there was centred upon us the most powerful light of an immeasurable lens.' Temperatures reached 50 degrees Celsius in the sun. Their skin reddened and cracked. They gasped for water. The horizon shimmered with illusory shadows that took on the appearance of lush vegetation. When they were forced to stop to let the radiator cool and restock it with water from their emergency reservoir, Guizzardi and Barzini couldn't help but press their lips to the siphon – the same one they used to refuel the car – and take long gulps. The water gushed out hot, tasting of petrol and varnish. Prince Borghese wet his lips but pulled the others away, insisting that the remaining water be conserved for the car.

Hours passed as they drove on, each man vacantly staring at the bleached landscape, the guiding line of telegraph poles – which Barzini found himself counting mechanically – under the towering cloche of the sky above them. The middle of the desert, a stretch of forty miles in which the bones of fallen animals lay especially

---

* Georges Cormier, a competitive man, none too fond of Godard and rather lacking in imagination, did not at first see anything wrong with having gone on without them. In an article sent to L'Auto from Urga on 23 June, he reported that seeing the car 'stopped in an immense plain as far as the eye can see' and hearing Godard confess that he was out of fuel was 'one of the funniest peculiarities of our trip'. He also presented the ongoing race as one between three cars: the Itala and the two De Dion-Boutons. 'The Spyker and the Contal Mototri are following. I would primarily like to pay tribute to the latter, whose driver Pons had to suffer more than anyone else from the ruts and rocks because of the alignment of his three wheels.'

thick, was 'a place of agonies. There is breathing about it an inde-scribable spirit of death.' Relief came only through movement, through the 'freshening breath which the speed of the car brought against our faces'.[15]

Those left far behind in the desert despaired. Pons and Foucault had, like Godard and du Taillis, run out of petrol. On the night of 18 June, they had slept beside their empty tricar within sight of a Mongolian encampment, but did not ask for help. The next day, sense deserted them completely. They spent the early morn-ing walking towards Pong-Kong with neither food nor water. By mid-afternoon they had reconsidered. 'Stunned by the sun, dazed by fatigue, tormented by thirst and hunger, we turn[ed] back and, hobbling and hopping, arrive[d] back with the Mototri at six o'clock in the evening.'

On the 20th, self-preservation finally overcame prejudice and they sought help from the nearby Mongolians. Their faith in their comrades and reluctance to engage with locals had very nearly proved fatal: a few more hours and they might well have suc-cumbed to heatstroke. 'It is a little humiliating to admit,' Pons would write for the French sporting magazine *La Vie au Grand Air* published weeks later, 'that one receives aid and assistance from savage tribes, when so-called civilised people have odiously abandoned you in the desert.' Forced to leave their equipment, their spares and the Contal where they lay, the pair were given milk and biscuits by their hosts, then they walked back to Kalgan with a passing camel caravan, a journey of five days.* Their dream of completing the Peking–Paris was, finally, at an end.[16]

This was not the only time the Europeans' assumptions about 'savage', 'uncivilised' or 'lazy' foreigners were disproved. To their evident surprise, whenever the motorists engaged with them, they found Mongolians to be kind, hard-working and honest. When a

---

* There are various legends concerning the fate of the Contal Mototri. Some believe that it rusted to nothingness where it lay. Others swear that it was later dragged to Urga as a gift or tribute and is either there still, hidden in some corner of the city, or has long since been broken up for parts.

group prevented him going down the wrong road, Cormier wrote, 'I must once again attest the intelligence of these people: the man was perfectly able to convey to me that the telegraph station was over there. He showed me Prince Borghese's wheel tracks in the sandy parts.' Later, he called them 'hospitable, selfless, highly intelligent'.

One Mongolian chief, 'clad in his gorgeous purple robe', invited Barzini, Borghese and Guizzardi to his village. ('He was so effusive and wholehearted that we could not disappoint him.') There, in the midst of a shower of hospitality, a younger Mongolian man suddenly turned to the guests and asked, '*Sprechen sie Deutsch?*' He explained he had spent time in Berlin several years previously while taking part in an exhibition about people of different races. 'There was an encampment of Mongolian *yurtas* with its horses, its dogs, and its women,' the man explained. 'And a great crowd came to see us every day and spoke to us, so I learnt German.'

On their first day in the desert, du Taillis and Barzini had been astonished when the former was flagged down by a group of men who wished to return a piece of Prince Borghese's luggage that had fallen off the Itala. Even to those hard-bitten commentators, the contrast between the behaviour of those they termed civilised and those they believed uncivilised was marked.

'Mongolians? Honest barbarians?' Barzini joked to du Taillis. 'Miserable people of the wild who give themselves the luxury of returning picked-up goods?'

'Yes, and without even asking for a tip!'

'But my dear friend, wherever have they gone to, the brigands of the prairies? Those men whose actual duty it was to assault us.'

'They are probably gone to Europe.'[17]

For du Taillis, the generosity of the Mongolians was no longer a laughing matter. When Cormier and Collignon had driven past on 19 June, the journalist had initially thought they were joking. He looked after them, expecting to see them stop at any moment. He turned to Godard, eyes wide, arms held out in mute appeal, but the driver was busy hurling a stream of invective and pebbles after the rapidly diminishing De Dion-Boutons, face red and puckered with rage. Du Taillis turned to watch the cars' progress.

Cormier had promised to send fuel from Udde when they arrived. But when would that be? What if they were delayed or had a breakdown on the way? Udde was well over 100 miles away. How long would aid and fuel take to reach them once it had been arranged?

The Gobi had sprung on them like a trap. It stretched as far as the eye could see in every direction. Nothing but sand and gravel, peppered here and there with clumps of tough grass. Du Taillis swung himself up into his seat in the Spyker, throwing his head back against the canvas stretched over their luggage. Closing his eyes, he ran through a mental tally of their supplies. Godard, the damned fool, had ditched almost everything they had back at Pong-Kong to make room for Pons's luggage. By way of food, they had part of a roast chicken, some military biscuits purloined from the troops stationed in Peking, some chocolate and two litres of water – enough to last a day and night if they carefully conserved their energy. But this, it was clear, Godard had no intention of doing. Instead, he paced and fumed, finally demanding that du Taillis get down and join him: they would walk towards Udde, pushing the car. The journalist ignored him.

The sun wheeled overhead. They spoke little, occasionally taking sips of water and nibbling biscuits. By that night, supplies of both had run out. Their clothes, patched dark with sweat during the day, chilled them as the temperature dropped precipitously, the salt from their sweat stiffening the fabric. By the following morning, their remaining food, the roast chicken, had turned putrid and crawled with insects. Around ten o'clock, through their binoculars they spotted a Mongolian woman riding a camel. She offered them a few sips of muddy liquid from her waterskin. She allowed them to harness her camel to the Spyker, but it was no use: the wheels sank in the sand. She shook her head, untied the ropes and departed.

By the second night, both men were showing symptoms of heatstroke and exhaustion. Unlike the Contal crew, whose vehicle had failed within sight of an encampment, the Spyker crew were utterly alone. They had no means of getting a message to their companions, or even to the staff at one of the telegraph stations who knew all about the race and would probably assist them. They had no way of knowing when the help that Cormier had mentioned might come, or if, indeed, it ever would: it seemed unlikely he would stick around to ensure that it did.

Feverish, but too dispirited to erect their tent, they stretched out on the ground beside the Spyker and barely moved for ten hours. The following morning, 21 June, they were awakened from their torpor by an incessant jingling. The sound came from the bells attached to the harnesses of a large camel caravan in single file, accompanied by a dozen men. The procession slowed down but would not stop, despite both men begging for water and food. When they, too, disappeared into the distance, despair began to take hold. Du Taillis, the less robust of the two, was now too fevered and weak to walk. Godard, however, announced that he would go and seek help or die in the attempt. He set off, walking slowly and sinking a little into the shifting pebbles with each step, growing smaller and paler until he was swallowed up by the heat haze. Du Taillis watched him leave.

Suddenly he jerked awake. A shout roused him. Blinking, du Taillis shifted his head to the side, noted from the movement of the shadows that some hours had passed and saw what he thought was a hallucination: Godard riding double behind an imposing horseman. Godard grinning, yelling, waving his hat above his head. It was really him, not some product of du Taillis's overwrought imagination. Behind the driver were a dozen more riders, dressed like the first in coats that reached mid-thigh, decorated with geometric beadwork, and pillbox hats made from velvet and gold cloth.

When they reached the brightly striped car, the small caravan dismounted, gravely formed a circle and got down to business. Godard bartered using dramatic hand gestures and the promise of money: since Cormier had taken charge of all the French group's finances, they had none on them. Du Taillis, somewhat revived thanks to the water offered by their rescuers, contributed little, but he didn't need to. Godard, du Taillis would later write in *Le Matin* with real warmth, had 'a talent for making himself understood'. Two hours later, he had arranged for a rider to deliver a letter to the telegraph operator explaining the situation and to return with a can of petrol. In the meantime, two camels would tow the car in the direction of Udde and the group would walk there together. Godard magnanimously made a present of his gold watch chain – the only valuable thing he possessed – to the group's leader. With that, the deal was struck.

As they had talked, darkness fell around them. Since it was the preference of the horsemen to travel at night, they set off towards the Udde telegraph station, the deep violet path lit only by moonlight. This left Godard and du Taillis to trail behind them at a distance, alone save for two camels they had begged from the Mongolians to tow their stricken car. Hunger and thirst tormented them. The only well they encountered was brackish and opaque. As the sun began to rise, they were driven to such extremes that, 'Lying on our stomachs, under the bonnet, we got two sips of water from the radiator, oily water, but priceless to us.'

At last, they saw a purple shadow moving towards them across the pallid ground: the horsemen had returned. Within moments they were surrounded by the smiling faces familiar from the day before. A couple brandished canisters of the promised petrol like trophies. Dizzy with relief, the Frenchmen fell upon the fuel, poured it carefully into their tank, started the engine with whoops of joy and set off once again towards Udde, their new friends riding alongside them in a yelling honour guard, the long red ribbons of their hats rippling out behind them like tongues of flame.

The Spyker team had spent four days in the Gobi. 'Four days,' du Taillis recounted in his telegram from Udde on 22 June, 'of thirst, hunger and mortal fear.' They did not, however, emerge unscathed. In du Taillis's book, *Pékin-Paris: Automobile en Quatre-vingts Jours*, written on his return to France, it is clear that at the beginning of the journey he took extensive notes: a single day could occupy several pages in which he would describe the moods of his companions, the landscape and the food. After his time in the Gobi, his narrative lost focus. His telegrams to *Le Matin* became shorter and more sporadic. He quoted more heavily from other sources, and there are confusions over dates. Later, he would have to spend extra days resting, sometimes taking the train to catch up with the other drivers rather than travelling with them. It is likely that du

Taillis was suffering the after-effects of severe heatstroke and had only providence – and Godard – to thank for his survival.*[18]

Ulaanbaatar, Mongolia's capital, has had many names. The literal translation of its present one is 'Red Hero' and was adopted in 1924 when the country became a Communist state. In 1907, Mongolians typically called it some variation of *Da Khüree* – literally, 'Great Encampment' – while to Westerners it was Urga, from the Mongolian word for 'palace'. The palace in question, according to de Windt, belonged to the *Kootookta*, or 'Living God', one of the two mortal deities in Mongolian Buddhism (the other being the Tibetan Dalai Lama), and the axis around which the city revolved. It was 'an imposing edifice of Tibetan architecture, all white, gold and vermilion, like an ornament off the top of a twelfth-cake'. The city huddled around it, 'a compact mass of red and white dwellings, golden domes, and snowy tents, surmounted and surrounded by thousands of brightly coloured prayer flags, which flashed and waved in the sun with every breath of the pure morning air'.[19]

Despite its initial beauty, the city had a sinister atmosphere.† The inhabitants – Mongolians, Chinese and Russians – seemed to meet there 'without blending or communing. There is between them the bulwark of an age-long hostility.' The Russian consulate was surrounded with 'trenches, ditches, wire nettings, loopholes, all the most modern and efficacious aids to defence'. The reigning *Kootookta* lived under Russian protection: none of his recent predecessors had survived past their twentieth birthdays. (After

---

* Du Taillis gives broad hints of his condition in his book. 'We would be almost fine if four days of suffering, deprivation and indescribable fear had barely touched us. Instead, we were carrying the memory of it very tangibly: Godard had a very high fever, and I was suffering from dysentery that was not to leave me much respite for days to come.' He does not admit, however, to the later absences from the group for days at a time that Cormier records.

† An oft-remarked feature of the city at this time was its funerary practices: the dead were not buried but left in the open. Harry de Windt rode out of the city to see the site for himself. In a 'cleft between two low green hillocks . . . [was] a valley literally crammed with corpses in every stage of decomposition, from the bleached bones of skeletons that had lain there for years, to the disfigured shapeless masses of flesh that had been living beings but a few days or hours ago . . . I can never think of the place even now without a shudder, and devoutly wish I had never seen it.'

all, Barzini explained, an 'intelligent, energetic and daring man at the head of the Mongolian people might become dangerous to Chinese sovereignty'.) The Qing dynasty was making clumsy efforts to strengthen its hold over Mongolia as a defence against Russian aggression. The Mongolians, for their part, resented their enforced role as China's political pawns: one American journalist notes that they were 'indifferent to their rulers and ready at any decent provocation to throw off their yoke'.[*20]

The motorists were oblivious to these machinations. For them Urga was merely somewhere to rest and recover. They stayed with the local manager of the Russo-Chinese Bank, Mr de Stepanoff, and were lavishly entertained. After the privations of the desert, days of being unable to bathe[†] and, in the case of du Taillis and Godard, having barely survived, the de Stepanoff home seemed like a mirage. 'Russian, French, and Italian flags adorned the stairway. In a great hall sparkled a long table spread with twenty or thirty covers, laden with sweetmeats, with spotless serviettes artistically folded, and superb crystal and porcelain.'[21]

The Itala had roared into Urga on 21 June, eleven days after the departure from Peking, and would depart two days later. The two De Dion-Boutons, driven by Cormier and Collignon, arrived on the 22nd. Godard, determined to catch up, embarked on a mad dash from the desert telegraph station at Udde, covering nearly 400 miles in twenty-three hours to arrive at dawn on 24 June. It was, considering the state both men must have been in, an incredible feat. Even the unsympathetic Georges Cormier noted that they arrived 'in a sorry state of fatigue and hunger . . . Godard is the more drained of the two; slumped over a pile of wood, he can no longer move.'

If recriminations – hushed or otherwise – passed between the crews of the Spyker and the two De Dion-Boutons in the decorous

---

[*] In 1911, a delegation of Mongolians would appeal to St Petersburg for support against China; this coincided with the collapse of the Qing dynasty, and, in December of that year, Mongolia declared its independence. This would prove the beginning of a new phase – rather than an end – of the political tussles between Russia and China over Mongolia.

[†] Not quite completely unable. Cormier, in a rare show of humour, captured one of his companions taking a bath at one of the desert wells, naked save for a pith helmet.

surroundings of the de Stepanoff home, none mentioned it.* Du Taillis's palpable resentment against Cormier seeped into his dispatches, but he largely held his tongue. The former racer was de facto leader of their group and in charge of their finances. He was relatively wealthy, as a car dealer and acquaintance of the Marquis de Dion: not someone a striving writer for *Le Matin* could cross lightly. Whatever took place behind closed doors, from this point on Cormier and du Taillis would be united in publicly blaming Prince Borghese for a new rupture of the Peking–Paris convoy. Du Taillis, in a curt line to *Le Matin*, wrote, 'Prince Borghese had set off for Kiakhta on Sunday, having declined to wait for us.'

Barzini, for his part, explained that they left promptly driven by fear of their vehicle not being able to cross the River Iro in northern Mongolia, something that had been troubling all the drivers since their departure from Peking. For Borghese, driving by far the heaviest vehicle in the group, rivers and mud were particularly challenging; he would do everything in his power to give himself the best chance of getting through. 'One downpour of rain suffices to change [the Iro] into an insurmountable obstacle . . . A Russian merchant coming from Kiakhta told us that the height of the water of the river was at that moment about four feet, a considerable height for a motor car to attempt to cross.' Still, the prince's decision was a crucial one and a bit of a gamble. Any pretence of the Peking–Paris being a collective endeavour had disintegrated entirely. From that moment on, the Italians were on their own.[22]

---

* Pons, once safely back in France, was not so reticent. In a piece for *Le Matin* he rejected du Taillis's suggestion 'that I had to give up the idea of continuing because of the difficulties encountered on the road and because of the capabilities of my vehicle. Here is the exact truth: I gave up because, on 18 June, my fellow travellers, failing to respect the laws of the most elementary humanity and not taking into account the promise that each one had made not to abandon each other . . . they did not wait for me and did not retrace their steps to make sure that I did not need their help.'

7

THE DIVIDED EMPIRE

Russia between Revolutions

[Are not] you too, Russia, speeding along like a spirited troika that nothing can overtake? . . . Everything on earth is flying past, and looking askance, other nations and states draw aside and make way.

Nikolai Gogol, *Dead Souls*, 1842

As dawn approached St Petersburg on 22 January 1905, it gilded the roofs and cupolas of a hushed city. Ivan Vasilev rose as quietly as he could and decided to take advantage of the hushed Sunday morning to attend a protest. Organised by a priest, Father Gapon, after months of strikes and economic unrest, the march aimed to present a petition to Tsar Nicholas II at the Winter Palace. Father Gapon, like many Russians, blamed the bureaucracy for the plight of ordinary people. They were certain – almost certain – that if they could only convey their distress to the tsar directly, he would accede to their demands to improve living and working conditions. Clearly not everyone was so sanguine. Just before he set off, Ivan Vasilev wrote a note to his sleeping wife and son. 'If I fail to return and am killed, Niusha, do not cry. You'll get along somehow to begin with, and then you'll find work at a factory. Bring up Vaniura and tell him I died a martyr for the people's freedom and happiness.'[1]

A few hours later, soldiers stationed outside the Winter Palace, spooked by the seemingly boundless sea of faces before them, fired into the crowd. Around forty people were killed and hundreds injured in the ensuing stampede; many were women and children. Later that day, reprisals and counter-reprisals by revolutionaries, police, soldiers and pro-tsarist, nationalist militias such as the Black Hundreds turned into full-scale rioting. In all, some thousand people were wounded or lost their lives. Ivan Vasilev never returned home.

Bloody Sunday in 1905 tarnished the image of the Tsar as a benevolent ruler and threw Russia's existing social problems into stark relief.

Another casualty that day was the myth of the 'Good Tsar', one of the last props shoring up the tottering edifice of Russia's autocracy. A Bolshevik in the crowd wrote that the 'reverend and almost prayerful expressions' in the faces of those around him 'were replaced by hostility and even hatred'. Maxim Gorky, triumphant in the face of impending revolution, wrote to his wife the next day, 'People have died – but do not let that trouble you – only blood can change the colour of history.' Sunday, 22 January, later known as Bloody Sunday, marked the start of the 1905 – or 'First' – Russian Revolution, a precursor to the October or 'Second' Russian Revolution of 1917, and would be pivotal in the downfall of the Romanovs, who had ruled Russia for three hundred years.[2]

Russia had entered the twentieth century on a cresting wave of change. After centuries of expansion, her territories were vast. The country spanned some 2,900 miles from north to south and 6,700 miles from east to west. Militarily, she had within living memory suffered two bruising defeats – in the Crimean (1853–56) and Russo-Turkish (1877–78) wars – and was soon to suffer a third humiliation at the hands of the Japanese (1904–05). Her military was underfunded and demoralised.

This was worrisome, because many of Russia's relationships abroad were strained. To the east, she had been an aggressor during China's Century of Humiliation; to the west, she was torn between alliances with France and Germany, and locked in a rivalry with the British Empire known as the Great Game. This confrontation, which lasted for most of the nineteenth and into the twentieth century, played out as a struggle for imperial conquests, particularly in central and south Asia. In this context, China was used by both as a counterweight and a source of valuable land.

During the 1850s and 1860s, a series of agreements, now known in China as 'unequal treaties', allowed Russia to annex swathes of valuable territory to the north of the Amur River, around Vladivostok and down the east coast as far as the Korean border. The two would continue to spar over territory into the 1900s. These skirmishes suited Britain: she dispatched military advisers to China to shore up the latter's defences and make her an effective bulwark against further Russian expansion.

Russia was suffering from substantial internal problems too. During the 1860s, Tsar Alexander II (born 1818; reigned 1855–81) emancipated serfs – a third of Russia's population – and enacted a series of legal, military, educational and administrative reforms to help modernise the empire. These initiatives stalled or were rolled back during the short reign of the reactionary Tsar Alexander III

(b. 1845; r. 1881–94), who came to power after his father's assassination. However, the disastrous handling of the 1891 famine and outbreaks of cholera and typhus – which affected nearly 40 million people and killed up to half a million – gave new impetus to reformers. So too did the escalating demographic and land crises.

The country's population surged to more than 150 million in 1900 and was continuing to rise: by 1913 it was over 175 million. This left many peasants (nearly 80 per cent of the population) with little or no land. What they did have was often meagre, and their farming methods were well behind those of their European contemporaries. Even good years saw them producing fewer crops, certainly insufficient to sustain a ballooning populace. In some regions, famine threatened almost annually.

In response, millions abandoned the villages of their birth to seek work in cities. These were overcrowded and dirty. Outbreaks of cholera and typhus recurred with alarming frequency. A 1904 survey found that sixteen people lived in the average St Petersburg apartment: six or so in every room. Wages were low; workers had few rights; unemployment and poverty were widespread. By 1913, there were an estimated 25,000 homeless people in St Petersburg alone.[3]

Russia's size and expansion over the previous centuries also meant that, as it struggled with urbanisation and the land crisis, sectarian and nationalistic conflicts were flaring too. An Englishman who had travelled up the Volga from the Caspian Sea in 1874 marvelled that 'There are no less than thirty-six different races included in the Russian dominions. Some of these, and the most interesting, are to be seen, not in a state of fusion with the others, but each living a life of its own, intermarrying only among its own, preserving its own peculiar institutions, manners, customs, language, and religion, apparently absolutely unaffected by the civilisation of the country in the midst of which it has pitched its camp.'

The latter half of the nineteenth century had seen increasingly draconian attempts to 'Russify' minorities. The empire's 5 million Jews were forbidden from owning land, serving in the army as officers or joining the civil service. In Poland, students had to study their native literature in Russian translation, and officials, down to and including railway porters – most of whom had never learnt Russian – were punished for speaking Polish. During an outbreak of cholera in 1907, a medical committee in Kiev refused to allow public health notices to be printed in Ukrainian, with the result that many unknowingly drank infected water and died within sight of warning signs written in Russian.

Yet, for all the empire's problems, by 1900 there had also been substantial modernisation. Sergei Witte, minister of finance during the 1890s, encouraged huge investment in her infrastructure and industry. The Trans-Siberian Railway, largely built during this decade, was transformative. It facilitated voluntary peasant migration eastwards into Siberia, where more land was available, and encouraged or increased the profitability of mining and industrial and agricultural endeavours, such as butter exports: big business around the turn of the century. By 1905, Russia had 40,000 miles of railway, major cities shone with electric street lamps, her oil industry matched that of America and her steel production surpassed that of France.[4]

Geographically, the Russian Empire of the nineteenth and early twentieth centuries was distinctly lopsided. The country's centre of political, cultural and social gravity lay in the far west, around Moscow and St Petersburg. The latter city – the country's imperial capital – had been built during the eighteenth century explicitly to showcase the country as a great European power. Its buildings were modelled after those in other western European cities, especially Amsterdam and Venice, and built by craftsmen, engineers and architects that Peter the Great (b. 1672; r. 1682–1725) had gathered on his travels through Europe. Even the city's abattoir was remodelled with rococo flourishes.[*][5]

The country's relationship with its eastern territories, however, was more contradictory. For centuries Siberia had been part of the Mongol Empire. It was only with the collapse of the Siberian khanate in 1598 that Russia began its expansion eastwards. At first, the

---

* So naked were the stylistic borrowings that one nineteenth-century writer commented that 'Petersburg differs from all other European towns by being like them all.'

lure was the fur trade; later the promise of gold, minerals, wood and land. During the reign of Catherine the Great (b. 1727; r. 1762–96), convicts and local peasants were induced to start building the Siberian Tract – by this time seen as a vital artery linking western Russia to its vast Siberian territories. Simply made and unpaved, it nevertheless represented a technological leap. By the end of the nineteenth century, the 6,000-mile journey from St Petersburg to Nerchinsk, just east of Lake Baikal in southern Siberia, could be made by a messenger on horseback in only eleven weeks.

During its heyday, the Tract hummed with traders, travellers and mail carriers. Travel, however, was understood by all to be laborious, frustrating and time-consuming. Post houses – small, government-funded hostels where travellers could stay and change horses – were established roughly every fifteen miles. All were built on a similar model, adding to the strange monotony of long-distance travel. The simple, four-room wooden houses all had black and white pillars by the door and the imperial arms over the gateway. Two rooms were reserved for the concierge ('a master of hospitality, a tyrant of the road') and their family; the other two were for travellers. They were cramped, hot, smelly and dirty. Many were still there – more dilapidated than ever – in 1907.

Post houses were also where travellers changed horses. Siberian post horses were spirited little beasts – they needed to be since it was not unheard of for teams to be brought down by packs of wolves in the forest. (Wolves were rarer by 1907, but the Itala team still surprised one in the forest near Kansk.) Horses were traditionally harnessed three abreast – a 'troika' – rather than in pairs, and could only travel a set distance from their home post house. At each stop, teams of horses were allocated according to the grade of travellers' papers (*podorojne*). For travellers with 'ordinary *podorojne*', this often meant long waits to get fresh horses. Days could pass before a team became available. And, no matter how long a traveller had waited, if the bearer of a precious '*podorojne de courier*' or 'crown *podorojne*' arrived at the post house, the team would be unhitched from the ordinary traveller's vehicle and attached to theirs.

The low design of the *telegas*, or carriages, and the horses' twelve hooves meant a pummelling for travellers. Dust in the summer; clay-rich mud in the spring and autumn; snow and ice during winter. One winter traveller recalled, 'Whenever our pace quickened to a trot or gallop, the larboard [left-hand side] horse threw a great many snowballs with his feet. He seemed to aim at my face, and every few minutes I received what the prize ring would call "plumbers in the peeper, and sockdolagers on the potato-trap".'[6]

The Tract and the journey east also had grimmer associations. Between 1801 and 1917, a million men and women were deported over the Urals by the tsarist justice system. (By way of comparison, Britain deported 160,000 people to Australia between 1787 and 1868.) Many were guilty only of 'vagrancy', an elastic condition criminalised in 1823, which in practice tended to sweep up those peasant communities deemed weak or troublesome, such as the disabled or mentally ill. Critics of the tsar, agitators, activists and revolutionaries were also exiled. In 1907, Siberia was bloated with political exiles: some 45,000 had been sent east in the six months following the issue of the October Manifesto of 1905.

This abusive system – Siberia as a destination of exile specifically, and the significance of the one-way journey east – loomed large in the Russian imagination.* It also cast a shadow over perceptions of Russia internationally: encounters with exiles were staples in any description of the country. Groups on the move were instantly recognisable. Once sentenced, they were issued with coarse grey and dark brown clothing, one side of their heads was shaved, and they would usually be put in chains. The journey was traditionally done on foot. Convoys, supervised by armed guards, would trudge eastwards, spending each night in cramped, unsanitary conditions with little oversight to prevent violence, which was endemic. The journey from Moscow to Irkutsk – roughly equivalent to the Pacific Crest Trail hiking route up the west coast of the

---

* The Vladimirka, a short stretch of road leading from the Central Forwarding Prison in Moscow to Nizhny Novgorod, was especially notorious. By the time Isaak Levitan, a landscape painter and friend of Anton Chekhov, painted an empty stretch of it in 1892, the artwork was seen as a political rebuke and explicit reference to the convict system, even though no exiles were visible in it.

United States or the distance from Geneva to Istanbul – would take thirty weeks.

Fyodor Dostoevsky, himself exiled for many years, drew on these experiences to write books including *The House of the Dead* and *Crime and Punishment*. Anton Chekhov, who spent eleven weeks as a free man on the Tract in 1890 to visit the convict outpost Sakhalin Island, composed an explosive exposé about his experience. 'The Siberian highway,' he wrote, 'is the longest and, I should think, the ugliest road on earth.'[7]

Once in Siberia, punishments varied. Some served out sentences living in remote villages; others would be sent on to prisons or labour camps, working in mines or mending roads. Scandals involving rapes, murders, epidemics, corruption and corporal punishments meted out by sadistic overseers were commonplace. John Dundas Cochrane, a Scottish traveller who visited the Nerchinsk mining region in the 1820s, wrote that it was 'impossible to conceive the haggard, worn down, wretched and half-starved appearance' of the forced labourers he saw.[8]

Arguments about the causes and scope of the challenges in Russia in the build-up to the 1905 and 1917 revolutions have raged ever since. One thing that nearly all factions agree on is that the man at the country's helm was hopelessly, tragically ill-suited to the role. His brother-in-law, Alexander Mikhailovich, wrote that when the future Tsar Nicholas II heard his bullying father had died in 1894, his initial reaction was abject terror: 'What am I going to do, what is going to happen to me, to you . . . to mother, to all of Russia? I am not prepared to be a Tsar. I never wanted to become one. I know nothing of the business of ruling.' Privately, Grand Duke Mikhailovich agreed.[9]

Nicholas was short but well muscled and lean, with a wide forehead and a dashing, horizontal moustache. His temperament was contrastingly flabby. Indecisive, weak and often obtuse, he preferred to imagine that things could and should continue as they always had, rather than confront the challenges of the turbulent time in which he found himself.

From the very moment he assumed power, it was obvious that

Nicholas would prove incapable of meeting the moment. At his own coronation, in May 1896, around 2,000 people were killed or wounded when a crowd stampeded at a fair being held to commemorate the event. Nicholas, perhaps not knowing what else to do and insensible to the impression this would create, carried on with the celebrations as if nothing had happened. Known as the Khodynka Tragedy, this set the tone for his rule.

Like his reactionary father, he mistrusted reforms and reformers, loathed his own bureaucracy and preferred to hark back to a lost golden age of seventeenth-century Muscovite Russia. He adored researching and even trying to revive the clothing, architecture and forms of address from this era. It was the theme of a costume ball held at the Winter Palace in 1903, at which the imperial family attended in bejewelled regalia designed by the artist Sergey Solomko in collaboration with prominent historians. His brother-in-law called it 'the last spectacular ball in the history of the empire . . . while we danced, the workers were striking and the clouds in the Far East were hanging dangerously low'. [10]

Part of the fantasy Nicholas indulged in was the unbounded love of the Russian people for their tsar and his divine right to rule his people as he saw fit. He was a poor judge of character, mistrustful and jealous of his most able ministers, and easily led by sycophants and members of his inner circle. This latter group included his wife, Tsarina Alexandra. In private the couple cultivated a stuffily cosy domesticity – they called each other 'Hubby' and 'Wifey' – but when it came to his role as a leader she was both myopic and implacable. Her measure can perhaps be taken from her response to a worried letter from her grandmother, Queen Victoria, urging her to 'win the love and respect' of her new subjects. 'You are mistaken, my dear grandmama; Russia is not England. Here we do not need to earn the love of the people. The Russian people revere their Tsars as divine beings, from whom all charity and fortune derive.'* She would continually exhort her husband to be

---

* Poignantly, given the brutal fate of her family in 1918, a painting of Marie Antoinette hung over the tsarina's writing desk.

'more autocratic than Peter the Great and sterner than Ivan the Terrible'.[11]

She was also largely responsible for her husband's connection with Rasputin, the notorious holy man. His style combined mysticism and eroticism, supreme confidence and a flair for the dramatic that was particularly effective in the wake of the late nineteenth-century vogue for spiritualism and seances. Physically, he was unprepossessing, with long, dirty hair, straggling moustaches and poor personal hygiene: people compared his body odour to that of a goat. Nevertheless, he had great magnetism, had many followers and was especially popular with aristocratic women. He had innumerable affairs. One lover reportedly told everyone that the orgasm he gave her was so powerful that she fainted.

In an age disillusioned with the corruption and inauthenticity of the Orthodox Church, Rasputin's coarse, peasant simplicity seemed like an antidote. By 1907, he was busy making himself indispensable to the distraught and superstitious tsarina as a healer for the couple's son, Alexei. The damage this relationship would inflict on their reputations – there were, inevitably, rumours that Rasputin and Alexandra were having an affair – was a contributing factor in the fall of the Romanovs a decade later.[12]

Tsar Nicholas II may have been insensible to it, but by the turn of the century it was obvious that revolution was in the air. Karl Marx's *Capital*, published in Russia in 1872, had helped foster a small but burgeoning group of Marxists intent on converting as many disaffected workers, intellectuals, students and soldiers to their cause as they could. The events of Bloody Sunday in January 1905 unleashed mayhem. The stoking of anti-Jewish sentiments by right-wing groups (with the tacit support of Tsar Nicholas II himself) resulted in pogroms.* The public were disgusted by defeat after humiliating defeat at the hands of the Japanese, and by the incompetence and corruption of the country's military leaders. A peace treaty, signed

---

* 'The impertinence of the socialists and revolutionaries had angered the people once more,' Tsar Nicholas II wrote in a letter to his mother on 27 October 1905. '[And] because nine-tenths of the trouble-makers are Jews, the people's whole anger turned against them. That's how the pogroms happened.'

in August, was not as costly as it might have been, but the damage had been done. Military discipline began breaking down. Aboard the battleship *Potemkin*, stationed at Odessa, a row over a flyblown piece of meat escalated to mutiny. Seven officers were murdered and the revolutionary red flag was flown.

By October 1905, anarchy reigned. Workers were walking out of factories in cities. A general strike organised by the Soviets of Workers' Deputies brought the country to a standstill. Railway trains and lines were attacked, even as they served to spread revolutionaries and their literature around Russia. In the countryside, peasants refused to pay rent and began seizing property, vandalising and setting fire to the large estates of their noble neighbours. Some 3,000 manors were destroyed between 1905 and 1906, 15 per cent of the total. Noblemen either fled or were pinned to their estates by terror, anxiously watching to see how brightly the night sky glowed as it reflected the embers of their neighbours' homes. To Grand Duke Mikhailovich it seemed, simply, as if 'Russia was on fire'.[13]

On being advised by the new minister of the interior, Alexander Bulygin, that political concessions needed to be made, Nicholas was both horrified and taken aback: 'One would think you are afraid a revolution will break out.' 'Your Majesty,' Bulygin replied, the revolution has already begun.'

With all other avenues closed off, in October 1905 the tsar finally and unhappily acceded to a raft of measures proposed by Sergei Witte to restore order. On the one hand, the October Manifesto guaranteed freedom of assembly and of the press, an end to religious discrimination and, most importantly, the creation of a legislative parliament, the Duma, elected on a democratic franchise. (Although the tsar had no real intention of sharing power and found any excuse to undercut the Duma: he peremptorily dissolved the first one on 8 July 1906 after just two months.) On the other, Pyotr Stolypin, who took over as prime minister in 1906, enacted a series of draconian measures to quell the unrest.* Regional leaders were empowered to declare martial law – by

---

* Even as Russia roiled, Stolypin followed the Peking–Paris – and wrote at least one congratulatory telegram to *Le Matin*.

January 1907, twenty-three provinces, twenty-five districts and nine cities had done so. Many tens of thousands of 'politicals' were executed or given long sentences. Between 1905 and 1910, the number of penal labourers rose from 6,100 to 28,500, while those exiled to Siberia nearly quintupled, from 6,500 to 30,000. News of the continual chaos, assassinations and executions damaged Russia's political interests abroad – not to mention her ability to secure much-needed loans.[14]

In January 1907, just before the announcement of the Peking–Paris, famine threatened large areas of European Russia, Poland and the Caucasus. A scandal erupted after it was discovered that a privy councillor had misappropriated £14,300 entrusted to him to buy grain, using it to pay off his gambling debts instead. Warsaw's chief of secret police was assassinated. While preparations for the race were under way, local officials interfered with elections and allowed the Black Hundreds to carry out violent attacks, particularly on Jews. On 11 February, ten days before the opening of the Second Duma, two clockwork bombs were discovered in the home of Sergei Witte. The new Duma proved more radical than its predecessor, its deputies frequently clashing with Stolypin and refusing to pass his reforms. On 24 April, a British diplomatic dispatch from St Petersburg gloomily reported 'no improvement . . . in the state of public order. The number of murders and robberies remains much as usual, and, in addition, cases of strikes accompanied by terrorist agitation in factories are reported with increased frequency.'

By late May, a month before the race started, the *New York Times* was reporting 'open revolution' in a dozen provinces. The St Petersburg correspondent of the *North-China Herald* responded a month later: 'The number of mysterious murders, strange suicides, corpses found in boxes, unaccountable disappearances that fall to the lot of the unappreciated and incompetent local reporter every week would make a New York yellow journalist insane with envy.' In rural areas, houses were burned, wells poisoned and cattle hamstrung; in cities, having someone murdered cost only forty or fifty kopecks. 'We are living in a society,' concluded the correspondent, 'where the father suspects his son and the brother his sister. Confidence, the basis of the social structure, is absolutely destroyed.' On Sunday, 16 June, while the Peking–Paris motorists were in Kalgan, mere days from crossing the Russian border, dozens of elected deputies were arrested on the tsar's orders, the electoral law rewritten and the Second Duma dissolved. This marked an end to the First Russian Revolution, and fired the starting pistol for the Second or 'October' Revolution of 1917.[15]

8

Urga – Irkutsk

23 June–1 July

For the first time since Peking, we could treat ourselves to the intoxication of speed. All the levers, all the gas, all the ignition was thrust forward. We were flying. It was a dream, soaring like a star through the vastness of azure and gold.

Jean du Taillis, Spyker team

Although the automobilists spent time each night poring over their maps of northern Mongolia and the Russian border, there was little they could glean from them. At best, they might be described as unreliable. Du Taillis wondered if the scale on his map might be too small to depict roads, since he could not see any indicated. Barzini, in better health and part of a well-provisioned team he had come to trust, was more optimistic. 'The roads beyond Kiakhta were marked by two lines instead of only one. Was not this the proof of some great change to come?' he wrote. 'Those two lines were refreshing to look upon. From time to time we spread out our map for no other purpose than to run along those lines with our eyes, and gain a foretaste of the joys of uninterrupted racing at forty miles an hour.'[1]

Until they reached the border, though, the going would be very tough indeed. From Urga, they had to travel 350 miles due north through land that was as sodden as the Gobi had been dry. The long-feared rain was on its way. The early part of the route was so hilly the vehicles could ascend only in their powerful reverse gears, or by putting the automobiles into a low gear, hopping down and steering while walking alongside to lighten the load. Later, they would need to ford large rivers, including the Chara and Iro, and pass through the moss-green floodplains of the Orkhon, Mongolia's longest river. They had no guides and could no longer follow the telegraph wires. The ground was too low-lying, and the water courses too restless, to allow one sure path to take root. Generally, travellers relied on experience and trial and error, doubling back if they hit too deep a river, forced to meander back and forth as the mutable flow of land and water dictated. When it became absolutely necessary to cross a waterway, at best there would be wooden rafts. More often, they would have to look for fording places – shallow spots with good footing, where vehicles might only wet their wheels – often only navigable when the waters were low.

Throughout the race, the vehicles often had to rely on animal power. River crossings were especially challenging.

Ignoring the reproaches of the other motorists, the Italians had set off from the Mongolian capital at dawn on the morning of Sunday, 23 June. Fifteen minutes later, in a valley bobbing with tufted grass and irises, and still within sight of the city, the car lurched to a stop. Perplexed, Borghese pressed firmly on the accelerator. The engine whined and the air around them was filled with spurts of gritty black spray. The Itala began to tip, as if the ground under the left-hand side was dissolving like sodden paper. The engine screeched with the sound of grinding metal and puffed out clouds of acrid white smoke. 'Stop! Stop!' Guizzardi yelled, beside himself. 'We are sinking deeper!' They had entered the first of the hundreds of miles of bog that lay between them and the border.[2]

An explorer who had earlier travelled the riverine expanse stretching north from Urga to Kiakhta on horseback compared it to an idyllic deer park, with its 'belts of copse-wood, of silver birch and hazel, the ground one blaze of hyacinths, wild roses, and pinks growing in the thick rich grass, with clear, narrow brooklets running here and there'. For the automobilists, the landscape took on a more sinister aspect. Barzini described it as 'traversing a true labyrinth, from time-to-time skirting stagnant pools, winding in and out among high rushes and tufts of iris plants in the vast and desolate swampy plain'. They navigated 'like sailors', using compasses, the sun in the day and stars at night. At one point Prince Borghese found himself driving through a sandstorm that had whipped up from nothing in minutes. Twice the heavy Itala sank in the mire. The Spyker – a few days behind and travelling in convoy with the two De Dion-Boutons – sank once. Du Taillis, seeing all four wheels disappear, bet ten bottles of champagne that the rest of the car would soon follow. This prospect roused his compatriots. After two hours of shovelling and levering, the ground released the Spyker with a final squelch. The Frenchmen were delighted to learn later that the prince, driving alone, had been forced to pay teams of Mongolians to pull his Itala free.

The Italians' fear about imminent rainfall proved well founded. When Borghese, Barzini and Guizzardi reached the Iro on 24 June, the river was 300 yards across and lapped at Guizzardi's waist when he waded through to check the depth. Two days later, when the two De Dion-Boutons and Spyker rolled up to its banks, the water reached Godard's chest. There was no ferry. The only way to cross was to unload the automobiles, remove the sensitive magneto ignitions, smear the engines with protective layers of grease and rags, and beg locals to harness their oxen to the cars to drag them from one bank to the other. Scores of local men and women joined in – on horseback and on foot – waving and pointing, singing and goading the oxen forward as they struggled against the current. The water gurgled over and through the sides of the cars that half floated, half bumped along the bottom. At last, each one emerged 'dewy and steaming, leaving a long track of water behind it'.[3]

The Itala reached Kiakhta on the afternoon of Monday, 24 June, two long days after setting off from Urga. The Frenchmen arrived three days later, by which time the Italians were nearing the shores of Lake Baikal, 240 miles to the north. After China and Mongolia, the border felt oddly familiar to the motorists. 'Here is Kiakhta!' du Taillis enthused. 'Here are its white churches, its white houses, its green roofs! The yellow world has disappeared! European civilisation reveals itself at last!' Barzini wrote that it was as if 'Europe had come to meet us at the gate of Mongolia'.

The formalities of the border crossing were not as onerous as they might have been. The motorists were warned against taking photographs – a stricture they ignored – and to watch out for revolutionary bandits.* They were also tipped off that they would be under surveillance: it was feared their party might contain Japanese spies. Still, the tsar had issued an *ukase*, or decree, ensuring the drivers special treatment at the border and official protection through his empire.† Each car was given a brass plaque with a registration number. Since they were the first over this border, the Itala was 1, Cormier's De Dion-Bouton 2, Collignon's 3 and the Spyker 4. Passports were checked, registration fees paid and glasses of champagne proffered and drunk.[4]

---

* Cormier, a keen photographer, was particularly incensed by the ban on photography. He rebelled against it at every opportunity, delighting in capturing the irritated faces of officials who were trying to prevent him from using his beloved camera.

† This *ukase* is mentioned by du Taillis in his contemporaneous account for *Le Matin* and in his book; it is also mentioned in Cormier's book. Barzini's book downgrades the *ukase* to a *podorojne*, or travel permit, from the minister of the interior. It is possible that the tsar's *ukase* was issued after the Itala crossed the border but before the arrival of the French drivers three days later, on Thursday, 27 June. If an *ukase* was issued, it indicates that the tsar took a personal interest in the Peking–Paris.

Thanks to its strategic position at the border of the Chinese and Russian empires, Kiakhta was built by and for generations of wealthy traders. Barzini called it 'a village of millionaires'. Many of the homes were palatial, surrounded by gardens and conservatories. In the streets, women walked along the wooden pavements in the latest French fashions. At their hotel, the drivers dined in a room illuminated by gold candelabras. The city's cathedral, financed by tea merchants, was the town's extravagant centrepiece, with doors of solid silver. 'The altar alone, of solid gold, silver and platinum, cost £30,000.' Inside, candlelight glinted on a 'huge candlestick and chandelier studded with emeralds, rubies, and diamonds' and many ikons 'thickly encrusted with precious stones'.[5]

The source of this wealth was a single commodity. 'Tea,' a foreign traveller observed, 'is the universal national beverage of the Russians. The moment a Russian awakes he swallows tea, and every act of daily routine is prefaced by libations of tea.' The raw material for all this consumption travelled in large chests, each carefully wrapped in raw hide to protect it from rain, snow or upset. Large caravans composed of two-wheeled carts in summer and sleds in winter were on the move for up to sixteen hours each day to ensure the tea reached its destination as fresh as possible.

'You have no idea what Kiakhta was seven or eight years ago,' a customs official confided to Barzini. 'As many as 5,000 cases of tea used to be unloaded here in a day; nearly fifty million pounds of tea passed this way every year.' Much of it headed to the annual fair at Nizhny Novgorod, through which the motorists would later pass. From there it would be traded throughout the empire and beyond.* Kiakhta's fortunes had begun to sour in 1856 when a treaty between the two empires opened up several Chinese ports to Russian ships, making the route through Kiakhta slower, unreliable and less popular. By the turn of the century the Trans-Siberian Railway, which bypassed the town completely, was threatening to deliver the *coup de grâce*.

---

* Neither the motorists nor the coverage of the race mentioned that while the route they sped along in their automobiles had been determined by the telegraph line, it was also the way that, for many thousands of years, chests of tea – among other goods – had been lugged by camel, mule and man. It is still in use, although diminished, today.

For the town's traders, the Peking–Paris automobiles looked like redemption. An association of tea merchants immediately formed to organise automobile caravans to claw back trade and profits to the overland route. 'They estimate,' du Taillis wrote, 'that while the camel caravans take twenty-two days, automobiles will be able to provide the same service in less than nine.'[6]

Heading towards Lake Baikal, the Italians passed into towns and cities through triumphal wooden arches, often painted to look as if they were hewn from marble. These, Barzini learnt, marked the route used fourteen years previously by the future Tsar Nicholas II, after a trip to lay the foundation stone for the Vladivostok station of the Trans-Siberian Railway. They passed through Verkhne-Udinsk (now Ulan-Ude), a large barracks town and got their first real glimpse of the political turmoil in the empire. Gone were the old-fashioned billowing blouses and long smocks Russian soldiers had worn during the Russo-Japanese War; now the city was peppered with khaki-clad men, as if occupied by an invading army. Bugles sounded, swords and spurs clanked in the streets. Sentries were posted at the doors of the banks and public buildings. 'Even the telegraph-office was occupied by the military: there were soldiers at the door, soldiers armed with rifles and bayonets in the little public waiting-room, and in the office, and before the safe. I felt,' Barzini wrote, 'as if I was writing my telegrams in the ante-room of a prison.'[7]

A methodical man, Prince Borghese had intended to drive along the southern shore of Lake Baikal rather than use the ferry crossing. His maps – the ones that he and Barzini had so enjoyed admiring in northern Mongolia – indicated the presence of a good road. What faced him once he reached the lake, however, was rather different: the overland route, rendered obsolete by the development of the railway, had been surrendered to nature. At best, it was weed-choked and scarcely distinguishable from the surrounding wood. Whole sections had been battered and washed away by storms. Those that remained were 'crossed by fallen trees and by timber brought down by the floods of the thaw'. Inns and other infrastructure that had once served travellers had been abandoned. They came across an old post office, uninhabited and eerie, the roof half tumbled in.

Prince Borghese, Guizzardi and Barzini found themselves measuring progress in feet. Hours passed as they tried to cut a path using axes and spades along the abandoned road. After fruitless hours, the prince straightened, wiped away the sweat cooling on his forehead and admitted defeat: they would turn the Itala around and return to Misovaya, where they had spent the previous night. This was a sad, run-down port in the south-east corner of Lake Baikal. Like the road, it had suffered since the building of the Trans-Siberian Railway: above the waterline lay a listing school of boats, hauled up out of the water to rot.*

The De Dion-Boutons and the Spyker, now only a day behind the Itala and closing in on Misovaya, found themselves in the woods amongst cyclamen, ragwort, foxgloves, ferns and wild strawberries. The track was, more or less, a continuation of the surrounding woodland: 'potholes, lethal and breakneck. In some places the route became unclear: you had to choose between swamps and inaccessible climbs.' The French motorists found their surroundings picturesque but the going monotonous. They took comfort in the sudden closeness of the Trans-Siberian Railway. The puffing columns of steam gave Cormier the impression they were 'less isolated, less cut off from civilisation. From now on,' he continued hopefully, 'we'll be able to find healthier, more varied food to eat.'

The sight of the railway also inspired the prince. While the lakeside track was unnavigable, crossing the hundreds of rivers – 175 by their count – was proving truly insurmountable. At one time there had been plenty of ferries and bridges; by 1907 few remained. Where possible, the prince preferred to ford the rivers rather than trust his car to the rotten, creaking bridges, which in any case were often half submerged in the still swollen waterways. One collapsed with a screeching groan right after they crossed it. Before long, they came across another so twisted that it resembled a distorted skeleton, completely impassable.

---

* The very same day the Peking–Paris motorists struggled and failed to make any headway at all, 29 June, Selwyn Edge set a twenty-four-hour distance record at the purpose-built racing track at Brooklands, a record that would stand for eighteen years.

A solution occurred to them as they sat around a sputtering sam-ovar drinking tea in the modest log-house home of a Misovayan grandee. All these rivers did, in fact, have new and intact bridges over them: those belonging to the railway line. The prince tele-graphed the regional governor general requesting permission for the Itala to drive along the railway tracks.[8]

Meanwhile, lost in the primeval forest on the way to Misovaya, the two De Dions and the Spyker had to be dragged through a river by teams of farmers' horses, like carts on their way to market. Cormier, Collignon and Godard looked on in disgust at this undignified spectacle that seemed to mock their very identities.

Over the following days, the fine drizzle that had greeted them at the Russian border on 27 June turned into persistent rainfall. Water pooled in the footwells of the cars and worked its way into their mackintoshes, so their woollen undergarments grew heavy and began to reek. They were tired and out of sorts. Russian hos-tels and post houses, it turned out, were basic in the extreme: 'Just a straw mattress, a single sheet and a blanket.' To make matters worse, on their first night in Russia, Collignon had woken with a shout to find someone trying to climb into bed with him. In a panic, he grabbed the intruder by the scruff of the neck and hurled him across the room. Bizac and Cormier sat up in bed, striking matches with shaking hands to try and locate the source of the ruckus. In the flickering light the pathetic figure of one of the hostel's staff was revealed, dead drunk, crumpled by the door.

Adding to their discomfort was the increasing physicality of their journey. The road had become so slippery that they had to unload their automobiles at the base of each incline. This took on a dis-piriting rhythm: once the vehicles had crested the summit, driv-er and passengers would trudge back down and carry everything – 'the exterior pieces, camping equipment, tents, beds, luggage, spare tyres' – back to the car, reload it, tie down the canvas over their already sodden possessions and set off towards the next hill. The final five versts to Misovaya took them two hours: a very slow walking pace.* They arrived wet, filthy, covered in mosquito bites and bad-tempered, exuding none of the glamour the townsfolk might have expected from motorists, and in far fouler moods than the prince and his companions had demonstrated over the preceding days.

With ill grace, the Frenchmen allowed a local policeman to guide them to where they would spend the night. They immediately set to ordering dinner: bread, eggs, roast beef, fried potatoes and beer. Later, they turned down a local delicacy, a special variety of fish. They complained loudly about the slowness of the service and the quality of the food, extolling the superiority of French restaurants. Further seasoning their discontent, a visitor dropped by and told them the prince had set off the previous day with the intention of following the train tracks. That settled it. Cormier decided, contrary to the rules, that they would all load their cars *onto* the train instead.† It did not improve their moods to discover the next morning, when they went to pay their bill, that what they believed to be a hotel had in fact been the private home of the mayor and it had been the cooking of his wife that they had been castigating.[9]

By now, Borghese, Barzini and Guizzardi had come to regret their intrepid attempts on Lake Baikal's southern shore. 'We are alive,' Barzini wrote from Irkutsk on 3 July, 'by a miracle!'

After days kicking their heels around the samovar in Misovaya, papers signed by the governor general permitting them to use the railway bridges had been delivered by two policemen on Saturday, 29 June. ('Such a visit,' Barzini noted ruefully, 'is not always a good augury in Siberia.') They had set off the following day at half

---

*Versts were a measure of distance in Russia at the time. One verst is equivalent to 0.66 miles or 1.07 kilometres.

† Now three weeks and several thousand miles into the race, the rules formulated months previously at the offices of *Le Matin* probably felt like episodes from another life. At a meeting in late February, it had been agreed that 'Under penalty of disqualification, the contestants will refrain from travelling by rail between Peking and Paris.' In du Taillis's book, he suggests that they did attempt to drive around Lake Baikal before deciding to take the train; his dispatch to *Le Matin* on 2 July suggests otherwise. Cormier's dispatch to *L'Auto* churlishly suggested that the prince had caught a boat from Misovaya, eliding the Italians' far more convincing attempt to complete this section by car.

past four – missing the Frenchmen by a whisker. As they left, Lake Baikal was as flat and soothing as a piece of sea glass. They hoped to reach Irkutsk – 200 miles away, beyond the western shore of Lake Baikal – on Monday, 1 July.

After three hours' rough lakeside driving, they cut across to the railway track by the River Mishikha. A policeman, thrilled with the novelty of directing an automobile and clutching a small red flag to signal to any trains they came across, hopped aboard the Itala and indicated the direction of the first railway bridge. Like all strategically significant bridges in the empire, guards had been stationed at each end to prevent revolutionaries from blowing it up. The first of many they would cross that day, it was 'scarcely wider than the sleepers', Barzini later reported, 'and without parapets, and suspended over deep ravines, the waters of which we can see foaming beneath us. The right wheels of the automobile, running outside the right rail, pass almost over the edge of an abyss.' The motion of the car over the sleepers was a juddering, tyre-and-nerve-shredding gallop. Barzini, ensconced in the back and determined not to look down, tried not to imagine the yawning gulf below them. The policeman, perched on the running board, squeezed his eyes shut, his knuckles turning white around the red flag.

At nine o'clock, a local stationmaster called out from a little booth by the track that a train was on its way. Rather than stopping to wait, the three men decided to head back to the abandoned road parallel to the tracks to try their luck until the train had passed. Half a mile further on, they reached a deep river, spanned by an old wooden bridge. Eyeing it, they hesitated. It was certainly rickety, but no more than others they had passed over, and it looked strong enough to risk. They began to inch over, Guizzardi at the wheel, the prince, leaning forward, tense beside him, with Barzini wedged behind.

The wooden structure, grey with age and patched here and there with moss, swayed and creaked as it bore their weight. They held their breath. Just as the front wheels approached the opposite bank, there was a terrifying crash beneath them. Time slowed and stretched. The machine lurched, rearing up as its tail end plunged through splintering planks towards the river below. Barzini found himself half submerged in the water, his legs pinioned by luggage and wooden beams, the car looming perpendicular over him, threatening to fall back and crush him beneath the water. Droplets of hot oil rained down from the engine. The prince's trousered legs, with their neat turn-ups, were suspended above him, kicking madly. Through the ringing in his ears, Barzini heard shouting in Russian. He turned his head: the policeman had jumped clear at the last moment and was running off in the direction of the railway line. Guizzardi, still in the driver's seat with his feet in the air, twisted himself free like a cat and limped round to help Barzini lift the debris off his throbbing back and legs.

Moments later the trio, shaking with adrenaline, stood looking at the wreckage. A small portion of the grille and the front wheels were poking forlornly through a ragged hole in the bridge, but the Itala appeared to be in one piece. Around them lay sodden luggage, bent tools and splintered wood. Their clothes were torn, wet with oil and river water and stained with engine grease and the green mosses that had covered the bridge. Miraculously, however, there was not a single broken bone between them.

The horrified policeman, still clinging to his flag, returned with a team of railway workmen. Barzini could barely move. He lay down and watched as the men, under the direction of the prince and Guizzardi, righted the car and built a ramp for it to ascend the bank. The prince swung himself behind the wheel while Guizzardi went round to the front of the car, his hand on the crank handle. At a nod from Borghese, Guizzardi plunged the handle round. The engine coughed, then caught and held at a steady thrum, as if nothing had happened.

By two o'clock in the afternoon, the three men were back driving along the tracks, this time accompanied by two railway engineers, tasked with gingerly walking out to test each and every bridge they came to. The prince, perhaps hoping to lighten the mood, shakily suggested that at least they had enjoyed their fair share of adventure for the day. Guizzardi and Barzini exchanged glances and said nothing.

Four miles from Tankhoy, at a small port a few miles east of Lake Baikal's southernmost point, a railway worker ran up to signal the Itala off the rails. The faint whistle of an approaching train could be heard, and a column of steam rose through the forest ahead. The prince, raising a hand in assent and lowering his goggles, turned the wheel to guide the heavy automobile off the tracks and down the embankment. The Itala jerked to a stop: the wheels were caught in the soft sand between the sleepers.

Borghese, Guizzardi and the engineers panicked. Scarcely risking a glance to judge the distance between them and the train, they jumped down, put their shoulders to the car and pushed. '*Vai! Vai!*' Guizzardi shouted, beckoning up to Barzini to come to help. The journalist could only gaze back at him in horror: his legs, which had been tingling unpleasantly ever since the accident, refused to move. He cast around. One man was trying to build a little ramp from a pile of wood nearby to help lever the car up and over the rails, hurriedly looking over his shoulder and up the track. Barzini was growing frantic, massaging his muscles and bracing his arms to pull himself clear of the spare tyre he was using as a seat. Beneath him, the metal tracks began to hum as the train emerged around the wooded bend a few hundred yards away. He dared not look up. Below him, he heard confused shouting; ahead, the train's screaming whistle. His companions ran to the back of the Itala to give it one last shove. This time, the wheels bit. With a sickening lunge, the car heaved off the track just as the train roared past with an indignant blast of its whistle.[10]

When they reached Tankhoy hours later, it was buzzing with the news of their near escape. The prince, still pale, found himself clapped on the back, cheered like a hero and ushered to the only place the village had to offer its guests: the stage of the little wooden theatre. They prepared for bed in silence, tired and sore, conscious of the armed guards with fixed bayonets that the authorities placed at the entrance to guard against attacks by revolutionaries and exiles. Dramatically spotlit, and in front of a set painted in readiness for an upcoming amateur production by some railway workers, they stiffly removed their clothes, examining their various scrapes and bruises before empty rows of seats. 'The curtain was raised,' Barzini wrote, 'the footlights flooded us with light (we had to hunt an hour before we could turn them off), and amid all that splendour we three undressed to go to bed, sighing and moaning . . . We seemed to be acting out some scene out of a farce.'[11]

9

SEPARATE SPHERES

Women and the Automobile

> The average woman is probably quicker than the average man in gathering from a road map the information which it has to offer.
>
> Dorothy Levitt, 1909

Dr Bill French did not trust horses. This was unfortunate: as a doctor in Washington DC in the late nineteenth century he had to travel a good deal and was therefore utterly dependent on his horses, which, he assured his family, were just waiting to rebel against him. It was with alacrity, therefore, that he visited a new showroom selling steam locomotives in 1899 with his daughter, Anne Rainsford French. For both father and daughter it was love at first sight. 'She stood there in sleek black dignity with her Stanhope body on four pneumatic bicycle tyres. Her dashboard was all patent leather and the whipsocket was still in place as insurance to the buyer who might have to hitch up his horse on occasion. Bill objected only to the whipsocket.'

Before long Anne was entrusted to make repairs, drive her father to appointments and keep abreast of the rules of the road – state laws varied, but where they lived the speed limit was nine miles per hour on main avenues, five at crossings and four when turning. Then, in early 1900, she became the first woman licensed to drive an automobile in America.

'I don't think the examiners were happy when I applied,' she later recalled. 'I was all prepared to take down the engine and put it together. They said it wouldn't be necessary and just asked me to name a couple of the parts.' Newspapers weren't quite sure what to make of her either. They reported her achievement, but seemed a lot more interested in her appearance. '[She] is plump and pretty, with a dazzling complexion and fathomless blue eyes. Her shoulders are absolutely flawless from an artistic point of view.' Her younger brothers called her 'Old Fathomless' thereafter. Perhaps most telling was the reaction of her husband a few years later: he would not let her behind the wheel at all. 'Driving is a man's business,' he told her firmly. 'Women shouldn't get soiled by machinery.'[1]

It is often assumed, since motor culture would later revolve around the needs and desires of men, that women were entirely absent from the automobile's salad days. In fact, there were a number of female motorists in 1900.* From the beginning, however, they were met with the widespread cultural assumption that they were interlopers in a male space, a belief that would stubbornly persist – even strengthen – over the following decades. Events like the Peking–Paris only helped cement the brawny, venturesome image of the automobile, making it still more difficult for the majority of women to access them.

Automobiles, especially ones that could travel great distances, represented a challenge to the status quo in an age when women were culturally aligned with the home. Women were either idealised as more delicate than men, less mentally capable and in need of protection, or traduced as 'fallen' and therefore unworthy. Riding in or driving an automobile put a woman's reputation at risk. As Miss Francis Thorton, a New York driving instructor in 1914, put it: 'In the early days . . . there was much opposition to the idea of women driving because it was thought that they were too frail to drive a "devil wagon".'

The term 'fast', applied to women of loose morals, irresistibly recalled speeding automobiles. A 1905 serial in *Motor* magazine portrayed the relationship between cars and women as distinctly

---

* Since automobiles were often bought by and registered to men, even if women drove them, it is difficult to get accurate numbers for women driving in the 1890s and 1900s. Some American cities and states that had registration records suggest female car ownership in America in 1907 was steadily climbing, but still only around 10 per cent.

sinful: 'To think of "The Monster" as she called it, was to long for it . . . She wanted to feel the throb of its quickening pulses; to lay her hand on lever and handle and thrill with the sense of mastery; to claim its power as her own – and feel its sullen-yielded obedience answer her will.'

Women's fashions of the era – voluminous skirts with tightly fitting waists, puffy sleeves and veiled hats – were decorative rather than practical and scarcely allowed women to clamber into an automobile, still less drive one. Those determined to do so had to jury-rig their attire so it would not get caught in levers or dragged into axles. Christabel Ellis, who took part in the July 1908 Ladies Bracelet Handicap race at Brooklands, for example, tied a cord around her skirts above her ankles to keep them in check.[2]

Since automobiles were usually open to the elements, women drivers were also conspicuous. A female columnist for *The Motor* wrote in 1904 that 'There is nothing secretive about the motor car. It says, in effect . . . look at me and look at my passengers.'

As well as extra scrutiny on the street, incidents and accidents involving female drivers held a disproportionate allure for the press. On a single day, 27 July 1907, the *Daily Telegraph* gleefully carried two such stories. One involved the German empress whose vehicle broke down in Braunsberg – a region famous for its horse-breeding – and sought refuge in a hotel. 'The consequence was that when the Empress entered, muffled up and completely disguised in waterproofings, she was greeted with remarks of a chaffing and anything but a respectful nature.' The other involved the death of an eighty-year-old pedestrian near Wimbledon, who, the paper announced, was hit by 'Mrs C. N. Williamson, part author of *The Car of Destiny*'.[3]

The most prominent female drivers around 1900, like their male counterparts, tended to be wealthy, which helped insulate them from public censure. The Gilded Age novelist Edith Wharton was a keen motorist. She was born into a prominent New York family, the Joneses, who are thought to have been the ones to be kept up with.* Others included Dame Ethel Locke King, Camille du

---

* The phrase 'keeping up with the Joneses' came from a 1913–38 comic strip of the same name that ran in the *New York Globe* and the *New York World*.

Gast, Hélène de Rothschild and Alice Roosevelt, who horrified her father, the American president, by travelling unchaperoned from Rhode Island to Boston in a Panhard racer at speeds of up to twenty-five miles per hour.[4]

For other women, risks to their reputation held less sway than the exhilaration and freedom motoring offered. Dorothy Levitt was an early racing driver who twice broke the women's world speed record and beat several men during a race at the 1905 Brighton Speed Trials.* She parlayed her driving prowess into modest celebrity, and proclaimed in the opening pages of her 1909 book, *The Woman and the Car*, that 'Motoring is a pastime for women ... the intense pleasure, the actual realisation of the pastime comes only when you drive your own car.' Her attitude was not that driving was safe or easy, but that women were more than capable of motoring anyway. She recommended that her fellow female motorists carry with them – in addition to pliers, hammer, grease injector, screwdrivers, inlet and exhaust valves and

Dorothy Levitt, record-breaking driver and inspiration for early female motorists, at the wheel of a Napier at the Brooklands racing circuit in 1908.

---

* In July 1905 Levitt raced an 80 hp Napier at 79.75 mph at the Brighton Speed Trials; the following year she broke her own record at Blackpool with a speed of 91 mph in a 90 hp Napier.

innumerable other tools and spares – chocolates, a hand mirror and a small revolver.*[5]

Even for women like Levitt, opportunities for racing were limited, and women were often excluded from national automobile clubs. In 1909, for example, the American Automobile Association banned women racing drivers altogether, right around the time Joan Newton Cuneo beat several leading male racers in New Orleans. After an experimental relaxation of the rules to allow the 1908 Ladies Bracelet Handicap, Brooklands also forbade women from official races, using the rationale that, since there were no female jockeys, it was hardly to be expected that there should be female racing drivers.

When they could find opportunities to race, women had to tread a fine line. Stories about their wins took care to emphasise how feminine they were, as if to balance out the fact of their driving prowess. The male editor of Levitt's book, for example, wrote in the introduction, 'Looking at Miss Levitt one can hardly imagine that she could drive a car at such terrific speed. The public, in its mind's eye, no doubt figures this motor champion as a big, strapping Amazon. Dorothy Levitt . . . is the most girlish of womanly women.'

Others were less accepting. A writer for the American magazine *The Outlook* insisted, 'No licence should be granted to anyone under eighteen . . . and never to a woman.' In 1908, the mayor of Cincinnati tried banning female drivers entirely after one drove over a city treasurer. 'The only proper machine for a woman to run,' he decreed, 'is a sewing machine.'[6]

Assumptions about who would drive automobiles and why had an impact on the development of the industry. For example, while it is often assumed that electric automobiles were abandoned because they were an inferior technology, in fact they were more reliable

---

* Chocolates she advocated as soothing. The mirror was useful not only for fixing the hats and scarves that women wore while driving, but also, as cars did not yet have rear-view mirrors, 'to occasionally hold up to see what is behind you'. When it came to firearms, Dorothy favoured a Colt, finding it 'very easy to handle as there is practically no recoil – a great consideration to a woman'.

and cheaper for short, intra-city journeys. As a result, in 1900 a little over a third of cars in New York, Chicago and Boston were electric and the market for them persisted into the 1920s. It is true that the technology stagnated, but one reason this happened was that electric vehicles became associated with women, while those with combustion engines were seen as more masculine.[7]

The 'femininity' of the electric car was already entrenched by 1900. Perhaps the most compelling reason for the association was their 'circumscribed radius', as one commentator put it: because the batteries only carried a certain amount of charge, electric vehicles could never stray too far from home. 'Has there ever been an invention of more solid comfort to the feminine half of humanity than the electric carriage?' the writer Carl H. Claudy asked in the pages of *Motor* in 1907. 'What a delight it is to have a machine which she can run herself with no loss of dignity, for making calls, for shopping, for a pleasurable right, for the paying back of some small social debt.'[8]

Electric vehicles were also seen as cleaner, more convenient, quieter and easier to run – all qualities that, it was assumed, suited them to women. Motoring companies of all kinds took note and began marketing features that made driving easier specifically for women. The Haynes-Apperson, an early American car, assured potential buyers in 1904 that it was 'the only powerful automobile simple enough for a lady to run *easily*'. Models with electric ignition systems – easier to use than crank handles – often used images of women or copy that referred to them. One from 1906 proclaimed that 'A lady at the wheel of the Harrison Motor Car can start and operate it – the motor is started from the seat . . . a crank is not necessary.' Model Ts, which remained crank-started until the 1920s, could be fitted with electric starters as an optional extra. Manufacturers of one such device reassured potential purchasers that theirs was so easy to use 'a woman or child can do it'. Another that 'Ford owners, particularly their wives and daughters' would henceforth be freed 'from all the trouble of driving a car' once their device was installed.[9]

This kind of rhetoric would develop over the century, with the automotive industry making assumptions about women being helpless, bad drivers and generally unsuited to the rough and tumble of the road. In 1964, an American Volkswagen Beetle ad showed a car with a rumpled bumper bearing the legend 'Sooner or later, your wife will drive home one of the best reasons for owning a Volkswagen'. Underneath it explained that while women were 'soft and gentle', they 'hit things'. Around the same time, Goodyear produced a series of ads showing various women stranded by the side of the road and staring in dismay at flat tyres. 'When there's no man around,' the copy read, 'Goodyear should be.'

A decade earlier, Chrysler produced a Dodge La Femme: an ill-advised salmon-pink and cream automobile upholstered in 'Orchid Jacquard', sold with a complimentary lipstick holder, umbrella, handbag, cigarette case, lighter and purse. All, naturally, pink. The exact number sold is not known, but it was probably small: the entire concept was dropped after just two years.[10]

What Chrysler's admen – like so many others – overlooked was that a woman's reasons for wanting a car were identical to those of a man: conspicuous display; convenience and sociability; newness; adventure; self-reliance; speed; participation in a modern, exciting technology; and, perhaps most potently, a sense of freedom. The latter motive was especially true for those who lived far away from amenities, family and friends, in rural areas or suburbs. During the Middletown studies conducted in Muncie, Indiana, during the late 1920s and early 1930s, women revealed they had quickly come to see the automobile as a necessity, even in lean financial times. Some admitted to putting the expense of running a car above the purchase of clothes or even food. 'The car,' one woman told the interviewers, 'is the only pleasure we have.' Another, who had prioritised buying an automobile over indoor plumbing, found the choice an obvious one: 'You can't go to town in a bathtub.'[11]

Edith Wharton wrote lyrically about the experience of car ownership and driving after a two-week, whistle-stop tour of France with her husband and brother in 1906.* Her series of essays for

---

* Henry James, hearing about her frenetic itinerary, could scarcely imagine anything worse, calling it 'a frustrated, fragmentary merely-motory time'.

The Atlantic about the journey – later published as *A Motor-Flight Through France* – opens with an illuminating *cri de cœur*:

The motor-car has restored the romance of travel. Freeing us from all the compulsions and contacts of the railway, the bondage to fixed hours and the beaten track, the approach to each town through the area of ugliness and desolation created by the railway itself, it has given us back the wonder, the adventure and the novelty which enlivened the way of our posting grandparents. Above all these recovered pleasures must be ranked the delight of taking a town unawares, stealing on it by back ways and unchronicled paths, and surprising in it some intimate aspect of past time, some silhouette hidden for half a century or more by the ugly mask of railway embankments and the iron bulk of a huge station.[12]

10

Irkutsk – Tomsk

1–11 July

They say that when you have had two misfortunes in one day, a third one is sure to come.

Prince Scipione Borghese, Itala team

Although its expanding suburbs had long since crossed the river, the ancient heart of Irkutsk lay inside a sharp curve in the Angara River, forty miles west of Lake Baikal. Reflections of the many green and gilt-domed churches shimmered in the fast-flowing waters. The city boasted a large railway station, a series of wide shopping streets and a theatre. An imposing white structure with high walls and barred windows served as the central forwarding prison, through which millions of men and women passed on their way to their assigned place of exile.

Jeremiah Curtin, a big-bearded American ethnographer and collector of folk tales who visited in 1900, found Irkutsk something of a curiosity. Unusually for a Siberian town, many of its buildings were made of stone or brick rather than wood. And, although it dated back to the seventeenth century, much of it seemed new. (In fact, the few older edifices, including the prison, were survivors of a fire that had consumed three-quarters of the city over two days in July 1879.) Close to, its hasty rebuilding gave Irkutsk an irregular, dishevelled appearance. Nevertheless, it was the administrative capital of Eastern Siberia, a principal node in the country's exile system and the home of many of the region's wealthiest men. Socially, it consisted of an uneasy mix of government officials, brash gold-mining millionaires and tradespeople, many of whom were also exiles.*[1]

Bruised and battered from the accidents and near misses of Sunday, 30 June, Prince Borghese, Barzini and Guizzardi drove into Irkutsk the following evening. By now, all three had their assigned roles. The prince would secure lodgings, glad-hand local dignitaries and Peking–Paris committee representatives if there were any, purchase supplies and check their onward routes; Guizzardi would tend to the Itala and refuel; Barzini would head to the telegraph office and send off sundry reports. From Irkutsk, he dutifully praised their Italian Pirelli tyres – they had 'stood the strain splendidly' – and boasted that even after 1,500 miles 'of the most impossible roads', the Itala did 'not require the least repair'. This was not entirely true. Earlier that day they had discovered that the brakes, damaged by the bridge fall, had engaged while they were driving, causing so much friction that the lubricating oil had caught fire. Flames were licking over the gearbox before the three men managed to douse them with water from a muddy ditch.

They allowed themselves a full day in the city to settle their nerves and repurchase anything that had not survived the fall through the bridge into the river or the drenching in motor oil. There was also – despite the natural inclinations of the prince, who rarely allowed others to call the shots – fun to be had. Hosted by Mr Radionoff, a wealthy merchant and enthusiastic member of the Russian Peking–Paris committee, Borghese and Barzini dined on caviar and champagne and toured the nightspots of the Paris of Siberia, staying out well past dawn.

Guizzardi, meanwhile, attended to the Itala. Excess baggage and equipment were discarded, finally clearing the third seat for Barzini. A fire hose sloughed off the accumulated mud from the first twenty days of travel and a painter was engaged to write 'Pechino–Parigi' in white letters along the car's weather-beaten paintwork.[2]

---

* The explorer Harry de Windt dined at one home where 'an enormous gold nugget was placed on the table, and used as an ashtray, our host announcing in a loud voice that its use in this capacity lost him annually £300 in interest!' This was obviously a favourite boast in the city to impress foreigners: the same line had been used on Victor Meignan several years previously.

On 3 July, after their first lie-in of the race, the Italians left at eleven o'clock in the morning, accompanied by Mr Radionoff (who had invited himself to travel with them), bearing an enormous parcel of sandwiches to assuage their hangovers.*

Seven hours later, Collignon, Cormier, Bizac, Godard and du Taillis rolled into Irkutsk train station from Misovaya.† The two De Dion-Boutons needed urgent repairs: water from the river crossings and rain had swollen the fibres of their clutch plates, which would now need replacing. But the Frenchmen also wanted to enjoy themselves, and the city wanted to make the most of its exotic and rather less formal visitors.

In addition to the city's other amusements – and more to the tastes of Cormier, Collignon and Godard than the theatre or folk-dancing exhibitions Borghese and Barzini had visited – Irkutsk boasted a lively velodrome. There the Frenchmen were feted for two days straight with a series of banquets, as well as impromptu bicycle and motorcycle races – won on two occasions by Godard, who was insufferable in victory. Applause thundered around the track and the local military band struck up tune after martial tune. 'People jostle to lift us up,' du Taillis informed readers. 'They want to carry us on their shoulders in triumph.' The climax of this spectacle was a women's motorcycle race. It was, according to du Taillis, 'the graceful crowning moment of the evening'.[3]

On 7 July, the Itala and her crew had reached Krasnoyarsk while the two French and single Dutch vehicles, having squandered the time they had made up since crossing the Russian border, were 570 miles behind them near Cheremkhovo, a village half a day's drive from Irkutsk. Already, the automobilists were beginning to

---

* Mr Radionoff, who Barzini evidently resented for immediately seizing the free third seat and relegating him to the running board, travelled with the Italians for some three hundred miles. He left them only at Nijni-Udinsk after contracting 'a sudden cold in the eye, which prevented him from appreciating any longer the pleasures of too long a motor journey'.

† Edgardo Longoni may also have been present: there is some confusion in the accounts about when exactly he had to return to Europe. He isn't mentioned in accounts often, but does crop up in Cormier's book from time to time. Du Taillis, however, reported to Le Matin that Longoni only travelled with the automobiles from Irkutsk to Krasnoyarsk.

feel heartily sick of Siberia, with its damp weather, bad tracks and dull food. Days melded into one another. Trees, tea, boiled eggs, dry biscuits, rain and mud. Du Taillis's latest dispatch, in contrast to the chatty epistles sent from Irkutsk, began brusquely: 'Progress is difficult.' Barzini likewise remembered these days as 'wet, dark, and sad. The hours went by and the miles went by, slowly . . . We were gloomy and almost resentful.'

Just outside Cheremkhovo, Godard and du Taillis, swaying along in their overladen, Harlequin-bright striped Spyker ahead of the two De Dion-Boutons, saw someone on the track motioning for them to stop. He was 'a great devil . . . huge, with green eyes, a sparse but dishevelled beard and bushy hair'. The man, who the motorists took to be a shaman, wore a loose tunic buttoned at the shoulder, coarse long Siberian boots and a dingy, discoloured cap on his head that had once been red velvet. At the sight of this apparition the two men got the giggles. Once they had begun, they laughed until tears rolled down their cheeks.

Clearly this was not the greeting the Siberian had been hoping for. Face darkening, he 'gesticulated like a man possessed, and advanced with threatening gestures'. Still chuckling, Godard reached down to engage the hand-operated clutch to drive off just as the man reached into the car and began yanking at levers.

'What an old soak!' Godard yelled. 'I'll put him in his place!'

Turning, he threw a wild punch. The shaman lost his balance and fell to his knees at the edge of the road. The next instant, the Spyker was back in gear and on the move, Godard's hands gripping the wheel. Du Taillis swung round in his seat to see the man, still prone, raise 'vengeful hands in our direction'. A few minutes later, seemingly without reason, the Spyker stopped dead without warning, as if its engine had been choked off by the shaman's gesture.[4]

Siberia covers an area of more than 5 million square miles – almost 9 per cent of the earth's land surface. It is twenty times the size of France; a third larger than the United States. In 1907 the population was miniscule, even after centuries of forced and voluntary migration. Samuel Turner, a British butter merchant

*Many of the tracks the vehicles had to use through Siberia were little more
than mud slicks winding through forests and fields.*

and climber who visited in 1905, noted that the population was
one-sixth that of Great Britain, or 'somewhat less than that of
Greater London'.[5]

It would take the Italian team nine days to drive the 1,000-
odd miles from Irkutsk to Tomsk through Nizhneudinsk and
Krasnoyarsk. The less powerful French cars took four days longer
to cover the same ground. The stricken Spyker did not make it at
all, since it had to be towed by two horses back into Cheremkhovo
and was diagnosed with a defective magneto, a vital part of the
ignition system. Godard conferred with Cormier and Collignon
and agreed that the two De Dion-Boutons should go on while
Godard would hang back to try and fix the magneto or, if this
proved impossible, telegraph Spyker and beg for a replacement.
Du Taillis would stay with him.

The route to Tomsk took the remaining cars west-north-west.
'The rough ascents enclosing the Baikal declined gradually until,
changing into a gentle undulation, they became extinguished in
the limitless plain of Central Siberia,' giving Barzini the impression of 'a vanishing storm'. They drove through hour upon hour
of Siberian taiga, blanketed with dense, resinous pine forests,
bisected by flashing rivers. The automobiles crossed the Yelovka,

the Belaya, the Oka, the Ili and more. Few had bridges; most were plied by small ferries.

Every few miles they came across a clearing cut into the forest to make way for a village. A notice on a black and white pole would grandly announce its name, distance from its nearest neighbour and the number of inhabitants. Other than that, they all seemed the same. Surrounded by spreading skirts of pasture, rye and barley fields, they huddled in valleys, sheltered from winds that swept down from the icy tundra to the north. Simple fences penned in geese, pigs, chickens and horses that otherwise wandered freely among the houses. Constructed of unpainted wood, the cottages seemed from a distance to blend in with the bark of the surrounding forests. In marshy areas, they slowly subsided into the ground, lolling like old tombstones.

Driving through each, the cars sank up to their hubcaps in the mire that accumulated between the houses. Odours of old skins, resin and turpentine assailed their noses, woodsmoke cut with a base note of sewage. Windowsills were crowded with potted red geraniums. Horseshoes hung above doors or were set into thresholds. Barefooted women in pale dresses stopped like actresses in a pastoral tableau – drawing water, picking mushrooms, scattering grain for hens – and stared at the automobiles. Some ran into their homes to hide. Others would slap their thighs and laugh until they cried at the sudden appearance of horseless carriages with their mud-caked crews in driving coats, hats and googles. Men would gather round and offer them vodka.* When Cormier, at his most schoolmasterly, refused on behalf of his compatriots, the villagers would shrug, taking a nip themselves.[6]

This was the height of Siberia's summer, usually a busy period for communities who depended on the land. This year, however, was a little different. The *rasputitsa*, a Russian term for the wet spring and autumn seasons when roads become impassable, was proving to be

---

* Vodka was little known outside Russia at the time. Collignon dubiously described it as an '*eau-de-vie* made from grain [that] gives them [the Russians] real pleasure'. The butter merchant Samuel Turner placed the word 'vodka' in inverted commas and provided the following helpful footnote for his readers: 'Brandy, usually made from barley or rye. A cheaper kind is made from potatoes.'

132

a drawn-out affair.* Over the first few days of July, it was unseasonably cold. The competitors, in their open cars, found themselves 'benumbed, although clad in furs', but worse was to come. For the first 400 miles or so from Irkutsk conditions were relatively dry. 'The roads are good,' Barzini telegraphed from Nizhneudinsk, 'and we are able to maintain an average speed of twenty miles an hour.' But by the end of the first week, rain began to fall in earnest and seemingly without end, turning the tracks to sludge.

Even if conditions had been dry, the quiet road would have been challenging. De Windt, who travelled along it by tarantass, a private horse-drawn carriage, was scathing. It 'would scarcely be called a road in any other country ... I often wondered our tarantass did not come to pieces altogether long before reaching Tomsk.' That had been before the railway had been built. A decade of neglect had not improved matters. In dry conditions, it was dusty, overgrown and riven with ruts. Cormier described it as being, at best, like a flowery meadow 'where the grass often grows taller than our cars' chassis. It is extremely unpleasant to drive in these conditions, without being able to see the ground, and feeling constantly apprehensive about falling into a hole.' Entire days were spent in first gear. 'We use second a few times,' he recorded morosely.

As torrential rain set in, the surface turned to sludge. The cars sulkily rebelled against the conditions. They slipped, balked and skidded, tyres throwing up fans of viscous brown mud that beat back down over the occupants. In some places, felled branches and tree trunks had been laid over the surface to keep peasants' carts from sinking. As Cormier's De Dion-Bouton drove over them, its 'axles knocked, the springs creaked and bent, the body of the car and its heavy load leapt from side to side. The engine snorted, slowed down, whinnied again.'

---

* *Rasputitsa* literally translates as 'disagreeable travel' and was so severe, even in normal years, that it had a noticeable impact on Russian society, economics and politics. It was credited with halting various invasions, including that of Napoleon in 1812, and would later hinder German advances during the Second World War. In 2022, when Russia invaded Ukraine, they themselves became casualties of 'General Mud', as convoys became stuck and advances stalled.

They were reduced to speeds that would have disgraced country carts hitched to the weariest of nags. Distances they expected to cover in a day took them three or four. 'It is impossible to give any idea of what a Siberian road is like with such rain,' Barzini wrote with fervour. 'This black earth is, as it were, a pulp, into which for thousands of years the dead grasses have been worked . . . It is saturated with organic material, soapy, thin; when it is wet there is no more difficult feat than to walk on it keeping one's balance. The car stopped on it and gave signs of the most deplorable rebellion against being guided. It turned round, it threw itself aside.'

After over an hour trying and failing to climb one incline slap-stick-slick with mud, the Itala's crew gave up and glumly sheltered from the rain in the home of a railway watchman. 'Is summer always like this here?' they asked. 'There has not been a summer like this here within the memory of man,' he replied as he calmly filled his pipe. 'Never have we had such cold and such rain in July. It is impossible to work the fields . . . We shall have a hungry winter in Siberia.'

Guizzardi, terrified by the very idea of finding himself marooned in this dismal wilderness during a famine, wound a length of chain around the left-hand driving wheel. This gave the necessary traction but had the effect of picking up extraordinary amounts of mud, bits of wood, slime and stones, only to deposit the entire mass back over driver and passengers. They did not care: they were advancing at last. 'The car speedily tackled that climb which had so often repelled us,' Barzini crowed. 'About halfway up the machine had a moment's hesitation, but it was only a moment: the chain was digging into the ground like a claw.'[7]

With the rain, the woods and villages through which they drove became still more desolate and forbidding. Mosquitoes, scourge of Siberian summer, hung around men and beasts like whining shadows. Taking their cue from the locals, the motorists acquired long black netted veils that hung from the crowns of their heads to their heels. These, too, were soon spattered with dark mud.

The first weekend of July found the motorists, each thickly shrouded like a village matron at a funeral, wending their way through interminable forests. 'Never,' Barzini wrote, 'even in the

desert, did we feel the solitude so much as between these interminable barriers of shade flanking us on either side, shade that
was made more dark and fearsome by the crepuscular light of the
threatening heavens.' Rain clouds sometimes plummeted to earth
around them, occluding their sight so that the trees loomed out
suddenly and 'assumed bizarre, spectral forms, looking sometimes
like the dim profile of a fantastic city, the sharp tops of conifers
resembling the pinnacles of some gothic edifice'.[8]

Adding to the automobilists' unease were warnings from officials they met and tales they heard whispered in the post houses:
escaped convicts haunted the forests between Irkutsk and Tomsk.
During the winter months they would try to survive living rough
in the taiga, setting off westwards in the spring and summer.
Over the second half of the nineteenth century, as the number
of convicts and escapees increased so had the violence. In June
1845, an entire group of gold prospectors had been hunted down,
robbed and slaughtered by four vagabonds. In another infamous
case, several merchants and their families had been murdered.
It was reported that each of the heads of the children had been
crushed under the wheels of the travellers' own carriages. When
de Windt had travelled this route, in the 1880s, a family had been
waylaid and shot just a few days previously. 'Most of the victims
are first stunned, and then their throats are cut. In no case of late
has a traveller's life been spared.' Even local villagers were not
immune. They habitually left food outside their houses at night to
appease any escaped convicts who might be lurking in the area.
Retribution if they failed to do so – usually in the form of arson –
was common.[9]

About seventy miles from Krasnoyarsk, the Italians were driving
between dense columns of trees when they saw a group of men in
the track, carrying guns. Not knowing whether these were 'sportsmen, runners of contraband, or perhaps the famous bandits', they
did not slow down to investigate, but floored the accelerator as
far as the terrain would allow, jerking over ruts and branches and
praying that their vehicle would keep moving despite the mud.
The sight of the growling machine bucking its way towards them
so unnerved the group that they 'fled and buried themselves in the

shrubs, whence they gazed at us with a terrified air. They were so completely stunned,' Barzini joked, 'that with perfect success we could have demanded their money or their lives.'[10]

If there had ever been any doubt, the section of road from Irkutsk to Tomsk drove the point home: the Peking–Paris was turning into a rout. Barzini's venturesome reports printed every two or three days in the *Corriere della Sera* and *Daily Telegraph* gave the impression of inexorable forward motion despite all obstacles, their path onward eased by the liberal application of the prince's fortune.

On 8 July, the Itala and her crew reached the banks of one river whose ferry had sunk in floods a few days previously. The prince, undaunted in his long, waxed driving coat and brimmed hat, called on the chief of the local village. Venerable and white-bearded, wearing a long jacket of embroidered velvet, the chief gravely studied the official *ukase* documents the prince presented to him. A crowd gathered around the men and began discussing the problem of getting the car across the river. 'Soon the village was mobilised, and the peasants, armed with axes, ropes, buckets, and poles, marched to the riverbank.' The ferry was hauled up from the depths, bailed out and a ramp up to it 'rapidly constructed with the extraordinary skill which the Russian peasants display in wood-working'. By noon they were crossing the river, the entire operation having lasted three hours.

A few days later, in a village on the final approach to Tomsk, the mud was so deep that the automobile had to be hitched to five draught horses. 'Two *moujiks* [peasants], formerly in the artillery, mounted the leaders, and we were carried along at a brisk trot for a good distance, the postilions using their *nagaikas* [short, thick whips] and the whole population witnessing our triumphal progress.' They reached Tomsk – the best part of 2,500 miles from Peking – on the evening of 11 July, a month after their departure.[11]

Du Taillis, meanwhile, remained ominously quiet. Telegrams sent on 8 July and published on the 10th confirmed that the Spyker was experiencing mechanical faults; that the weather over the previous few days had made driving supremely difficult; and that the two De Dion-Boutons were going on ahead, the small French flags attached to their bonnets now tattered, muddy and forlorn. The telegrams were sent from Nizhneudinsk, a village halfway between Irkutsk and Krasnoyarsk and fully 600 miles behind the Italians.

Four days later, a place-holding article from an anonymous journalist vainly tried to distract from the silence with banalities: 'They have covered 3,222 kilometres [2,000 miles] in thirty-one days; they still have 8,058 to go.' Intriguingly, this article also included a brief telegram from Cormier, nominally the correspondent for *L'Auto*: 'We are being devoured by mosquitoes. We are continuing on to Krasnoyarsk, which is 300 kilometres [186 miles] from Kansk.'[12]

This was hardly edge-of-your-seat reporting, nor was it the easy French triumph envisaged by *Le Matin* and other Peking–Paris backers, especially the powerful Marquis de Dion. Behind the scenes, the organising committee scrambled to regain control of the narrative and prevent national humiliation. This latter emotion was being stoked by the Italian racing driver Felice Nazzaro, who was in the process of demolishing the French – until now 'the aristocracy of the world of automobilism' – in the 1907 Grand Prix season. Nazzaro's most recent triumph had been a particularly sore one, coming as it had in front of a sulky French audience at Dieppe on 2 July. The band at the event vented their feelings by quietly sagging through the Italian national anthem during the prize-giving ceremony and then – fervently, defiantly – blasting out the Marseillaise. 'If there were any further need,' a French journalist fumed, 'to demonstrate the uselessness and absurdity of speed races, the results [at Dieppe] provide ample evidence. An Italian driver, Nazzaro, in an Italian Fiat car, won the race.'[13]

Back in Russia, a rapid fire of telegrams lit up the network of wires between Paris and Siberia. Du Taillis – the Parisian organising committee insisted – must abandon the Dutch Spyker and her ne'er-do-well driver and immediately take the train to Krasnoyarsk to take up a place with the two De Dion-Boutons in order to continue reporting. This would mean the end of the road for Italian sportswriter Edgardo Longoni, who would now be out of a seat, but too bad. He had not made much of an impression and, since the Parisian newspaper was the instigator of the Peking–Paris, their journalist must take precedence. Besides, the

De Dion-Boutons were now the only French cars in the race: France's honour depended on them.* [14]

As a sop to his bosses, du Taillis's next dispatch, sent from Krasnoyarsk on 13 July, was longer than usual and full of lively colour. He told the story about the mysterious 'shaman' and the consequences for the Spyker, making a visible effort to paint Godard in a flattering, even heroic light. The reason for his efforts to defend the wayward driver soon became clear. On hearing of Godard's plan to take the car by train to Tomsk to get the magneto fixed, a stern message was sent by Georges Bourcier Saint-Chaffray, head of the Peking–Paris committee: 'Tell the competitors that anyone taking the train would risk absolute disqualification.'† Du Taillis, keenly aware that he owed Godard his life after what had happened in the Gobi, covered for him. 'Godard is willing to go back. He will return by train to the stopping point . . . and will repeat the journey by road as soon as the magneto has been repaired.' Despite the journalist's efforts, as mid-July approached, Godard was on the brink of being thrown out of the race. [15]

---

* How seriously French national pride was intertwined with her reputation as home of the automobile can perhaps be judged from an article responding to Nazzaro's win at the French Grand Prix in July: 'French firms have spent French money, used French materials, spilled French blood and risked human lives in order to put an Italian-made automobile in the limelight.'

† Of course, this technically meant that all the French competitors were already disqualified, since the two De Dion-Boutons and the Spyker had all taken the train from Misovaya to Irkutsk on the evening of 2 July.

11

HORSE POWER

Before the Automobile

He works with the precision of a machine. A machine that never breaks down.

British Pathé, 'Barge Horses' (1947)

There is probably nothing more remarkable in the history of locomotion, and we may add in the history of commerce, than the persistence of the horse.

The Economist, September 1907

Before the automobile, horses reigned supreme. In 1900 there were between 3 million and 3.5 million horses in America's cities. Chicago alone had well over 83,000; 1907 Milwaukee had a human population of 350,000 and an equine one of 12,500. William J. Gordon, author of the 1893 book *The Horse-World of London*, estimated his city's equine herd to be around 72,000. The situation was the same the world over. In 1907, city streets from Irkutsk to Anchorage and from Turin to Auckland would have been rich with the sights, sounds and odours of horse – from the mist rising from their coats and dung on cold mornings to the clatter of their hooves and the baritone trumps of their farts.[1]

Horses were part of the *lingua franca* of nineteenth-century cities. At one end of the social scale, in parkways and along fashionable thoroughfares, elites had judged one another for centuries by

Horses were integral to the urban areas of the early 1900s. It would have been near impossible to imagine cities functioning without them.

the quality of their horses, stables, carriages and horsemanship. Many chose to have themselves commemorated in portraiture or sculpture astride glossy steeds, while particularly cherished horses almost became extensions of the family. John Boyle, the fifth Earl of Cork, erected a monument on his Somerset estate to one beloved equine:

> Under this Urn are interred the bones of KING NOBBY; a Horse, who was superlatively beautiful in his kind. He loved his master with an affection far exceeding the love of brutes. He had sense, courage, strength, majesty, spirit, and obedience. He never started, he never tript, he never stumbled.[2]

Horses could be humdrum too. They were essential to agriculture, the economy and transportation and performed so many quotidian roles that they could become almost invisible. In Moscow, for example, locals could immediately identify which district a

140

fire engine was from by the colour of the horses harnessed to the hook-and-ladder wagon, water pump and crates that they pulled to fires. Golden pintos for Tverskaya; black pintos for Rogozhskaya; bays in Arbat; cream with black tails and manes for Khamovniki; red chestnut for Myasnitskaya; dappled greys for Yakimana. In Moscow, as elsewhere, city dwellers would have understood that these animals had been acquired young. It was essential to train them to respond to the fire bell so they could be readied as fast as possible; they also needed to be trained to run *towards* fire and hold their nerve even in close proximity to the flames.

As well as pulling fire engines, horses were employed in haulage, in public and private transport, in mines – there was, thus, a market for blind horses – assisting municipal road sweepers and at breweries. They became part of urban folklore. Cab drivers were advised never to buy former fire horses, because some remembered their training rather too well and, whenever they heard a fire bell, would suddenly take off in the direction of their old station. Others were pranksters. One belonging to the Henry Meux & Co. brewery in London disliked the pigs the brewery also kept. The horse would take some of its feed and drop it near the water trough. If a pig came snuffling up for the dropped feed, the horse would grab it squealing by the tail and drop it in the trough before 'caper[ing] about the yard, seemingly delighted with the frolic'.[3]

Although we tend to think of the nineteenth century as the steam age, the advent of the railway had not lessened the need for horses. Quite the opposite, in fact. The 84 million tons of goods the railways put into circulation in Great Britain in 1890 (setting to one side, for a moment, the cargoes of ships and barges) all needed ferrying from station to depot and from depot to warehouse, shop or home. There was great demand for all sorts of specialised horse-drawn vehicles – hearses, bakers' carts, tree transplanting and ice delivery wagons – each instantly recognisable, often brightly painted to act as a mobile billboard for the company or service to which it belonged. As well as favouring teams of roughly equal strength, which increased efficiency, consumer vehicles preferred matched pairs for cosmetic reasons. Undertakers chose blacks, for example,

141

while London cab horses were usually bay; greys were shunned, perhaps because their pale hair clung more visibly to clothing.[*4]

As well as goods, horses moved people. Because they were expensive to buy, stable, maintain and feed, carriages and horses of one's own were indicators of social rank. In 'The Adventure of the Solitary Cyclist', a Sherlock Holmes story written in 1903, the detective makes a crucial deduction the moment it is revealed Mr Carruthers, who claims to be a wealthy man, possesses no coach and horses. Those of more limited means had other options. Private vehicles and horses could be hired for days or seasons. Then there were cabs, trams and omnibuses. London in the 1890s was home to 11,297 hansom cabs. Cab horses were renowned for their idiosyncratic names: Scorch, Blaze and Blister were acquired on a hot day; Slush, Gaiters, Sou'-Wester and Puddle on a wet one.

Omnibuses were an innovation of the 1830s. These large, horse-drawn vehicles carried a dozen or more people in an enclosed cabin with wooden benches and were operated on fixed schedules and routes across cities. They were also an immediate success with the urban poor and middle classes: in 1854 34 million Parisians rode omnibus lines. By 1890 each of the horses used to draw London's 2,210 omnibuses could expect to draw an average of a ton and a quarter of weight, twelve miles per day at around five miles per hour.

The prevalence and popularity of cheap public transport in cities had social implications. Representatives of both sexes and varying social stations found themselves crammed together, posing dangers both real and imagined. According to the *Los Angeles Record*, the air inside a crowded streetcar 'was a pestilence; it was heavy with disease and the emanations from many bodies ... A bishop embraced a stout grandmother, a tender girl touched limbs with a city sport, refined women's faces burned with shame and indignation – but there was no relief.' In 1907, a group of women would form the Society for the Protection of Passenger

---

[*] 'Why clubland should object to grey horses is not known, but the fact remains that a man with a grey horse will get fewer fares,' wrote William J. Gordon in *The Horse-World of London*, 1893.

Rights, which criticised the subway as 'crowded to the point of indecency' and complained about the 'rowdyism' of 'young ruffians'. Moral and epidemiological concerns aside, these and other more regular and affordable forms of public transport allowed people to live farther away from their place of work, reshaping cities by creating new suburbs and slums. Although by 1900 many European and North American metropolises were beginning to experiment with electric, steam or combustion engine transportation, the vast majority were still horse-drawn.[5]

Horses created their own economy. Twelve million acres were required to keep America's herds fed around the turn of the twentieth century, while 368,000 Americans were employed as teamsters (the driver of a team), 54,036 as grooms and 26,737 as livery stable keepers. Horses needed to be bred, transported, sold, healed when they came up lame and destroyed when they were too sick or injured to work. To accommodate its horses, cities needed large stables, each of which employed yet more people, such as farriers, groomers, stable-hands and vets. London's Road Car Company, based at Farm Lane in Fulham, had two-storey stables with space for a herd of 700; Great Western's South Wharf Road stables were four storeys high. Tack, saddles, whips, liveries and blankets needed making, buying, mending and, after a day in the driving rain, drying.* William Gordon noted with approval that Great Western had a special drying room for their horses' collars, 'whence it comes in the morning ready for wear, warm and comfortable as a clean pair of socks'.[6]

Urban working horses – as a result of heavy loads, diseases spreading in close quarters, the constant stopping and starting, and the risk of accidents on slippery cobbles – died young. Epidemics were not infrequent. During North America's Great Epizootic of 1872, thirty-six horses were succumbing to disease

---

* One horse goods supplier, C. M. Moseman & Brother, owned a showroom near Broadway in New York, five storeys high, which contained no fewer than fifty kinds of saddle, 200 varieties of whip and 275 harness styles. They also had branches in Berlin, Paris, Moscow and Vienna. There were even fashions in horse clothing: straw hats came into fashion for first humans and then horses in the early twentieth century.

and being removed from Manhattan's streets daily.* Even in a regular year, dead horses were hardly an unusual sight in cities. In 1912 Chicago had to deal with 10,000 carcasses annually.

Although the nineteenth century saw the publication of litanies of hand-wringing editorials and books on the moral implications of horse cruelty, corrective measures remained half-hearted. For one thing, there was still money to be made from the carcasses. Hair was sold for cushions and as a stiffener for plaster; hides for leather; hooves for glue; bones for knife handles, phosphorus (for matchsticks) or bootblack. The flesh ended up as food for pets, or even – 'lost in the oblivion of sausage meat and labelled "made in Germany"' – humans.†7

Despite the loyal service and entrenched interests urban horses represented, by 1900 city dwellers were increasingly preoccupied with the problems they posed. Gordon was forced to admit that 'there can be no doubt that from an aesthetic point of view [London's] streets would be considerably improved' if horses were removed. A 1908 article entitled 'The Horse vs. Health' charged New York's 120,000 beasts with being 'an economic burden, an affront to cleanliness, and a terrible tax on human life'.

Iron-shod hooves and cartwheels ringing against cobbles were noisy, and in cities the din started before dawn. Benjamin Franklin, one of the America's Founding Fathers, was complaining of the 'thundering of coaches, chariots, chaises, wagons, drays and the whole fraternity of noise' in the late eighteenth century. People also abhorred the accidents and traffic that horse-drawn vehicles created. Horses spook easily and their instinct is to rear and run away, highly undesirable traits when the horses in question are hitched to passenger or goods vehicles on busy roads. In New

---

* An epizootic, as the name suggests, is an outbreak of disease in an animal population. The Great Epizootic of 1872 was caused by a strain of equine influenza, and it spread across North America with astonishing rapidity. First detected in Ontario, Canada, in the early autumn, by 25 October the *New York Times* was reporting that 15,000 horses in their own city were afflicted, and travel in the state was suspended a few days later.
† Some countries, like France, were culturally accepting of hippophagy (horse-meat consumption) and needed no sausage-meat subterfuge.

York during the nineteenth century, there were more accidents per horse-drawn vehicle than motor vehicles would later cause.

They were also expensive. Feeding 10,000, roughly the number engaged in London's omnibus service, required 25,000 tons of maize, 2,700 tons of beans and peas, a little over 1,000 tons each of oats and bran, as well as 20,000 loads of hay-and-straw mixture. It cost £20,000 a year to shoe them all and, since two out of every three died in service, their bodies had to be removed from the streets (at a cost of over £1,300 annually).[8]

The most frequent complaints against urban horses, however, concerned dirt, disease and smell. In the early part of the nineteenth century, horse dung could be swept up and sold for a handsome profit to farmers as manure. However, as urban horse populations exploded, so too did dung piles and street sweepings. By the 1890s the economics had flipped and owners and city councils were having to pay to get dung removed, making horsepower less economical.

In some places, particularly in poor areas of large cities, horse droppings simply accumulated faster than they could be removed. Health officials in Rochester, New York, calculated in 1900 that the city's 15,000 horses were producing enough manure annually to create a pile 175 feet high and an acre wide: a breeding ground, they estimated, for 16 billion flies.

In New York itself, carts on Ludlow Street on Manhattan's Lower East Side 'were standing in filth of every kind from one to two feet deep'. On the west side of Thomson Street there was 'snow, ice, mud, garbage and general filth from three to four feet high'. Although streets piled knee- or even thigh-high with waste were rare, they did exist. In dry weather, the dung turned to dust and blew into people's faces; when it rained, it became a malodorous sludge. In Moscow, the yard at Tverskaya fire station was heaped high with manure. 'From under [it], especially after rain, a brown, fetid liquid streamed straight across the yard, seeped under the closed gates to the lane, and then flowed along the pavement toward Petrovka Street.'[9]

Dirty streets and fly-blown dung raised the spectre of disease, a potent fear at a time when cities were vulnerable to outbreaks of cholera, malaria, smallpox, yellow fever and typhoid. In London in 1873, a 'serious outbreak of typhus fever' near Drury Lane was blamed on 'an accumulation of filth'. A *New York Times* article from 1871 warned of 'a reign of the cholera plague, now striding westward'. The author suggested the arrival of the disease in the city was all but inevitable, given 'pest-breeding nuisances, smelling rank and offensive in the nostrils', such as 'decaying vegetables, putrid meat, stale fish, and abominable nastiness of every description . . . tons of manure and refuse'. A decade later, a Dr J. C. Peters was still railing that 'typhus fever, diseases of the eye, ear, and throat, diphtheria, pneumonia, and consumption are in a great measure due to the filthy condition of the streets and the bad ventilation and drainage'.[10]

Although cities employed teams of street cleaners and sweepers – this work took on extra urgency during epidemics – by the mid-1890s it was getting more expensive to manage the refuse.* Paris was employing nearly 3,500 men and women to sweep its streets at a cost of around 8 million francs, paid for by a special property tax. By 1895 New York was spending some 3 million dollars a year on street cleaning. Nevertheless, photographs of some residential streets show them waist-deep in dung. Juvenile street cleaning leagues were formed to encourage civic mindedness in the young through meetings, special titles and certificates and songs. 'And We Will Keep Right On' was one of these:

> There's a change within our city, great improvements in
>   our day;
> The streets' untidy litter with the dirt has passed away.
> We children pick up papers, even while we are at play;
> And we will keep right on.

---

* Cities were highly resourceful when it came to disposing of waste. The river embankment for London's Battersea Park was largely built on compacted horse dung. 'For more than two months past,' the *Illustrated London News* reported in December 1848, 'about three barges have been daily employed in conveying the sweepings of the streets of the metropolis to Battersea . . . The soil is particularly soft and slushy when first deposited, but soon consolidates, and forms a fine, firm, substantial stratum.' New York, meanwhile, acquired barges to take the street sweepings out to sea so they could be dumped; one of these barges was rather delightfully named the *Cinderella*.

George E. Waring Jr, New York's commissioner of street cleaning, claimed that his wife begged him not to accept the job.* 'It is utterly hopeless,' she cried. 'You surely can never clean Elizabeth Street; you will only disgrace yourself trying to do so.'[11]

Although the problems horse-powered cities presented were increasingly obvious, the idea that automobiles and other similar machines would upend the status quo remained outlandish in 1907. Much of the world was geared towards the use of animal power, particularly that of horses. Transitioning to machines and automobiles would require astronomical upheaval and outlay. At stake were the livelihoods of millions of people who reared, cared for, grew food for, drove or made goods for horses. Roads smooth enough to make automobiles practicable would need to be built and maintained, something few countries had the stomach for after nearly a century of heavy investment in railways.

Horse partisans could also point to the automobile's own flaws. They were expensive compared to both horses and trains, faced hostility from other road users and seemed likely to depend on uncertain fuel supplies. Indeed, commercial operations that had tried replacing horses with motors were not guaranteed success. The London Power Omnibus Company, registered in 1905, collapsed in July 1907. Motorised public transport in Berlin, introduced in 1903, quickly went through several crises. During the summer of 1907, nearly half of the city's fleet of 700 motor cabs disappeared from the streets, because small proprietors could not make ends meet 'in consequence of the high price of petrol and rubber'.

The belief in the primacy of the horse even carried into the automotive industry itself. 'Horsepower', a measure of the power of machinery, was coined in the late eighteenth century by James Watt, who wanted a way of simply comparing his steam engines

---

* Waring's 1898 book *Street-Cleaning* is a very earnest exploration of his work in New York, followed by a review of the methods and street cleanliness of various other global cities, from Berlin to Boston via Genoa and Paris. When I checked it out from the London Library in August 2019, it was the first time the book had been issued since September 1991.

to horses in a way that felt familiar to his customers. Other habits and nomenclature transferred over too. At Brooklands motor racing circuit there were 'paddocks' and an 'enclosure'; cars were weighed by a Clerk of the Scales so they could be handicapped; and drivers were distinguished by wearing coloured silks, like jockeys. In France, only a week after the start of the Peking–Paris, *Le Matin* saw nothing contradictory in their sponsoring a Festival of the Horse at Chantilly – a week-long event 'showing off the horse in all its beauty, power and usefulness'. Likewise *The Economist*, evaluating the prospect of mechanisation in September 1907, confidently predicted 'the triumph of the horse'.[12]

12

Tomsk – Tyumen

11–19 July

> A Cossack horseman, with no saddle or stirrups, gallops along-
> side us for about ten kilometres.
>
> Georges Cormier, De Dion-Bouton team

Tomsk was an impressive, modern city. Although the streets were
clogged with mud, they were broad and imposing, flanked by handsome public stone buildings. Aside from the governor's palace, in
which Prince Borghese, Guizzardi and Barzini stayed during their
time in the city, it had a famous university, giving it a reputation as
the intellectual capital of Siberia. The streets thronged with well-
dressed women, and there were no fewer than three hotels.

When the Itala left early on the morning of Friday, 12 July,
it was preceded by an honour guard consisting of a pair of galloping Cossacks and a group of cyclists and motorcyclists. As
the road dipped towards the River Tom, the bugling horns and
chiming bicycle bells became muffled by dense mist rising off the
river. Within moments the entire convoy had been blotted out like
something from a fairy tale.[1]

From Tomsk, they were heading 600 miles due west through
Kolyvan to the city of Omsk.* As they continued, firs gave way

---

* Due to its position midway between Tomsk and Omsk on the Siberian Tract, historically Kolyvan was an important trading town. It was still prominently marked on maps
around 1900, even though the roads had largely been abandoned in favour of the railway, which ran twenty miles further south. Newer roads have also bypassed Kolyvan,

to birches. For hours they wound their way through an unaltering maze of pale, speckled trunks. After a few hundred miles, the bunched hills of wooded taiga relaxed into a treeless plain. These steppes were settled by nomadic, yurt-dwelling Kyrgyz people, similar in their customs to Mongolians. 'They have,' Barzini marvelled, 'the same dress, the same appearance, the same mode of riding, almost on their horses' necks.'

Near Kolyvan, the ground softened to a marsh furred with tall grasses and cleaved by the River Om. Here the Itala was guided by a troika sent on the orders of the governor of Tomsk, Baron von Nolken. Barzini, already holding on to the straps securing the luggage, grimly braced his feet under the seats in front of him to avoid being tossed out of the bucking automobile, which at every moment seemed in danger of either overturning or sinking into the mire:

> It was a desperate undertaking, driving in the midst of high grass, amongst which shone large stretches of water. The troika made the most fantastic turns, and sometimes disappeared behind clumps of marsh plants, dwarf willows, and rushes. We were guided by the tinkling of the bells suspended from the curved yoke, beneath which the middle horse strained his neck in the rapid flight.[2]

The quiet road – still faintly imprinted here and there with the vestiges of 'an extinct traffic' – was a continuation of the one they had been following since Irkutsk. This was the eastern part of the great Siberian Tract, the historic overland route connecting European Russia to Siberia. Cormier, deeply contemptuous of this unpaved road – 'nothing but a deep rut' – noted that villagers 'seem to view it as the highway of the world. When you ask them if it is the right road to take to the next town, they reply with "It is the Tract", as if that makes clear that there is nothing more to add.'[3]

Sunday, 14 July, found the Itala and her crew speeding over a plain towards Omsk at the long-dreamed-of speed of thirty miles

and it has stagnated. All that remains of its former glory are some beautiful old houses and a gold, onion-domed church.

an hour. Crops of wooden windmills, each of a different design, sprang up around them like fantastical mushrooms. The trail was relatively good, which was fortunate because the chain they had wound around the rear left wheel days earlier to counter difficulties with the mud had not been without its drawbacks. As Guizzardi anxiously explained to his companions, it had damaged the tyre and strained the wheel's wooden rim. He pointed out the cracks that had formed where the spokes slotted in. All three tried to ignore the ominous creaks it was making on each revolution; they did not have a spare. With their minds engaged alternately with congratulating themselves on their pace and worrying about the wheel, they were slow to pick up on the cloud of acrid smoke billowing from beneath the automobile.

The brake had once again engaged while they were driving and the heat from the friction had set fire to the lubricant. Scrambling down, the three men saw, to their horror, yellow flames hungrily attacking the wooden carriage work, gnawing their way towards the fuel tanks and catastrophe. They dashed to the ditches on either side of the road to fetch water but found only mud. Guizzardi, practical as ever, hurled his waterproof overall onto the fire to suffocate the flames. The prince's ankle-length fur coat flew through the air to join it. They grabbed axes, spades, penknives: anything to cut away sections of still-smouldering wood. After a tense half hour, during which time Guizzardi completely detached the foot-brake, insisting they rely solely on the handbrake from then on, the Itala limped on to Omsk.[4]

Lying at the junction of the Om and Irtysh rivers in the midst of Siberia's dairy country, Omsk was a dull, well-to-do regional capital.* One foreign visitor dismissed it as a city 'in which the largest building is a military academy and the most picturesque building a police station; in which there is neither a newspaper nor a public library, and in which one-half the population wears the Tsar's uniform and makes a business of governing the other half.'

* British butter merchant Samuel Turner earnestly described the local product as having 'some very fine qualities . . . with a good, waxy body'.

The Italians arrived to find the city engaged in its Sunday promenade. 'Along the wooden pavements moved the peaceful crowd of citizens, walking with the peculiar gait of people wearing their best clothes and anxious not to spoil them.' The town buzzed with the news that Leo Tolstoy had not, in fact, been killed in a peasant arson attack – as had previously been reported – and was still defiantly alive. (His nearest neighbours had not been so lucky.) The motorists spent two days in the city resting, repairing the motor's charred carriage work and meeting local representatives of the Russian Peking–Paris committee.*[5]

The committee, headquartered in St Petersburg, had been formed immediately after the announcement of the race and contained some of the most prestigious names in Russian society. Members included the wealthy auto-enthusiast Sergei N. Koribut-Kubitovich, cousin of Sergei Diaghilev – already precipitating a craze for all things Russian in Paris, but still two years from producing the *Ballet Russes* – Marquis Eugene da Passano and Ludvig Alfred Nobel, member of one of Russia's richest families and nephew of Alfred, patentor of dynamite and creator of the Nobel Prizes. They were clearly also an independent-minded group. With neither the knowledge nor agreement of the original committee in Paris, the Russians had decided to amend the route over the Urals.

The plan formulated in Paris had the motorists heading due west from Omsk, through Petropavl and Kurgan, crossing over the Southern Urals between Chelyabinsk and Birsk, before fording

* Du Taillis later implied the Italians had cheated by having the Itala rebuilt or replaced entirely during this time. 'Workers and police officers claimed that bulky crates, whose customs clearance costs amounted to over seven hundred roubles, disgorged a completely new Itala,' he wrote. 'I cannot be the judge of these matters, my role being simply to relate the rumours and to state that in Siberia it is perfectly useless to seek the truth.' Ettore Guizzardi was still outraged by this allegation when interviewed in 1963 and many Peking–Paris enthusiasts still believe it to be true. It is, as du Taillis knew full well, highly unlikely. It would have taken more than two days to transport an entire vehicle – or even significant new parts – to Omsk. (Transporting a new magneto from Amsterdam to Tomsk, a few hundred miles further down the same line, took nearly a fortnight.) Another reason to dismiss this tale is the left driving wheel. If the Italians *had* been asking for new parts, getting hold of a replacement for that would have been a priority, but the original, damaged wheel remained.

the Kama River at Yelabuga. The Russian committee advocated a longer, more northerly approach that looped north through Tyumen, Yekaterinburg and Perm before veering south-west to rejoin the original route at Kazan. They argued that this course more than made up for its greater length in practicality: it would take them along a better road and traversed the Urals at their lowest point. The motorists would also be able to take advantage of detailed road maps that had been commissioned for them – neatly annotated with the altitude of various points and the distances between villages and fuel depots – as well as a chain of organised hospitality along the way. 'We accepted with enthusiasm,' Barzini wrote from Omsk, 'and are ready to double the distance if thereby we could find a better course.'*

Back in Paris, the organisers were caught unaware. The Peking–Paris, which had been envisaged as a demonstration of France's automotive superiority, was now literally off track. The automobiles were dispersed over the Siberian landscape, like apples tumbled from a basket. There were too few journalists between the vehicles in the rear to coherently cover the action. The methods the impecunious Godard had deployed to enter the race were causing a steady seepage of embarrassing press reports. (In an interview with a Dutch newspaper around this time, Jacobus Spijker raged, 'Godard is an incredibly cunning man who lied his way into receiving a Spyker.') To cap it all off, the entire enterprise seemed likely to become a transcontinental repeat of the French humiliation at the Grand Prix. With the world's gaze fixed upon them, an Italian car with an all-Italian team had been out in front from the first. Despite a two-day halt in Omsk, they were now 640 miles clear of the De Dion-Boutons, with the precise whereabouts of the Spyker unknown.

It was too much. Attempting to massage a more flattering narrative from events, Georges Bourcier Saint-Chaffray, on behalf of the organising committee, penned an addendum to one of du Taillis's reports in *Le Matin*. Conveniently forgetting earlier

---

* They also had little choice: Borghese had employed the Nobel firm to supply the Itala with its fuel. Since Ludvig Nobel was among those responsible for the change in itinerary, Borghese's fuel had in all likelihood been redirected long before he reached Omsk.

assumptions that the Italian car's weight and power would be a drawback on Siberia's terrible roads, he argued that there had never been any doubt of the Itala's insurmountable advantage. No, the truly important thing being demonstrated, he concluded, was that economical French cars like the De Dion-Boutons 'can withstand such efforts, bear them, and even carry out the feats that were only expected of 40-horsepower vehicles'. To hammer home the point, this was all printed under the bold subheading 'This Is Not a Race'.

Blithely unaware of this contretemps, at three o'clock in the morning of 17 July, two fellow motoring enthusiasts piloted the Itala out of Omsk and onto the northern road towards the Urals. They drove an idiosyncratic old motor car, 'nearly as small as a baby carriage', that had been designed and built nearby. *Sotto voce*, Barzini joked to Borghese that the passenger, 'a pleasant Swedish man of colossal stature' who wore a white waterproof cloak with a voluminous ruched hood, resembled a prima donna returning from a late night at the opera. The illusion was rudely shattered when the tiny antique drew aside to let the Itala pass and the Swede, overcome by the moment of farewell, pulled a pistol from inside his cloak and sent a volley of bullets up into the predawn sky.[6]

By mid-July, the race had really taken hold of the public imagination. Newspapers that had previously only mentioned it casually began printing regular updates and opining that automobiles were tools of adventure, freedom and escape. To the *South China Morning Post*, the motorists were 'plucky motor car racers'. 'The cars,' reported the *Sydney Morning Herald*, 'had to cross ravines where there were no bridges, and mountains where there were no defined roads.' The *Illustrated London News* printed eight large photographs of the race bordered with stylised dragons breathing curlicues of smoke. 'Great interest is being taken throughout the civilised world – and a part, at least, of the other – in the motor journey from Peking to Paris,' the accompanying text assured readers. 'Sometimes the cars have gone at slow pace through the Siberian forests, startling the wolves that hide there ... Bandits have fled in terror before the cars.' Another article described one of the motorists' automobiles as their 'faithful mount. They camp next to it. They use it as a tent, as a pantry, as a bulwark against the dangers of deserts and strange plains. Better than a familiar horse, it carries them, at full speed, through fabulous spaces. Without a care, without rest.'[7]

Such a narrative suited everyone.* It bathed the motorists, their companions and scribes in the flattering golden glow of heroism and acclaim. So long as the machines kept running, the manufacturers were engaged in a near-mythic quest to advance global civilisation to a gloriously speedy future. Many of the brands involved, including Itala, Dunlop, Spyker and De Dion, used the publicity to market their products to potential customers. It was flattering to Western audiences and governments since a tenet of the colonial mythos was that modern technologies – guns, trains, electric lights, the telegraph – were militarily essential in conquering benighted regions and thereafter a blessing bestowed on the inhabitants.

Newspapers and magazines could leverage the immediacy of the instalments sent in by the participants to shift more copies. In fact, by the middle of July, a few days without an update heightened the drama. 'There is no news of Prince Borghese who has been on the steppe for two days,' *Le Matin* solemnly informed readers in 'Their Adventures'. Despite their efforts to set the race up to allow for near-daily updates, newspapers were learning that the trope of adventure allowed for some flexibility when it came to determining the precise whereabouts of the drivers on any given day. This was to prove especially useful, because on 19 July one car would drop off the map entirely.[8]

Once past Omsk, the Italians found themselves in a featureless terrain which undulated around them like 'a calm green sea'. There was little to distinguish one mile from another, one hour from the next. They passed few settlements. Since there was little wood

---

* Or nearly everyone. The host countries and their inhabitants were often not depicted in the most flattering light: adventures required rough-and-ready backdrops to amplify the danger and excitement of passing through them.

with which to build them, the cottages they did see were made of slender, interwoven branches, like giant upturned baskets, their roofs covered in turf and flowers. Still, the weather and road were good enough to move into third and then – greatly daring – fourth gear. It was loud in the open-topped Itala. Conversation, when it took place, was conducted in an uncomfortable half-shout. When Borghese and Guizzardi spoke to one another it was nearly impossible for Barzini, from his vantage in the back, to join in. There was little to discuss anyway. Borghese refused to look too far ahead. A disciplined man, he preferred to think only of reaching the next village, and then the next, as if each were their final goal.

He also insisted on a punishing regimen. He decreed that they set off every morning before dawn, driving through the day and often into the night, and once they had arrived there was still much to be done before they could turn in. The prince would track down the cache of fuel that had been arranged and sent out for them weeks before, often with the help of the local police. Guizzardi saw to the motor, tightening screws that had shaken loose, replenishing the fuel for the next morning. Barzini wrote up his reports and sent them off by telegram. This meant that they usually got only four hours' sleep a night; after five rough weeks on the road, this was taking a toll. Hours in the car passed in weary silence. When Barzini fainted from exhaustion on a street in Omsk, passers-by stepped over him, mistaking him for a drunk.

The prince alone seemed all but immune to the strain. Barzini wrote of Borghese's 'power of resistance; a resistance which was physical and, to an even greater extent, moral. He was tired, but he knew how not to look it. He took special pride in never appearing tired to strangers ... He fortified himself in the smiling impassibility of a diplomat, and stood the test for an indefinite length of time.' He only allowed himself to unbend a little when the three men were alone, passing his hand over his eyes and admitting he felt done in. Even so, the next morning he would get up and ensure that they were on the road early once again.[9]

Barzini would often dutifully enquire at the telegraph offices for news of the other Peking–Paris participants and update the prince

*Prince Luigi Marcantonio Francesco Rodolfo Scipione Borghese was a
forceful character: proud, aloof and determined.*

and Guizzardi. Truthfully, however, their lead and long separation
had given them a feeling of total independence. The ambitions of
the Parisian organising committee, the fate of Pons and Foucault
or the whereabouts of Godard, du Taillis, Collignon and Cormier
had receded into irrelevance. Instead, it was as if the world had
shrunk around them, until all that mattered was the small section
of track before them that day, the rhythm of the Itala's engine,
what they would eat and where they would rest their heads that
night.

During the afternoon of 17 July, a column of mole-grey smoke
reared up, splitting the sky before them in two. Half an hour later
they drove up to the cause: the steppe just to their right, baked
tinder-dry by days of sunshine, was on fire, the flames rapidly
blown eastwards by the prevailing wind. The nearest village was
busy deploying all the defensive strategies at its disposal. It was
too small to have a fire station, too remote to call one. 'Outside

every house have been placed all sorts of receptacles full of water, and groups of people stand ready around the wells . . . Men are at work digging trenches, and numerous *telegas* from the surrounding neighbourhood arrive with peasants, all provided with poles, spades, and other implements.' The Italians kept driving.[10]

The following day, a Thursday, they were invited to dine at an open-air banquet near Zavodoukovsk, a village sixty miles southeast of Tyumen. The landowner sent servants to intercept them with invitations and carriages to take them to his estate. Having accepted, the three Italians, still grimy and numb from the road, enjoyed a surreal meal. They ate in the shade of trees surrounded by women with ringlets, dressed in the fashions of the 1860s, and men in regional dress. 'Our host's brother, a gigantic-looking man, was dressed in the old Siberian costume, with a silk shirt covered with embroideries, and a lovely waistband covered with silver.'

This otherworldly experience notwithstanding, Barzini and his two companions were increasingly aware of the diminishing distance – both culturally and temporally – between themselves and western Europe. Newspaper correspondents and official photographers hired by the Peking–Paris committee swarmed them at Tyumen. As they passed through towns and villages they began to hear cries of 'From Peking?' 'Yes!' they shouted back.[11]

On Thursday, 18 July, the two De Dion-Boutons arrived in Tomsk, a week and nearly 700 miles behind the Italians and in immeasurably worse humour. It was, according to du Taillis, 'one of the most difficult days of our trek'. Rain tipped from the sky as if from a bucket. The cars drove through serried birches between which there was no discernible road, only paths that criss-crossed, maze-like, through the trees. 'Imagine this track which is impossible to follow,' he groused, 'hilly, full of gullies, covered with stagnant water, with mud.' The De Dion-Boutons juddered and lurched, skidding in the soft ground and jack-knifing over 'the gaping ravines of ruts'. Considerable time was spent knee-deep in bogs, shovelling mud free from the wheels. After fourteen hours, they caught sight of the domes of Tomsk's churches rising like moons over the final summit. 'At last,' Cormier, Collignon, Bizac and du Taillis half sobbed in unison. 'At last.'

As no one had been able to bathe for days, their clothes had begun to itch and stink. Dirt was embedded under their nails and in the calluses on their hands, and their moustaches were mis-shapen and straggling. As they crested the hill and descended into the city beyond, their minds fixed on the delights of a hot meal and a Russian bath, they found themselves surrounded by several hundred distraught cattle. Luck, it seemed, was not on their side.[12]

The Frenchmen spent two nights in a hotel with a lively dance hall and did not sleep a wink. (Du Taillis noted with some bitter-ness that the 'Maxixe', a jauntily refined two-step from Brazil, was extremely popular in Russia in 1907.) Cormier carefully photo-graphed the remnants of Tomsk's theatre, razed in pogroms that had consumed the city in the autumn of 1905.* The sight of the city made du Taillis almost emotional. 'It had been a month since we had seen so many beautiful houses, so many buildings built on architectural lines, laid out in beautiful order. We had forgotten the very concept of decently paved streets.'

It was while lost in the delights of city life that du Taillis ran into Godard. The driver had loaded the broken-down Spyker onto a train a week previously. He was now working on the magneto with a professor of electrotechnology from the local university and waiting for a representative from the Spyker firm to travel east with a replacement. His announced intention was to return with the repaired vehicle to Cheremkhovo, where it had broken down, to ensure that the race was fair and to appease the organising com-mittee back in Paris. Godard may have let a trace of a wink slip as he said this, but it was with sufficient subtlety that du Taillis could almost have sworn he imagined it.[13]

---

* Du Taillis, in his contemporary report, wrote: 'How pretty and dashing it seemed to us, like a peasant woman in her Sunday best, this excellent city of Tomsk, where last year's play programmes set the theatre facades on fire.' In his book, 'programmes' was amended to 'pogroms'. This may have been an error, or possibly circumspection on his part: the automobilists were continually shadowed by police and the telegraph clerks were frequently hostile.

13

TAKING UP SPACE

The Battle for the Streets

The roads were made for me; years ago they were made. Wise rulers saw me coming and made roads. Now that I am come they go on making roads – making them up. For I break things. Roads I break and Rules of the Road. Statutory limits were made for me. I break them. I break the dull silence of the country. Sometimes I break down, and thousands flock round me, so that I dislocate traffic. But I am the Traffic.

'The Motocrat', *Punch*, 1907

It is often imagined that the history of urban transport is Darwinian. In this view, speed and convenience are evolutionary advantages and so slower, inefficient means of locomotion cede their place to general acclaim. Thus carriages replaced carts and litters; bicycles replaced walking; railways and streetcars or trams replaced horse-drawn carriages. By this logic, cars are the well-adapted apex species. However, a look back to 1907 reveals a far messier reality, one in which the needs and desires of motorists – the few – would have an outsized impact on the many.

Tensions over the right to space on roads existed before cars arrived. The spring of 1840, for example, found the jobbing writer Edgar Allan Poe witnessing pandemonium in Philadelphia. The trouble centred on the Kensington portion of Front Street. This was the city's historic nucleus: dating back to the seventeenth century, it ran north–south and abutted the reeking Delaware River ports. Poe, then a resident of the city, knew the inhabitants of the area by reputation. Kensington was a working-class suburb dominated by immigrants: German fishermen, Irish textile workers, Polish labourers and factory hands. Normally hard-working and sober, they had been whipped into a frenzy.

The issue at stake was both the physical fabric and indeed the very soul of Front Street itself. A few months previously, a railroad company had applied for – and been granted – permission to lay rails into the street and the work had duly begun. But the Front Street residents believed the thoroughfare belonged to them: it was where they socialised, played, courted, showed off their Sunday best on their way to church, drove their carts, and where they cooled off in the evenings after hot summer days. Not one of these activities would be improved by regular, fast-paced conveyances ferrying passengers along the rails between Philadelphia and Trenton, so the Front Streeters began making their feelings known.

On the Monday of the second week of March 1840, a mob of men, women and children 'surrounded the laborers at the rails', wrote Poe, 'replacing the paving stones which had been displaced, and otherwise interrupting the work.' Sheriffs and their deputies were called, rioters seized, and fulminations thundered down from behind judicial benches. The protesters remained undeterred.

> On Thursday the disorders increased. [One man] was violently assaulted with paving stones discharged from the fair hands of the damsels of Kensington, who also led away in triumph a wagon containing iron rails for the road, the laborers being fairly driven off the ground. Many arrests were made, but with no good effect. In the afternoon the Sheriff and his whole posse were routed, and the rioters, having beaten them off, proceeded to tear up that portion of the road which was the nearest to completion; disengaging not only the rails but the wooden frames, and filling up the excavations with dirt and stones.[1]

The Front Streeters were by no means unique in their understanding of their rights of access to pleasant, safe streets, nor in their protestations when these rights were infringed.* Before the mid-nineteenth century, the spaces between houses may have been used thoroughfares, but they were also – by design as well as by custom – extensions of homes, shops and factories in which urban residents lived, shopped and worked. An 1896 petition from the residents of a Brooklyn street begging that their street be cobbled rather than asphalted perfectly illuminates the relationship. Their complaints about the latter – considered far superior by municipal engineers – included increased traffic and noise, the lack of privacy, danger to their children, and an estimated 35 per cent reduction in house prices.[2]

As well as acting as impromptu playgrounds, meeting- and marketplaces, the public nature of streets influenced architecture and social custom. Roads provided open space and relief from

Before cars became widespread, roads were commonly used as spaces to socialise with friends, family and neighbours in otherwise crowded cities.

* William Penn, the seventeenth-century English Quaker and idealist who designed Philadelphia in the wake of the Great Fire of London, laid out the gridiron of broad streets precisely to create a spacious, green urban utopia for residents. Each of the dwellings in his city – whose name translates as 'Penn's Woods' – would have 'ground on each side for gardens or orchards or fields'. Indeed, he boasted that it would be less a city and more 'a green country town, which may never be burnt, and always be wholesome'.

closely built, densely tenanted dwellings. People would take afternoon walks along thoroughfares, socialising or discussing business. Balconies and verandas allowed people to partake in and enjoy public street life from the semi-private confines of their own homes.

Transport also determined the shape of cities. Before the advent of affordable mass transport, inner cities had poor, overcrowded areas where labourers, domestic servants and other workers lived within walking distance of the wealthy households, shops and factories where they worked. The dirt, pollution and noise of cities following the Industrial Revolution, as well as the proximity of slums, meant affluent upper and middle classes began viewing them as undesirable. They and their families began decamping to nearby villages, moves possible only because they owned or were able to afford the necessary means of transport.

Within cities, vehicles were expected to temper their speed to that of pedestrians. Where greater speed was desired, either private, turnpike roads or raised walkways or pavements along the edge of existing roads were built to separate road users.* Although, as photographs and early footage of city streets show, pedestrians felt entitled to walk in the roadways even where there were pavements. In Chicago as late as 1926, nothing in the law prohibited pedestrians 'from using any part of the roadway of any street or highway, at any time or at any place as he may desire'.

Even as new, speedier means of transport were developed, their access to residential streets was not guaranteed if others took against them. This was the case in 1840 in Poe's Philadelphia. It was also true in the United Kingdom, where countless omnibus and horse-tram services failed thanks to opposition from residents. Kensington and Chelsea, Harrow School and the University of Glasgow all fought proposed trams in their localities, ostensibly because the noise and vibrations would disturb students and

---

* With the development and rapid expansion of the railways in the mid to late nineteenth century, the coaching industry was eclipsed and inter-city roads began to fall into disrepair. The idea of road travel even began to seem like something from a bygone era. This trend was swiftly reversed by first the bicycle and then the automobile. Cycling organisations, in fact, literally paved the way for automobilists. The Cyclists Touring Club in the UK and the League of American Wheelmen (LAW) in America had been lobbying politicians for a generation to repair roadways and replace cobbles with smooth, tyre-friendly surfaces.

residents, although there was also a class dimension to the opposition.* The 1870 Tramways Act formalised such resistance, giving local authorities powers to quash new schemes and banning them entirely in the City of London and the West End.[3]

Steam-powered and electric urban transport also faced resistance. While they were faster and more modern, they were regarded as more dangerous – exploding engines and rail-jumping were not unheard of – noisy and, in the case of steam, polluting.

Nor were cyclists immune. In 1894, the *Sheffield Weekly Telegraph* reminisced about the character of the cyclist in the previous decade:

> An entirely unaggressive person, who had to endure much contumely in print, and a great deal of rough usage in rural parts. Showers of stones greeted him from public-house corners; and it was regarded as the height of facetiousness to capsize him by a host of cunning devices.

Fear of attack was such that guns were marketed to cyclists for their protection. Smith & Wesson produced a hammerless, short-barrelled model 'specially adapted for Bicyclists', while in France 'Le Revolver-Cycle', invented by a Monsieur Joubert, fitted neatly into one end of the handlebars, ready to be whipped out at a moment's notice. By the 1890s, however, 'scorchers' – speedy, reckless cyclists – were popular villains. In London, a cyclist was fined for knocking an elderly man under an omnibus. The *New York Times* noted the death of a cyclist 'while racing with a trolley car'. The journalist called it a warning to other malefactors and argued that 'Reckless riding by wheelmen would be minimised were all offenders severely dealt with by authorities.'†[4]

Although the law was not always applied, authorities did enforce

---

* Trams, because they were relatively cheap, allowed working-class people greater freedom in where they lived by freeing them from the necessity of being able to walk to work. In addition to safety and aesthetic concerns, wealthy residents and businesses often objected to trams in their area, fearing an influx of poorer people and cheaper housing.
† The reporter was particularly regretful that 'one of the latest arrests for "scorching" was of a woman who frequents the Boulevard'.

the rights of pedestrians and homeowners against upstart forms of transport. A Philadelphia judge in 1840, for example, found that the railroad company had failed to secure the consent of the Front Street residents. Three years later, New York State's Supreme Court ruled that 'a train of cars impelled by force of steam power through a populous city may expose the inhabitants, and all who resort there for business or pleasure, to unreasonable perils; so much so that unless conducted with more than human watchfulness, the running of the cars may be regarded as a public nuisance'.[5]

Motordom, as automotive companies and users began dubbing themselves, faced similar hurdles. At the beginning of the twentieth century, they were a tiny minority, unpopular with other road users and in direct competition with cheaper alternatives, including electric streetcars or trams. Automobilists, however, were wealthy, well organised and wise to the fates of previous new modes of transportation. By taking a very different tack to their predecessors and unashamedly asserting their right to roads above that of others, they would be extraordinarily successful. In a few decades, they had laid claim to vast swathes of urban space, prising ownership of streets away from residents and other road users in order to make them better serve the automobile and her drivers.

Until the 1920s, pedestrians could cross the road wherever and whenever they chose. A long *Literary Digest* article on road deaths in 1920 found that it was 'a matter of both law and morals' that pedestrians were under no obligation to exercise 'all possible care' on roads. 'It is a fact that people afoot do not incur blame or deserve criticism quite as soon as do the manipulators of motor-cars, and that they have rights in the streets, even tho they choose to cross elsewhere than at the appointed places.' The *St Louis Star* bemoaned the fact that others were being 'forced to submit to the tyranny of the automobilist'. A judge in Philadelphia fulminated that 'It won't be long before children won't have any rights at all in the streets,' while another ruled that 'the streets of Chicago belong to the city, not to the automobilists'.[6]

However, from the 1920s, automotive interests were at pains to redefine roads as places for cars, with other road users cast as interlopers. Such efforts were particularly successful in America, where the concept of 'jaywalking' – crossing the road at a time or

166

place other than those officially sanctioned – was first concocted by motordom and then criminalised. Automotive firms employed clowns at fairs to mock simpleton hillbillies crossing roads willy-nilly, printed out cards that could be handed out in cities illustrating 'correct' and 'incorrect' road-crossing and deployed a barrage of advertisements in newspapers. The Packard Motor Company even erected an enormous imitation tombstone in Detroit in 1922 engraved with the legend 'Erected to the Memory of Mr J. Walker: He stepped from the Curb Without Looking'.[7]

Where new infrastructure was constructed, urban planners – often lobbied by automotive firms – put the needs of drivers front and centre. As early as 1924, a consultant for the Los Angeles Traffic Commission argued that 'The old common law that every person, whether on foot or driving, has equal rights in all parts of the roadway must give way before the requirements of modern transportation.' Thus pedestrian crossings were spaced so as to allow the traffic to flow more freely, rather than put where it would be most natural for a walker to want to cross. Petrol stations, motels, roadhouses – establishments serving food and drink built along highways – garages, car parks and service stations sprang up. In Washington DC, it was estimated in 1928 that 29 per cent of the surface area of its streets was occupied by parked cars. Motorways and highways spread across the surface of the globe like mycelia. Where they became congested with traffic, they were widened and then widened again, despite evidence from the 1960s onwards that this was counterproductive. In Italy, Fiat, Pirelli and Michelin funded major roads, including the A6 Turin–Savona motorway.* In America, motoring magnates including Henry Ford, Frank Seiberling (of Goodyear), Carl Fisher (of the Indianapolis Motor Speedway and Indianapolis 500) and Henry Joy (Packard Motor Car Company) backed road-building projects, kickstarting what would become the federal highway system.[8]

Speed was a particular bugbear. In the UK between 1865 and 1896 speed limits were two miles per hour in populated areas and four miles per hour elsewhere. In America in 1906 it was kept around ten miles per hour, around the speed of a trotting horse. Motorists,

* This road would earn the dubious nickname *l'Autostrada della Morte* ('the Highway of Death') because of its design, which had a central lane for cars overtaking in both directions.

however, detested and largely ignored limits, leading countries to raise them rather than waste police time trying to enforce them, even though it was plain that increased speeds made roads far more dangerous. As one disgusted British lawmaker explained in 1930, speed limits were not struck down because 'anybody thought the abolition would tend to the greater security of foot passengers', but because 'the existing speed limit was so universally disobeyed that its maintenance brought the law into contempt'.[9]

Mechanical alternatives to speed limits were also considered. In the early 1920s there were arguments that automobiles should be fitted with a 'governor' that would limit the maximum speed. When surveyed, America's exhausted police were all for the idea: two-thirds of police chiefs agreed that governors should be required in their cities. In Cincinnati 42,000 people – more than 10 per cent of the population – signed petitions in 1923 to introduce ordinances requiring cars to be governed to twenty-five miles per hour. Since the automobile's capacity for high speeds was a crucial selling point, the proposal horrified manufacturers. After funding a huge campaign, the motion was successfully defeated.

This experience, however, left American motordom even more determined to ensure that the streets belonged to them. In addition to poster and newspaper campaigns, a safety committee was formed in Detroit that would write reports for local newspapers on traffic in their city, emphasising the culpability of other road users for any accidents or jams. The lobby funded road safety education for children, telling them to stay out of streets and emphasising their responsibility for their own safety. Motor clubs and the American Automobile Association also sponsored children's clubs known as 'safety patrols'. These deployed peer pressure to great effect, holding humiliating mock trials of child jaywalkers: if convicted, the guilty had to wash blackboards or write essays with the subject 'Why I Should Not Jay Walk'.[10]

Cars also benefited from the struggles of alternative means of transport. Around 1900, mass transport was enjoying a boom. American cities including Atlanta, Washington and Los Angeles had developed electric streetcar and trolley systems in the 1880s and 1890s. Prosperous suburbs grew up along the tracks and around stations, including Highland Park in Dallas, West Hollywood and Angelino Heights in Los Angeles and Grosse Pointe in Detroit. By 1917, millions of people daily rode the country's 45,000 miles of streetcar rails. By the 1920s, however, just as automobiles were becoming ever cheaper, many streetcar systems were old, over-crowded, shabby and locked into agreements with cities to keep fares low, meaning that they were no longer profitable to run. They were also getting slower since streetcars often did not have dedicat-ed lanes, meaning they were increasingly enmeshed in auto traffic.

The motor lobby was also not above directly precipitating the decline of electric public streetcars if the opportunity arose. In 1974 attorney Bradford C. Snell argued in the US Senate that 'General Motors [GM] and allied highway interests acquired local transit companies, scrapped the pollution-free electric trains, tore down the power transmission lines, ripped up the tracks, and placed GM motor buses on already congested streets.' This was hyperbole. But it certainly was true that a group of companies – including GM, Firestone Tires, Mack Trucks, Standard Oil and Phillips Petroleum – were behind National City Lines (NCL), a holding company that bought up nearly fifty transit networks across the country between 1938 and 1950. Given that NCL's investors had a significant commercial interest in promoting transport based on combustion engines and tyres rather than electricity and rails, it is perhaps not surprising that NCL would rip up rails and replace electric trolleys with motorised bus services in more than twenty-five cities.*[11]

'America is a great place to be,' one professor of urban planning has quipped, 'if you're a car.' There are signs, however, that the auto-mobile is beginning to lose its century-long grip on urban space the

* Although streetcars and trolleys were declining even where the NCL had no involve-ment, the Federal District Court of Southern California would convict the companies of monopolistic practices in 1949. This struggle inspired the plot of *Who Framed Roger Rabbit?* In this 1988 film that mixed film and animation, Eddie Valiant, a private detect-ive in LA, is pitted against the villainous Judge Doom, who is plotting to replace Red Car, the city's streetcars, with 'eight lanes of shimmering cement running from here to Pasadena'. Asked why anyone would buy a car and drive themselves when they could ride a world-class transportation system for five cents, Doom replies, 'Oh, they'll drive. They'll have to. You see, I bought the Red Car so I could dismantle it.'

world over. Residents and urban planners are less reflexively well disposed towards automobiles than they once were. This is unsurprising: over the past century, cars have proven fickle friends to cities. Originally promising a modern, speedy solution to the snarl of cyclists, walkers, streetcars, carts and carriages, they instead promoted sprawl and congestion and encouraged inefficient urban design that made dwellers increasingly car-dependent as other forms of transport either withered or became more dangerous.

Today, the costs at both the individual level – hours spent in traffic, health issues triggered by pollution and the rising costs of insurance, parking and fuel – and the macro level are impossible to ignore. Data suggests that each American household spends over two and a half hours each day in a car. In a city, it is likely that for a good deal of this time the car will be almost stationary, surrounded by a sea of other cars also heading nowhere fast. More recently, traffic has been found to have a hefty economic impact: $461 billion between Britain, Germany and America annually. It is now common to read opinion pieces – such as one from the *Washington Post* – arguing that cars 'were never necessary in cities, and in many respects they worked against the fundamental purpose of cities ... Removing cars ... would help to improve the quality of urban life'. [12]

Cities have responded by making themselves less car-friendly, to encourage the use of alternatives. Some local governments have implemented congestion charges, limited traffic zones and taken space away from cars to use for dedicated bus and cycle lanes. Others have been more creative. In the United Kingdom, Bristol has banned diesel vehicles during rush hour and, in the Netherlands, Amsterdam has promised to banish all vehicles with internal-combustion engines by 2030. The Italian city of Bologna developed an app to gamify the active mobility of its citizens, with points awarded for walking, cycling and the use of public transport. A deputy mayor in Paris has vowed to axe 70,000 parking spaces in the city. 'We can no longer use 50 per cent of the capital for cars when they represent only 13 per cent of people's journeys,' he said. In their place, the city will plant greenery 'to adapt to the acceleration of climate change'.

Heidelberg, in Germany, has taken perhaps the most radical approach. In the past decade, the mayor has introduced cycle 'superhighways' and year-long financial incentives for those willing to give up their cars. They have also pedestrianised large sections of the city and, elsewhere, built Bahnstadt, a new development in which most streets are dead ends, the speed limit is under twenty miles per hour, cyclists have right of way and buildings are grouped around playgrounds, courtyards and amenities.[13]

Meanwhile, in America, where an estimated 5,500 square miles – slightly larger than the state of Connecticut – is currently allocated to parking lots, there has been a push to claw some of this back. Washington DC is planning to charge owners of especially large vehicles an annual registration fee of $500. A 2022 California bill forbids local governments from mandating parking spaces for new developments near transport hubs, while similar legislation has been adopted in Anchorage, Alaska; Cambridge, Massachusetts; and Nashville, Tennessee. During the Covid-19 pandemic, New York and Oakland closed down miles of roads to car traffic and Kansas City turned parking spaces into mini parks.[14]

Automotive firms often provide their own answers to these issues. Over the past century, for example, many lobbied for the building of new roads as a means of easing congestion. Unfortunately, this has long been found to make the problem worse.* More recently – and perhaps predictably – their answers have involved the sale of different kinds of cars. Some announced plans to abandon or curtail fossil fuels, to focus on electric vehicles to ease pollution; others promote ride-sharing and 'driverless' or autonomous vehicles as curatives for congestion.

Perhaps most notoriously, Elon Musk, the co-founder of Tesla, the electric car company, took aim at traffic as 'soul-destroying'. 'It's like acid on the soul. It's horrible. It must go away.' First suggested in a jokey tweet, Musk's Boring Company proposed digging a network of tunnels under and between major cities that

---

* The explanation for this is what planners and economists call 'induced demand'. When travelling by car is made easier, cheaper or both, more people want to travel by car. The addition of extra lanes on the infamously congested roads in Los Angeles, for example, has repeatedly been shown to produce another lane of nose-to-bumper traffic.

would whisk people to their destinations at speeds of up to 150 miles per hour, avoiding above-ground congestion entirely. So far, however, prototypes have been underwhelming: bumpy, a third of the promised speed (at best), catering only to modified Teslas with hired drivers and, inevitably, susceptible to traffic jams. In many cases, the Boring Company has failed to follow up on proposals. The author of an article for the *Wall Street Journal* that accused the firm of 'ghosting' cities describes the entire debacle as 'a collision of marketing with reality'.[15]

Automobiles will not be banished in the immediate future. Like fond parents, many urban residents will continue to love them no matter what their faults are. It will also take many years and a lot of money to provide practical, reliable alternatives. Still, cities will continue redesigning themselves to promote alternatives to mass personal automobile use. As this happens, cars will surrender space, making them less attractive as the default option. In time, we may return to a world rather like that of 1907, in which a scrappy multiplicity of transport options coexist as equals, and where streets are once again shared spaces.

14

Tyumen – Kazan

19–24 July

Fancy driving an automobile across a ploughed field ... The body of the car begins to creak and shake as though it is about to fall to pieces.

Luigi Barzini, Itala team

The steppes began petering out between Tyumen and Yekaterinburg. Barzini, Prince Borghese and Guizzardi were simultaneously disappointed and relieved. The expanse of the steppes meant speed – they had covered 200 miles on 18 July, 220 miles the day before – but it was unrelentingly tedious. The landscape numbed them. All they could talk about was how many miles they had travelled that day and how many more lay ahead. At least the balance had finally tipped: they were closer to Paris than they were to Peking.

As the road uncoiled beneath their wheels, the green plain around them furred with shrubs. These clumped and expanded. Trees appeared: stunted at first, then taller and broader. Soon the road was bordered with a double row of birches, outriders of a forest that would soon overwhelm them.[1] The region was a sparsely populated no-man's-land. Birches gave way to densely packed evergreens that soared overhead. Stretching north to south and across the path before them lay the Ural Mountains. Fir-clad and remote, they were, nevertheless, peppered with mines – iron, coal, platinum, silver, aluminium and diamonds – and factories whose industry relied on these raw materials.*

As they drove, they saw pillars of smoke rising from the chimneys of smelting works tucked into the surrounding valleys. The earth that squelched beneath their wheels and splashed up onto their faces shifted colour. Gone was the soupy black of the steppes; now they saw iron-laden reds, chestnuts, duns and turmeric-bright yellows. At one point, they slowed to a crawl to allow an enormous caravan of hundreds of *telegas* loaded with iron, coal and crude oil to pass by. 'The fruits of the Urals,' as Barzini put it. Yekaterinburg, ahead of them, owed its wealth to such goods. Tyumen, where the motorists had spent the previous night, had been founded in the sixteenth century to help facilitate trade in the region: it was strategically located amid a network of navigable rivers and on the railway line. Even so, one visitor thought it a 'bare and unfinished-looking place'.[2]

Villages were scarce, but the Italians noticed that when they drove through them, the women made 'strange signs of exorcism and spat towards us, as though to ward off some evil influence'. Although they did not know it, an advocate of this region's visceral mysticism was, by the summer of 1907, making perilous inroads into the seat of Russia's political power. Fifty miles east of Tyumen, along the curlicue scrawl of the Tura River, lay the village of Pokrovskoye, the birthplace, some thirty-eight years previously, of Grigori Rasputin.†

In the evening of 19 July, the Italians approached Yekaterinburg at speed. Ahead of them, over the mountains, the sky was dark and roiling. The track beneath them unfurled westwards over the hills like a ribbon. 'Our car made and left behind it in the still, heavy air a thick cloud of dust which did not settle for miles behind us. We could see that streak from the top of heights now and then, still there in the distance, as if it had been the smoke of some fire we

---

* When the two De Dions passed through weeks later, they happened across a sword factory in Zlataoust large enough to employ 8,000 workers.

† This region was also where Tsar Nicholas II, his wife, children and household were sent after the Romanovs were deposed in 1917. They were smuggled first to Tobolsk, 150 miles north-west of Tyumen, then to Yekaterinburg, where they were murdered on 17 July 1918.

had caused in passing.' A phalanx of bicycles and carriages greeted the motorists at the city limits. 'When we reach them,' Barzini
wrote, 'they surround us, and shout "Hurrah!"'[3]

The Urals stretch nearly 2,000 miles, from the polar shores of the
Arctic Ocean down into what is now north-western Kazakhstan,
petering out in scattered hills to the north of the Caspian Sea.* To
early inhabitants of the region they appeared simultaneously tantalising and terrifying, a locus of hardship and profit. The *Primary
Chronicle*, a history of the Kievan Rus' peoples from the mid-ninth
century to the early twelfth century, described them with some awe:

> There are certain mountains which slope down to an arm of the
> sea, and their height reaches to the heavens. Within these moun-
> tains are heard great cries and the sound of voices; those within
> are cutting their way out. In that mountain, a small opening has
> been pierced through which they converse, but their language
> is unintelligible. They point, however, at iron objects, and make
> gestures as if to ask for them. If given a knife or an axe, they
> supply furs in return. The road to these mountains is impassa-
> ble with precipices, snow, and forests. Hence we do not always
> reach them, and they are also far to the north.

Subsequent centuries, bringing with them more sophisticated
technology, hardened cartographers and determined travellers
and traders, diminished the Urals. By the twentieth century one
adventurer scoffed that their 'highest peak does not attain to more
than 6,000 feet' and that 'No part of the Urals is permanently covered with snow.' By the time the Peking–Paris motorists reached
them, Barzini was able to assure readers of the *Corriere* and the
Daily Telegraph: 'The Itala slips quickly down the easy gradients of
the gentle hillocks which usurp the name of mountains.'[4]

But if time had made the Urals porous – from geographical full
stop to mere comma – it had not divested them of their symbolic

---

* The southern Urals are now a principal petroleum-producing region and home to
many oil refineries.

power. For one thing, they marked the boundary between the continents of Europe and Asia. At 5.17 a.m. on 20 July the Itala roared past a little clearing by the side of the road in which stood a large marble obelisk. Although they had planned to, by unspoken consent they did not stop. If they had, and if they had climbed out of the car to stretch their legs, toast the milestone and examine the obelisk more closely, they might have admired the imposing letters carved into its sides: 'EUROPE' facing west, 'ASIA' facing east. They might also have noticed, all around the imposing furrows of stonemasons' chisels, a cacophony of shallower, yet more poignant scratches made by the exile convoys that had passed by: names, dates, messages to loved ones they would never see again.

It was obvious to Barzini why the prince and his mechanic felt no desire to stop and mark their progress. They were wholly preoccupied by something else: the Itala's rear left wheel. It was once again creaking and cracking, the spokes threatening to break free of the rim. Since it was largely made of wood, the best option seemed to soak it, so that the wood might expand and hold together more securely. Their only difficulty was finding a vessel large enough for the job. In Perm, their stopping place that night, Barzini reported that they had taken 'the practical, if rather comical, step of hiring . . . a bath at the bathing establishment here, and the wheel is now there undergoing its hydropathic cure'.*⁵

Russian periodicals and newspapers were now covering the race as eagerly as their French, Italian and British counterparts. Journalists and photographers began tailing the motorists. Censorship was routine at the time, with journalists and editors fined or even imprisoned for negative or unflattering reporting. (One unfortunate paper in Kazan was so heavily redacted in late

---

* Although there is no record of the motorists partaking in a bath themselves, it would be odd if they had not. Baths were a quintessential part of Russian culture, and of the traveller's experience there. Thomas Knox, who travelled through the region and did visit one, minutely reported his experience. He noted how gregarious Russians were while bathing, the tools deployed – 'a bucket, two or three basins, a bar of soap, a switch of birch boughs, and a bunch of matting' – and the intense heat when water is poured over the hot stones. There 'is a rush of blistering steam. It catches you on the platform and you think how unfortunate is a lobster when he goes to the pot and exchanges his green for scarlet.'

July that a news item on sunspots and an update on the Peking–Paris were the *only* articles deemed fit to print.) But under the guise of reporting on the race, even the frankest descriptions of Russian infrastructure could be sneaked into print: 'The roads in Siberia were horrendous. The abandoned great Siberian Road [Tract] had become overgrown with thick grass, which concealed both deep ruts and hazardous, sharp stones,' read one Russian article. 'Most rivers had to be crossed by ferry. Any Russian is well aware of the experience of boarding and disembarking from these vessels.'

The race provided a vehicle for Russians to discuss their own nascent automotive culture and imagine the future of travel from their own perspective. An article in Russia's oldest newspaper, St Petersburg's *Vedomosti*, assured readers that the Peking–Paris 'has worldwide economic and social significance ... Although automobiles have been considered solely luxury conveyances up to now, they have clearly demonstrated their actual suitability for covering great distances along poor dirt-track roads.' The article continued that this was 'particularly significant for Russia, with our railway network which is utterly inadequate for the satisfaction of the population's economic and transportation needs'. Rather than wasting money investing in new railway lines, the author argued, they could be repairing roads instead.[6]

This opinion was shared by Prince Mikhail Ivanovich Khilkov, a railway expert and a former Minister for Roads and Communications. 'Thanks to *Le Matin*,' he told one interviewer, 'the myth of the automobile as a mere distraction for sportsmen is over. The little vehicle that was able to travel the ghastly roads from Peking to Paris is more than just an expensive toy. This is what frenzied speeding on paved roads, fenced off like a racetrack, has never been able to prove.'[7]

The two De Dion-Boutons were making slow progress. Quitting Tomsk on 20 July – the same day that the Itala, over 1,000 miles ahead of them, crossed the border into Europe – they reached Omsk on the evening of the 24th, covering 550 miles in five days. The Italians had covered the same ground in three.

Collignon and Cormier took turns driving out in front, with du Taillis always in the lead vehicle while Bizac, the mechanic, rode in the rear. Bizac, habitually taciturn, was becoming sharp with du Taillis, with whom he had little in common, considering him dead weight. (It is possible he was also annoyed at being the team's *mangeur de poussière* – 'dust eater' – the consequence of always being relegated to the rear vehicle.)

Du Taillis, admiring of Collignon's mechanical knowledge and 'admirable evenness of temper', had little good to say about Cormier. He had always poked fun at the older driver's fustiness, but, since the Gobi, his comments had acquired a sharper edge. The journalist griped that Cormier 'understands nothing of the duties of the French abroad' when he 'refused to toast the glory of the Republic' on Bastille Day. In his dispatch to *Le Matin* on 22 July, du Taillis included an odd little anecdote that he claimed would help the reader 'recognise Cormier among a thousand drivers', but instead made him look obtuse and a little mean:

Georges Cormier, a well-connected figure in the automobile industry, was the de facto leader of the French automobilists.

When someone, no matter what their nationality, is introduced or introduces themselves to him, he will answer . . . with a sly smile: 'Hello, André.' Why André? That must be quite the secret. I was surprised by this constant familiarity with the most diverse characters. I questioned Cormier, who replied only with his sly smile.[8]

Collignon, who was described by nearly everyone as a simple, cheerful and straightforward soul,* seemed to be the only one of the motorists who liked and respected their de facto leader. On his return, he touchingly wrote that Cormier was 'an excellent companion' and that the two were 'dear friends'. Cormier did not return the sentiment. While his regular dispatches were generally short, focused to the point of mania on the quality of the roads and the weather, Cormier did indulge a taste for criticising his fellow automobilists. He mentioned that everyone called Collignon 'Pivo', from the Russian for 'beer', 'for he seems to have acquired a fondness for this brew – a fondness that grow[s] stronger by the day'. (Du Taillis, conversely, never mentioned this nickname and indeed called Collignon 'the epitome of sobriety'.) Cormier also informed his readers that Collignon was penniless and that his clothes, towards the end of July, became infested with lice.

In a longer article, sent from Urga on 23 June but published on 16 July, Cormier wrote that, unlike Borghese, he and Collignon had waited for two whole days for the Spyker in the Mongolian city. As his readers were by then aware, those few days in the Gobi had very nearly killed Godard and du Taillis. Cormier evidently saw things differently. The cause of the Spyker's delay, he wrote, was 'one of the funniest peculiarities of our trip: I still see Godard stopped in an immense plain as far as the eye can see – not the slightest ripple, giving one the sensation of infinity – and telling Collignon and me: *I have no more fuel.*'[9]

As if this ill feeling was not enough to ensure the De Dion team

---

* The account of 'S.A.P.', who wrote the foreword to Collignon's book about the Peking–Paris, is representative: 'Collignon is twenty-eight years old and in good humour . . .' it begins. 'He has no enemies. He is a happy citizen . . . Cormier, the other driver, wrote a beautiful book about his experiences. Jean du Taillis, a journalist, is working on another sparkling account. Collignon is content with the pamphlet you have here.'

lacked cohesion, Cormier was also probably responsible for the decision, made at Omsk on or around 24 July, that would slow them down further. This was the choice not to follow the suggestion of the local organising committee to take a more northerly route over the Urals – as Borghese had done – but to stick with the route sketched out in Paris during the planning meetings. Both Cormier and du Taillis were keen to present the prince's route over the Urals as a deviation, if not an absolute breaking of the rules. After implying, without evidence, that the prince had secretly imported an entirely new Itala, du Taillis wrote in his 1908 book that the Italians 'followed a different itinerary from the one we had to take, *in order to comply strictly with the conditions set* by the rally organisers. It so happened, as if by chance, that the one taken by the Itala, a little longer, it is true, avoided the major problems that the Ural Mountains had in store for us.' (italics in the original). Cormier gave the same explanation. The Itala's 'rule-breaking route' has become part of the race's mythology.

An exchange in the pages of *L'Auto*, published during the race, gives a different impression. Directly underneath a telegram from Cormier stating that Prince Borghese was 'no longer following the route indicated by *Le Matin*', the magazine printed a message from Georges Bourcier Saint-Chaffray, head of the organising committee:

As a follow-up to Cormier's telegram, dated Petropowlosk, 26 July, at 8 a.m., which you communicated to us, I must inform you – so that there is no unnecessary controversy or errors committed to print that could harm Cormier by wrongly representing him as a man who takes issue with and seeks to diminish the performance of one of his rivals – that, in the Peking–Paris, the choice of route is completely open. This is a challenge [*un défi*] and not a race [*une course*]. Prince Borghese took a longer route than the one Cormier takes. He did so at his own risk.[10]

Blithely unaware of backbiting, the Italians were busy grappling with problems of their own. Now firmly in European Russia, they hoped for more speed. As they set off from Perm towards Kazan on Sunday, 21 July, their route took them in a broad arc north of the Kama River to its junction with the Volga, which they would then follow upstream as far as Nizhny Novgorod. An hour out of Perm, however, the skies opened. 'Thunder roared incessantly,' Barzini wrote, 'and the rain fell with the violence of a cataract, inundating everything.' Water sluiced over the track, wrinkling and splashing against the rolling tyres. They were reduced to walking pace, but even so the automobile pranced and shied like a restive horse. After four hours, the rain stopped, the water subsided and they put on a burst of speed. Above the roar of the engine the three men heard a loud crack, then a crash. Peering down over the side of the automobile, they diagnosed the problem instantly: the damaged wheel had shattered. 'It's all over,' the prince called tonelessly up to his companions, squatting low to inspect the damage. 'We can't go another yard.'

Still, they could hardly remain in the middle of the road. To get them rolling as far as the next village, they set about a makeshift repair. Hacking twigs from surrounding trees and bushes, they secured them in bundles around the wheel until it resembled tumbleweed. As they worked, an elderly peasant materialised and stood watching them.

'There is a man here who can make you a new wheel,' he said, conversationally. 'He is the most capable maker of sleighs and *telegas* in the whole district.'

Relying on their smattering of Russian, Barzini and the prince established the whereabouts of the *telega*-maker and, once Guizzardi deemed the wheel sufficiently secure, proceeded the four miles in the direction indicated. It was approaching midday when they reached the spot. They found a remote farmstead, made of well-planed timber. It ranged around a forge and a yard hosting some curved sledge beams and a tarantass, which was held up on a stand to allow a fresh coat of varnish to dry.

Hearing the noise of the engine and the babble of Italian voices, the craftsman, a tall, imposing man of around fifty, walked into his yard followed by his apprentices. They were large men, with rough hands that they wiped on rags tucked into their leather aprons. They all had long hair and beards that flowed down over red shirts. They looked at the automobile and then at the Italians.

Indicating the twiggy bundle that had once been an automobile wheel, Borghese tried to explain the problem. Understanding shimmered out of reach temporarily. '*Latine intelligis?*' ('Do you understand Latin?') one of the apprentices asked.

In a chimeric gabber composed of two living tongues and a scholarly one – learnt, the apprentice said, while at home during the long winter evenings – their predicament was explained. The wheel was examined and a price arranged for the construction of new spokes to fit inside the battered rim.

Over the following seven hours, the small group in the yard coalesced around a common purpose. Guizzardi flapped anxiously about the Itala, like a moth around a lamp, muttering under his breath, pointing and cajoling, while Barzini looked on in fascination. 'No other tool is used than the axe, managed with marvellous dexterity. Soon, from great logs of pine, new spokes are fashioned, the chips flying from the axes like sparks from an anvil . . . [They] are not elegant. They are thick and rough, but they will withstand any shock.' The Russians, sweating with the effort and with red shirt cuffs rolled up over bulging arms, watched as their spokes were fitted into the rim, and the wheel reassembled and reaffixed to the automobile by Guizzardi. He cautiously turned the starter, climbed into the driver's seat and inched forward. It rolled. '*Grazie mille*,' the relieved Italians called out. '*Arrivederci! Do svidaniya! Salve!*' The Itala and its occupants bounded out of the yard, around a corner and out of sight.[11]

15

POWER PLAY

Automotive Fuel

A drop of oil is worth a drop of blood.

Georges Clemenceau (1841–1929)

Our vast industries have their root in the geologic history of the globe as in no other past age. We delve for our power, and it is all barbarous and unhandsome.

John Burroughs, *Under the Maples*, 1921

As the summer of 1859 drew to a close, it seemed as if the last of Edwin Drake's dreams might expire in the Pennsylvanian fug. Forty-one years old, he looked older. Deep grooves lined his mouth and forehead: these, coupled with his long, ascetic face, beard and heavy-lidded eyes, lent him the look of a saint from a medieval triptych. Raised in New York, he had worked on America's railroads, usually as clerk or conductor, until a muscular disorder forced his early departure. With a wife and children to support, he had continued to find odd jobs, most recently for a rackety firm called the Seneca Oil Company. They had promptly sent him to Titusville in Pennsylvania to hunt for oil, armed with only some letters of introduction that gave him the fictitious rank of 'Colonel' to excuse his limp.

That there was oil in the area was beyond doubt. Pools of it sporadically seeped up through rocks, especially along the banks of the narrow creek that meandered through the woods and fields around Titusville. Locals mopped up the dark liquid with blankets and used it as a remedy for scurvy, kidney stones and gout, as a local anaesthetic, or to remove stains from textiles – although they stank afterwards. They also sold it for a decent price, but demand was low. This had begun to change in the mid-nineteenth century. From 1846, cleaner, more effective methods of extracting kerosene or paraffin from oil were developed. In the era before electricity, when people relied instead on tallow and wax candles, or else expensive whale-oil lamps, kerosene provided cheap, clean light.[1]

This was how Drake found himself, in August 1859, overseeing what increasingly seemed like a doomed enterprise: drilling America's first commercial oil well.* Work had commenced on 20 May and there was nothing to show for it. A day's work would deepen the well by just three feet. Weeks and then months dragged by without success. Neighbours sniggeringly nicknamed the project Drake's Folly. The Seneca Oil Company stopped paying his wages, so he took out a 500-dollar personal loan. On Saturday, 27 August, the drill slipped into a crevice sixty-nine and a half feet below the surface. Dispirited, Drake's team downed tools for the day. When they returned, the well was surrounded by a bubbling black pool. Utterly unprepared for success, they desperately ran about scooping the precious liquid into whatever receptacles they could find, including, according to local lore, a bathtub and all the whiskey barrels in Titusville.[2]

Despite the bathos of the moment, Drake and his crew had changed the world for ever. By the turn of the century, his well had kickstarted an American oil rush and business was booming. Prospectors – known as 'wild-cats' – descended on Pennsylvania, conjuring evocatively named shanty towns from thin air: Petroleum Center, Bonanza Flats, Red Hot, Pithole and Wild Cat Hollow. Other oil-rich regions began cashing in too. Oil was discovered and drilled in Texas, Kansas, California, Baku (now part of Azerbaijan), the Dutch East Indies (Indonesia) and Persia.

---

* Edwin Drake is also notable for inventing the 'drive pipe': sections of piping used to line the hole being drilled to stop water overwhelming the works. Unfortunately, he did not patent this invention.

Companies formed to both extract and promote the use of oil. The Standard Oil Company was founded in Ohio in 1870 by a group led by John D. Rockefeller; in Europe, Royal Dutch (later Royal Dutch Shell) came together in 1890. Even as these two titans clashed over market share, they were astonishingly effective at developing new markets for their oil and securing profits. Standard Oil, once it began producing more oil than it could dispose of domestically, set its sights on China. The firm cheaply distributed kerosene-burning lamps inscribed with the words '*Mei Foo*', which roughly translated as 'beautiful and trustworthy'. Thus, around the turn of the century, this became the default lighting fuel for a country with a population of 400 million.[3]

These firms – especially after the spread of electric light in America and Europe began eating into their profits – were, of course, enthusiastic promoters of petrol-engine automobiles. After the 1906 San Francisco earthquake, for example, Standard Oil donated 15,000 gallons of fuel so that private automobiles could take part in rescue efforts.[4]

In Baku, then part of the Russian Empire, oil was extracted by hand until the early 1800s. In the 1870s, the tsar's government – and the ambitious Count Witte in particular – began encouraging investment in the area. Among those to do so were the three Swedish Nobel brothers: Robert, Alfred and Ludvig Emmanuel.* From the first they perceived a bright future for petroleum and set to work introducing modern, efficient methods of oil extraction.† By 1900 their firm was drawing 10 per cent of the world's oil

---

* Prince Borghese had tasked the Nobel firm with supplying his fuel through Russia, which they did by setting up depots every 150 miles. The French, meanwhile, relied on Mr Neuville, from a firm called Schreter. Judging from the number of times they complained that their fuel had failed to materialise, they would have been better advised to choose the larger and better-known Nobel Oil.

† 'Modern' and 'efficient' are relative terms. To get an idea of the technology being used at the turn of the century, watch 'Oil Wells of Baku: Close View' (1896). This thirty-six-second burst captured a year after the Lumière brothers created a working Cinématographe camera, shows several of Baku's wooden well structures in the foreground, while behind towering oil fires and smoke billow into the sky. The Lumières enjoyed distributing films depicting industrial landscapes and technologies, perhaps because it aligned their work with progress and the future.

from Baku. Record profits spurred on other investors, including the Rothschilds, to invest in Russian oilfields. Five years later, Russian oil accounted for nearly half the world's supply: 84 million barrels.

Being lighter, less bulky and easier to store and transport than coal, oil was an attractive energy source, particularly in countries that had both domestic demand and abundant supplies, notably Russia and America. Russian trains began switching from coal to fuel oil in 1889; soon after, the world's navies followed suit.[5]

In 2020, road transportation accounted for just under half of the oil consumed in OECD nations. In fact, the internal-combustion engine and automobiles in particular are largely responsible for propelling oil consumption over the century and a half since Drake sank his well into the Pennsylvanian rock. Not only do we merrily pump it into our cars, vans and trucks as fuel, but oil by-products have been used to build and repair the roads on which we drive and, in the form of plastics, are essential components in building cars themselves. The ready availability of oil means that whole sectors of other economies have come to rely on it too. Private automobiles and cheap petrol made suburbanisation, out-of-town retail and office parks possible; fuel is necessary for aviation and shipping; oil provides feedstock for paints, plastics, pharmaceuticals and even textiles; without oil to provide fuel for machinery and chemicals for pesticides and fertilisers, modern agriculture as we know it would grind to a halt.

So dependent on oil have we become that our ability to access it plays an outsized role in global diplomacy. In 1997, the American General J. H. Binford Peay III bluntly informed Congress that 'the international community must have free and unfettered access' to the oil reserves in the Gulf. Protecting the Strait of Hormuz, through which some 18 million barrels of oil pass every single day – a fifth of global supply – has for the past half-century been a shibboleth. It precipitated the Iran–Iraq War in the 1980s and influenced the actions of America and its allies in pursuing the Iraq War of 2003–11.

For the sake of oil, Western democracies have held their noses and cultivated strategic relationships with regimes and individuals credibly accused of human rights abuses. One of the very first things the British prime minister did after Russia attacked

Ukraine in February 2022 was to arrange visits to Saudi Arabia and the United Arab Emirates, both large oil producers. During the same conflict, many European countries found themselves in the uncomfortable position of placing tough sanctions on Russia while simultaneously writing large cheques to ensure uninterrupted energy supplies from the same source.[6]

The seeds for much of this were being sown at the time of the Peking–Paris, but our total – and in hindsight rather naive – dependence on oil was far from inevitable. At the turn of the twentieth century around 40 per cent of American automobiles were powered by steam, 38 per cent by electricity and just 22 per cent by petrol, which had a reputation for being smellier, noisier, less reliable and prone to price fluctuations. The general consensus was that different fuels were suited to different uses. Electricity, for example, was perfect for short, personal journeys or delivering goods from railheads to shops and warehouses. Steam was believed to be best for heavy haulage, while petrol was perfect for long-distance touring.

Indeed, the perils of over-reliance on oil as a fuel were already being debated. France, Germany and Great Britain (all still at the forefront of the automobile industry) had no oil reserves of their own, and very little in their empires.* They had also recently been shaken by a price shock caused by the political turmoil in Russia.[7]

The oil-producing region in the Caucasus between the Black and Caspian Seas was ripe for rebellion in 1905. The need for cheap labour for the oilfields had turned many urban areas into slums. The streets were 'littered with decaying rubbish, disembowelled dogs, rotten meat, faeces'. Wages were low and the work hard: in Batum, the Black Sea port, fourteen-hour days with two hours' compulsory overtime were not unusual. Sectarian violence between Armenians and Tartars was common. It was a hot spot for revolutionaries. 'Nina', the codename for Lenin's large

---

* All three were, however, making efforts to gain influence over oil supplies. For example, the British government took a great deal of interest in Royal Dutch Shell, and from 1914 would be the largest stakeholder in the Anglo-Persian Oil Company. Likewise, in 1906 the German government was involved in financing the European Oil Trust, which it was hoped would be a rival to Standard Oil.

printing operation, was run out of a cellar in Baku and many future Bolshevik leaders, including Joseph Stalin, cut their teeth in the region.* Alongside agitation stirred up by Communists, locals were resentful of the tsar's rule from faraway St Petersburg.

By early September, riots were endemic. Telegraph, telephone and railway lines were severed and oil works set alight. 'Great fires are still raging at Sabunto and Nomani,' read one bulletin, 'fed by immense tanks of naphtha, of which there are 1,600 in that region.' In Baku itself, people were fighting in the streets, much of the city was in flames and a hurricane was blowing. Authorities despaired of quenching the fire. At a stroke, half of Baku's oil industry was wiped out and oilmen gloomily predicted years of difficulty and expense before it could be returned to full strength.† As a result, fuel prices in Great Britain increased by 85 per cent between November 1904 and December 1906.⁸

One person preoccupied with the fuel question was the man the press had dubbed 'The Wizard of Menlo Park' or 'The Napoleon of Invention': Thomas Edison. An accomplished businessman and indefatigable self-promoter, Edison gave interview after interview about his inventions, the future and everything in between.‡ He repeatedly returned to the issues which today we would call fuel efficiency and renewable energy. In an article published on

---

* Stalin was the chief socialist organiser in Batumi, Georgia, from 1901 to 1902, when he planned numerous strikes and demonstrations. In 1907, his return to the region would coincide with yet more strikes. His role, in his words, was to foment 'unlimited distrust of the oil industrialists' amongst the region's workers.

† A visitor to Baku in January 1906 was appalled by both the evidence of past destruction and the continued violence. 'Riot, destruction of property, bloodshed, murder, were all a part of each day's work in Baku . . . Near to the station as we alighted from the train a murdered Armenian was lying in the gutter. Blood still oozed from his head. What immediately struck me was that no one gave him the slightest heed. Passers-by stepped over the corpse as if it were the carcass of a dog.' When he asked an Armenian companion about it, the man told him that the only solution would be to arm the Armenians. 'Give the Armenians guns, leave them alone, and in ten days there would not be a Tatar left north of the Persian frontier.'

‡ Many are as gauche as the modern celebrity interview. One began, 'The man whose masterly genius has conjured out of electricity and the kindred forces of nature a host of marvels beyond the dream of poet or prophet, whose work, more than the work of any other, has helped to characterise this as the "Age of Invention".'

7 January 1906, he described 'the control of the energy stored in coal, directly and without waste' to be one of the 'more vital and immediate discoveries' he was concerned with. 'Ninety per cent of the energy stored in coal is now lost,' he said. 'If, therefore, a means can be devised by which this enormous waste is saved, it will naturally revolutionise and vastly cheapen the production of power. The result will have an incalculable influence upon the material progress of civilisation.'[9]

An interview the following year found him looking to the sun and sea as energy sources when American coal supplies were exhausted (an eventuality he predicted would occur 'within a hundred years'). 'In Arizona, where there is a maximum amount of sunshine, engines generating power by the use of mirrors focusing on a copper boiler are in actual use . . . As for the sea, it is a vast storehouse of unutilised power.'

Many assumed that electricity, intertwined as it was in the

American inventor Thomas Edison (left) *was a strong advocate for electric cars.*

189

popular imagination with progress, would sooner or later power the world.* In 1899, *Scientific American* characterised electric cars as 'clean, silent, free from vibrations, thoroughly reliable, easy of control, [producing] no dirt or odour'. They were especially favoured in urban environments, where most journeys were short. They were more expensive than steam and combustion vehicles, but cities allowed for fleets of hired vehicles and there was no reason, given that the technology was clearly working, why this should not have continued to be the case.†

In 1900, the biggest practical disadvantage of electric cars was the need for an extensive refuelling infrastructure, in contrast to petrol, which could be carried by a driver in cans. Still, for the vast majority of drivers, this did not matter, since most journeys – then as now – were short and in urban environments where roads were good and electricity was close at hand.‡ At times and for specific tasks – short-haul deliveries or commutes, for example – they were more reliable and cheaper than alternatives. Electric cars accounted for a third of all vehicles on America's roads in 1900, and New York had a fleet of sixty electric taxis. The market for electric automobiles would persist into the 1920s. However, they were slower and, because of their smaller range, felt less adventurous than fuel-powered ones and so were often marketed to women.

Companies took note. General Electric ads from the era show women charging their motors at home, with copy that contrasted the convenience of electric cars with filthy, brawny, noisy, and rebellious petrol vehicles. 'There are no tiresome trips to a public garage, no waiting – the car is always at home, ready when you are.' This idea was so entrenched that Clara Ford, Henry's wife, would, from 1908, the same year the Model T was released,

---

* See, for example, the sci-fi novels of French writer Albert Robida, most notably *Le Vingtième Siècle: La vie électrique* (translated as *Electric Life*) (1890).

† The automobiles of one hiring scheme in London in 1898 were nicknamed, rather wonderfully, hummingbirds.

‡ In America, according to the 2017 National Household Travel Survey, 60 per cent of car journeys were under six miles and three-quarters were ten or less, well within the range of even a small battery.

drive a neat, forest-green electric automobile made by Ford's rival, Detroit Electric.

Overtly gendered marketing contributed to the downfall of the early electric car, but another issue was batteries. While most journeys were within the scope of existing batteries, the technology around 1900 did not allow people to travel long distances. Edison was working on a new, improved kind – 'on the market within six months' – 'which will enable every man to travel in his own private carriage' at a reasonable price. He predicted each would cost 200 dollars and would, on a single charge, carry a vehicle at twenty miles per hour for 150 miles and would 'travel a hundred thousand miles before it is worn out'. The *New York Times* characterised it as Edison's 'greatest wonder', capable of solving 'the problem of congested traffic in the big cities of the world'.[10]

Edison was not alone in his dreams that cars would run on fuels other than oil. At this time there was no broad understanding about what oil was or how it was made. While it was generally agreed that it was of 'organic origin', no consensus existed on whether it derived from animal or vegetable matter. And although it was still possible in 1900 to find scientists willing to prophesy that reserves were inexhaustible,* most had their doubts. 'With the constant withdrawal of large supplies of crude oil from Mother Earth,' wrote one expert, 'Nature's stores must be growing less.' In the motoring world, such anxieties found expression in an anguished casting about for alternatives and a marked reluctance to commit to petrol as *the* fuel for cars.[11]

Electricity and steam still had their advocates – and indeed vehicles fuelled by both would continue to be made and sold commercially for years – but others turned to alcohol. On 31 July, an article in *The Bystander* (opposite an advertisement for 'The famous ITALA CAR now running in the PEKIN to PARIS Race') concluded that 'the best fuel of all for motor purposes, and one, moreover, unlimited in quantity, is alcohol.' Distilled from beet, peat or potatoes, and more efficient than petrol, alcohol – the writer argued

---

* Dmitri Mendeleev, the Russian chemist responsible for developing the periodic table and who died on 2 February 1907, was among their number.

– was the surest guard against the 'petrol famine [that] appears to be inevitable in the near future, as the demand is increasing at a rate far in excess of the supply'.

A series of tests conducted in Berlin in the winter of 1902/03 indicated that alcohol-powered motors could be up to twice as efficient as those using petrol. Alcohol 'is to be the fuel of the future', it was reported at the time. 'It can be distilled from any form of vegetable matter, even such as potato parings and other refuse, and can be manufactured everywhere . . . Count René de Knyff, who was the leader in the recent Paris to Vienna race until his car broke down, used a mixture of alcohol and gasoline.'

As might be expected, countries that did not have oil supplies of their own were the most effusive about alternatives. Several races were held in France to promote the use of alcohol-powered automobiles. In 1899, four such vehicles zipped from Paris to Chantilly; two years later, fifty more took part in a Paris–Roubaix run; and a third such race was held over two days in 1902. A 'Coupe de l'Alcool' was offered by Prince Pierre d'Arenberg for the most impressive specimen. Unfortunately, however, while there was plenty of evidence that alcohol was *theoretically* the better fuel, in practice few could be persuaded to work on producing efficient alcohol-powered automobiles, since duties on alcohol made it prohibitively expensive. As an exasperated editorial in *The Autocar* put it in August 1907: 'It is useless for any car manufacturer to waste his time on perfecting an alcohol motor when petrol is the fuel of the day, and no one would look at alcohol. He might as well devote his attention to constructing a car to run on port wine.'[12]

Despite misgivings and the lure of alternatives, the trajectory towards a future reliant on oil would be set by 1920. There were two principal reasons for this. The first was the shift of the centre of the automotive world across the Atlantic from western Europe to America, where oil was cheap and plentiful. In this respect, the 1907 Peking–Paris race was both the apex and the end of an era. It was proposed and championed by a Parisian newspaper; the majority of those taking part were Frenchmen, who set out confident in their dominance; and the race was arranged so that the denouement would feature a triumphal entry of a presumably

French car into Paris, the spiritual home of the automobile. The very next year would see the release of Henry Ford's Model T and the well-publicised successor to the Peking–Paris: the 1908 New York–Paris race. Again, many of the cars taking part were French, but the winners would be the sole American team, driving a Thomas Flyer, an American car.

The second reason was the First World War. Although, at the outbreak of war, military establishments were suspicious of motorised vehicles and preferred a combination of rail and animal power, by 1916 the utility of motorised vehicles was becoming obvious. Railway lines were slow to build, vulnerable to attack and difficult to repair, while vehicles reliant on steam or electricity were cumbersome and difficult to refuel on battlefields. By 1917, Allied armies were consuming a million tons of petrol and their navies 8 million tons of heavy oil, consumption which drained French reserves to the dregs. A telegram from the French Prime Minister, Georges Clemenceau, to President Wilson makes the point with almost painful clarity:

> At the decisive moment of this War, when the year 1918 will see military operations of the first importance begun on the French front, the French army must not be exposed for a single moment to a scarcity of the petrol necessary for its motor-lorries, aeroplanes, and the transport of its artillery . . . A failure in the supply of petrol would cause the immediate paralysis of our armies, and might compel us to a peace unfavourable to the Allies . . . If the Allies do not wish to lose the War, then, at the moment of the great German offensive, they must not let France lack the petrol which is as necessary as blood in the battles of tomorrow.[13]

The Age of Oil has transformed the world. It has enabled the production of vast quantities of plastic goods, from water bottles to hip replacements; encouraged globalisation and greater mobility; realigned global power dynamics and alliances; revolutionised military engagements; remodelled urban layouts; and drastically increased pollution, impacting the health of ecosystems and the human population.

193

Predictions of how and when the sun will set on the Age of Oil are numerous. It is possible that oil production will fail to keep pace with continuing high demand – a scenario often shorthanded to 'running out of oil' – probably entailing a ruinous scrabble over the precious remaining barrels. It is also possible that demand will fall precipitously, thanks to public pressure, geopolitical considerations and changes in the automobile industry, such as more ride-sharing and a shift to electric vehicles or other fuels. In this scenario, oil prices would drop and it would no longer be cost-effective to produce it in large quantities. Sales of electric vehicles have certainly risen rapidly – doubling in 2021 to 6.6 million – from a near standing start a decade before. This trend is likely to accelerate, since the EU, Norway, China, California, India and Britain are among those banning fossil fuel cars between now and 2040.[14]

It is fascinating to imagine what might have happened if, long after 1907, multiple fuel technologies for automotives had continued to be pursued. In such a world, electric vehicles might have continued to dominate cities, while alcohol or some other fuel technology competed with fossil fuels for longer-distance journeys. Fuel options only now being focused on – such as hydrogen – might have attracted attention and proved financially viable decades earlier.

16

Kazan – Moscow

24–27 July

This is the truth that *Le Matin* wanted to demonstrate by launching the Peking–Paris challenge. The car is not only for sport. Above all, it is a practical vehicle, one of which much hard work can be expected. The proof is now well established.

Prince Scipione Borghese, Itala team

The prince, his chauffeur and the journalist awoke on the morning of Wednesday, 24 July, in Kazan. Two days – lengthy, trying ones – had elapsed since their departure from their saviour, the *telega*-maker, and his Latin-speaking apprentice.

It had been seven o'clock in the evening by the time the Itala and its hand-hewn wheel had re-emerged on the road some sixty miles south-east of Perm, heading towards the Volga and the setting sun. The roads had dried, and they were at times able to drive at a respectable twenty miles per hour. As they passed through the scattered villages, they amused themselves with a game devised by Barzini: guessing from the reaction of the inhabitants whether the settlement possessed a telegraph. Where there was none, it was not uncommon for villagers in Russian dress to cross themselves, burst into hysterical laughter or flee, yelling about infernal locomotives escaped from the train tracks. Where there was a telegraph, the motorists were expected and cheered.

Night and hunger surprised them in one of the former. At a loss as to what to do, and too stiff, sore and cold to contemplate a night in the open, Borghese won over scared prospective hosts by offering rides in the Itala. Two were coaxed up, and the prince set off as smoothly as he could. After a few moments their terror evaporated and, to Barzini's amusement, they became so enthusiastic they proved difficult to dislodge. A queue formed, and the Italian prince found himself shuttling villagers in slow, genteel circles around the streets accompanied by squeals of delight. Bolstered by this goodwill, the Italians were provided with a fulsome meal of tea, eggs, milk, bread and butter and a portion of bare floor on which to sleep.[1]

Over the previous two days, woods had given way to cultivated fields criss-crossed by waterways – the Vala, the Kunzhek, the Vyatka – and peppered with villages alternately Christian and Muslim. 'Minarets and delicate towers,' as Barzini put it in one of his dispatches, 'crescents and crosses, all mingled in the peace of the fields.' Villages and people had taken on a sleeker, wealthier aspect. Figures scattered over the landscape in loose garments of red, yellow or white unbent at their approach. Bodies swathed in silver braid, silk sashes and golden waistbands leant out of windows. Faces crowned with headdresses like embroidered mitres turned to watch them pass.

The three men had ample time to admire their surroundings. This was a region of river travel. The road was so pitted and rough that for long stretches it proved easier and more comfortable to drive over the ploughed furrows of fields parallel to the road. Progress was won at a jouncing eight miles an hour. The chassis creaked. Leaf springs began to give way. Having disabled the brake after the fires it had caused, they were reliant on the handbrake, which screeched and gnashed its teeth down steeper slopes.

While by now the three Italians knew each other well enough to be comfortable in silence, the atmosphere inside the car soured after hours of tension, tedium and discomfort that even the relative richness and variety around them could not relieve. They were also feeling homesick. The same landscape that had delighted Barzini earlier now depressed him. 'We crossed little tiny [sic], quiet, sad, solitary villages,' Barzini wrote, 'with little wooden houses just like those in the fields, except for a gay coat of white paint; we felt oppressed by a great melancholy as we thought of their monotonous, silent life.'[2]

At three o'clock on the afternoon of 23 July, forty-three days since they had left Peking, the automobile nosed along the valley of the Kazanka River and the Italians caught sight of the bustling blue line of the Volga. Right beside it was a city coroneted with towers and cupolas: Kazan. As they approached, a well-appointed horse-drawn carriage drew up alongside them. 'Are you from Peking?' the driver asked, around a cigarette. 'Yes, madame,' the prince replied to the dashing apparition in a man's hat and gold-rimmed spectacles. Graciously offering to show them the way, she whipped up her horses and led them past churches, lines of electric trams and cheering onlookers, finally leaving them with a salute at the doors to the Hotel d'Europe.

Kazan occupies rising ground where the Volga and Kazanka rivers meet, some thirty miles upstream of the Volga's junction with the Kama. A 'Moscow in miniature' was the verdict of one nineteenth-century visitor; in 1889, Harry de Windt called it 'the true boundary between European and Asiatic Russia'. Electric tramways and gas-lit streets rubbed shoulders with old-fashioned wooden houses. What they saw there led de Windt to indulge in a flight of orientalism and wonder whether 'the long journey from China had not turned one's head and indelibly mixed Europe and Asia in our minds':

> The veiled faces of the women, the fierce, swarthy Tartars in wild, barbaric costume, bristling with daggers and cartridge-belts, the mosques, minarets, and oriental-looking houses, mingling in strange incongruity with the modern stone houses of the Russian population, *à la mode de Paris*, four stories high, with balconies, *porte cochères*, and carved façades.[3]

During the first half of the fifteenth century, Kazan had been the capital of a great khanate, a fragment of the Golden Horde that had, at its fourteenth-century height, stretched from the Danube in the west to the Altai Mountains in the east. A century later, in

197

1552, it was captured after a protracted siege by the armies of Ivan the Terrible, Russia's first tsar. This tumultuous past had imprinted itself on the city. It was evident in the high, white crenellated walls of Kazan's Kremlin (the word literally translates as 'citadel' or 'fortress within a city'), which backed up the slopes away from the river, a memorial commemorating the siege and its large Tatar population.*

There was turmoil in the city's recent past too. Kellogg Durland, an American journalist and social reformer who travelled around Russia in 1906, reported that Kazan, on the edge of the 'famine belt', was ripe for insurrection. 'So complete was the failure of crops in some counties of Kazan this year that the harvest would not suffice for a single month.' The government relief needed for 1906 alone, he estimated, would amount to 32 million roubles, well over what the country could afford.† Prince Ukhtomsky, a loyal tsarist, gloomily predicted that the peasants were so utterly disillusioned that there was 'no hope of the present Tsar ever regaining their confidence'.[4]

Barzini's newspaper reports, bypassing social unrest completely, gave only an impressionistic account of their eighteen hours in the city – wide boulevards, busy population, an air of modernity. The three men were distracted and anxious. Superficially, their position was unassailable: they had completed nearly two-thirds of the race; a little over 1,000 miles separated them from the two De Dion-Bouton crews, still bogged down in the taiga to the east of Omsk; and they were 2,600 miles clear of Godard and his Spyker, last heard of marooned in Cheremkhovo, near Irkutsk.

Nevertheless, the preceding days had driven them to the limit of their nerves. It had taken them three days to cover the 360 miles between Perm and Kazan, a distance they might have expected to cover in one day on good roads and in a fully functional vehicle. Trouble with the wheel had drastically slowed their progress; the footbrake remained out of commission, the charred sections of the chassis testament to near disaster. Now, thanks to their bouncing

---

* On the night of 15 December 2014, this monument was vandalised with the slogan '1552 – Nobody and Nothing is Forgotten', spelt out in red letters.
† In the region of £341,000,000 today.

progress over the rutted ground, several of the leaves in the Itala's springs had broken. They had managed to find someone to repair them overnight in Kazan, but it was clear that the Itala was beginning to falter. Thus far their luck – not to mention Borghese's deep pockets – had held. They were only four days away from Moscow. From there, the route laid out by *Le Matin* headed due west out of Russia, across Germany and into France. But should something more fundamental give way, as it had on the Spyker, it might mean the end of the race, even as the finish line came within their grasp.

By nine o'clock on the morning of 24 July, the springs had been patched up. Guizzardi, who had come to identify with the car as if it were a beloved horse, minutely examined the work and sighed, looking up at the prince with an air of resignation. The prince, hands on hips, shrugged and moved to the front of the automobile to start the ignition. The Itala moved slowly through the city to the landing stages at the river's edge, weaving among crowds and piles of goods towards the small craft that would ferry them across the Volga.

'The greatest river of Europe,' Barzini wrote, 'vast, slow, majestic, proud [and] covered with ships.'

Since it connects inland central Russia with Iran, then known as Persia, the Caucasus and large portions of central Asia, it was always busy with traffic, but the hubbub was particularly pronounced just then. The Great Fair at Nizhny Novgorod was about to begin: goods and people from across the continent were being drawn to the spot like iron filings to a magnet.* The Italians were heading there too. Once the Itala had been stowed, the boat shoved off into the eddying traffic. Guizzardi stayed with the car while Barzini and the prince walked to the front of the deck, watching the river traffic slide by and setting their faces towards Nizhny Novgorod and Paris.[5]

---

* Nizhny Novgorod would become an automotive manufacturing hub under Stalin. It was renamed Gorky in 1932 after Maxim Gorky, the Russian writer who was born there. In 1959, despite its long history as one of the country's most international cities, it was closed to foreigners because several of its factories were considered vital to the security of the USSR. It resumed its former name in 1990 and was reopened soon after.

Nizhny Novgorod was an important regional capital and the site of a great annual fair, which the motorists did not stop to explore.

Nizhny Novgorod lies on the bank of the Oka where it turns and is swallowed by the Volga. Most who visited before the twentieth century approached it by water, from which vantage point it creeps 'up the precipitous slope of the hill, flanked here and there by small square minaret-shaped towers, with the old town reposing under the shadow of its fortress and looking down serenely on the busy scene below'.

If the oldest parts of the city were explicitly defensive, the site where the Great Fair was held was spread out as invitingly as a picnic on a blanket. A large annual market attracting traders from across the continent had been held on the Volga since the Middle Ages. Crowds jostled amid high-piled goods. Tea from Kiakhta; cotton from Khiva, now in Uzbekistan; iron from the Urals; jars of crude oil and sulphuric acid; cowhides, from which the leggings most Russian peasants wore were made; casks of dried fruit from the Caucasus. The French motorists would not arrive until mid-August, by which time the fair was in full swing. The

200

Italians, who passed through the city two days before its official start, experienced it as a brief, cosmopolitan whirl. 'Persians, who descend the Volga from Astrakhan; Armenians, severe of countenance; Circassians, armed to the teeth; Kirghese, coming from their steppes, leading over thousands of miles droves of horses, tied to their *telegas* by ropes of straw.'[6]

The middle group of Peking–Paris automobilists, composed of the two De Dion-Boutons and their crew of four, left Omsk on 25 July, sticking to what Cormier mulishly deemed the official route. This more southerly course led due west through Petropavl (now in northern Kazakhstan), Kurgan, Chelyabinsk, Miass and Kazan, where they would rejoin the direction taken by the prince. As the Russian subcommittee had explained, although more direct, this meant crossing the Urals where they were higher and taking a less-frequented road. (It would also mean the detailed maps the Russians had commissioned would be useless and the drivers would not be able to count on the network of support enjoyed by Borghese.) Still, Cormier had forcefully proclaimed his decision in the French press; there could be no turning back.

The day they set off, the two De Dion-Boutons found themselves caught in a large fire sweeping over the parched grasslands beyond Omsk. (The rain that had dogged the drivers since their departure from Peking had been absent from this region for over six weeks. Since this was a farming region, the drought was causing considerable distress. 'The animals died or had to be slaughtered. The wheat, which had not yet formed ears, was drying out on the stalks.') Taking a detour to avoid the flames, they got lost in the network of informal little tracks worn by feet and hooves that coiled over the landscape. On the 27th the rain recommenced its escort, just in time to welcome the Frenchmen over the Urals and into Europe. 'Never before have we encountered such a deluge,' du Taillis wrote gloomily. 'While the soil, which has become slippery, still bears our weight, it is like an ice rink. The wheels are spinning as if they don't want to leave Asia.'[7]

Meanwhile, well in the lead and gathering speed despite mechanical and structural complaints, were the Itala and her three-man crew. Borghese zipped from Kazan, through Nizhny Novgorod to Vladimir and on to Moscow in four days.* Nizhny's elite, including the region's governor, feted them at an al fresco banquet. 'The air was milk and the sky serene. Winding in the valley we saw the wide and languid Volga, blazing with the myriad lights of anchored ships. It seemed like the Milky Way stretched over the earth instead of the sky.' When they returned to the Itala the next morning, someone had garlanded it with flowers.[8]

The relief and sudden smooth speed once they approached Moscow made the three men almost giddy: this, at long last, was proper motoring. Barzini's triumphal dispatches were a world away from the gloom of the previous weeks: 'The road is straight as though cut by a cannonball through woods, fields, rivers, and lakes. In this rigidity of direction there is something grandiose, something hieratic.'

Moscow was also where the intensity of the fascination with the race would be revealed to the racers for the first time. True, they'd come across villages where people had known who they were, but nothing could have prepared them for the interest the world had begun to take in their journey. Preparations for their arrival in Moscow had been under way for days. Consuls from France, Holland and Italy, as well as Mr Perelman, president of the Russian Peking–Paris committee, and Mr Girault, representing the Automobile Club of Moscow, had formed yet another committee. This one was dedicated to ensuring the Peking–Paris motorists received a reception befitting the Russian capital.

Twenty miles outside Moscow, the prince spotted a pair of mounted Cossacks on either side of the road, wearing long black

---

* Barzini later took care to backfill these sections in his book, adding in details such as their stay on the night of 24 July with a poor miller family, the most piteous and moving passage of the account. Having been paid five roubles to help shift the Itala from a bog, their hosts proceeded to get so drunk they repeatedly professed their love for Borghese, Barzini and Guizzardi, and, indeed, for Italians in general. Hours passed, the men sobered and returned to work, the women and children made up cots in the corridor, while the Italians slept in the room where the vodka was. Barzini, unable to sleep, saw the miller's young wife stealthily enter the room. 'I heard, by a slight sound of glass, that she was taking the bottle. After a moment I heard the faint gurgling of liquor, long, interrupted by sighs . . . She was drinking.'

dress uniforms with red flashes at the shoulder and throat and daggers at their waists. Imagining they might have stumbled across a piece of pageantry intended for a member of the imperial family, he took little notice other than slowing down to avoid spooking the horses. As the Itala drove by, however, the horsemen fell in behind them at a gallop. A fresh pair was stationed every hundred yards; one by one they joined the procession. All at once, it dawned on the motorists that the Cossacks were there for *them*. Soon the mud-encrusted Itala, with the prince at the wheel, Guizzardi beside him and Barzini perched amidst the luggage at the back, was followed by a prancing, galloping escort a hundred strong. The road had been emptied of traffic; carts and *telegas* hustled aside to make way. 'Strangely enough,' Barzini marvelled, 'the drivers are not angry. On the contrary, they salute us effusively.'

A cavalcade of smartly polished and primped vehicles of all shapes and sizes had gathered to greet them at Goudenki, a small village just outside the city limits.* This delegation included members of the Russian Peking–Paris organising committee, several foreign consuls, well-wishers from the city's Italian populations and a murmuration of foreign correspondents, eager to finally be able to contribute their own reporting to a news story that had become a global phenomenon. As the Itala appeared on the horizon, a colourful crowd began to line the road. Chatter swelled to cheers.[9]

When the Italian automobile drew up, the contrast between travellers and greeters could not have been more evident. The Itala, charred and stripped down to help alleviate pressure on the weakened springs, was rough and rusted, its paint chipped and blistered. Its occupants – 'all begrimed with dust, mud, and oil' – felt like barbarians amidst the smartly dressed crowd.

An image taken on the road by a Russian photographer and later printed in several different magazines captured this moment for posterity.† In it, Borghese is at the wheel, his left arm resting on his

---

* The description of the scene in *Le Matin*, perhaps mindful of how far behind the French cars were, pointedly described this group as including '40 hp automobiles, with powerful and raucous exhausts' and 'small 10 hps trying to keep up with them'.

† The image was retouched and distributed (and may have even been taken) by Grigori Petrovich Goldstein. He would himself later achieve tangential renown by capturing

lap so that his hand curls loosely around the base of the steering wheel; his right hand clutches a lever. Beside him sits Guizzardi, wearing a natty little vertically striped hat. Both wear long coats – the prince's paler, with baggy sleeves nipped in at the wrist; the mechanic's double-breasted with deep cuffs and a rather prim collar with rounded points. Behind and a little above them, sitting at an angle so his legs dangle over the Itala's side, is Barzini, wearing a hip-length coat and waistcoat, incongruously starched white collar, breeches, puttees and ankle boots. All three gaze directly down at the eye of the camera, suspended between pride and discomfiture.

Borghese, noticeably thinner than he had been seven weeks previously, stepped down from the Itala looking to the gathered reporters nothing like a prince at all. 'He wore an enormous cape that covered him from head to toe. It was the colour of earth,' one scoffed. He was also, a local correspondent for *Le Matin* noted, ferociously burnt. 'The sun, the wind, the rain, the mud had tanned, browned, cracked, peeled and tinted [his face] this incredible brick colour.'[10]

Cameras flashed and whirred like earthbound fireworks; journalists readied their notebooks. The prince and Barzini were presented with the insignia of Moscow's Automobile Club. Guizzardi, who had shared the driving but would not be recognised, looked on proudly.

'Shouts of "Hurrah!" are raised, and we are surrounded. We clasp hundreds of outstretched hands. It is indescribable.'

Just at the moment when the great and the good were celebrating Borghese's seemingly unassailable lead, word of the Spyker and her maverick driver crashed the party. Godard, preternaturally silent for nearly three weeks, blinked back onto the map on 26 July. He made up for the duration of his absence with news of a truly astonishing run from Nizhneudinsk.

two iconic images of Lenin giving speeches in Red Square in 1919 and 1920, the latter of which was famous in Russia throughout the twentieth century, often cropped – after Stalin's rise to power – to remove Trotsky and Kamenev from the frame.

'The Spyker,' he telegraphed grandly to *Le Matin*, 'has covered 580 kilometres [360 miles],* having left at three o'clock in the morning and arrived here in Kansk at eight o'clock in the evening.'

He had, in one day's extended dash on the 15 hp Spyker, travelled ground the 35/45 hp Itala had covered in a little over two. True, he was more than 2,700 miles behind the Italians, but if he could continue at that rate and if they were further delayed . . . Suddenly, it seemed that the race was perhaps not quite a *fait accompli* after all.[11]

---

* He had, if anything, underestimated the distance: it is 425 miles [684 kilometres] from Cheremkhovo to Kansk.

17

TRANSFORMERS

Cars, Capitalism and Change

There be three things which make a nation great and prosperous:
a fertile soil, busy workshops, easy conveyance for men and goods
from place to place.

Attributed to Francis Bacon (1561–1626)

Why on earth do you need to study what's changing this country?
I can tell you what's happening in just four letters: A-U-T-O!

Quoted in Middletown, by Robert and Helen Lynd, 1929

The gleaming New York premises of the Automobile Club of
America opened on 17 April 1907. An imposing terracotta,
Indiana limestone and marble frontage extended 131 feet along
West 54th Street between Broadway and Eighth Avenue. Eight
storeys tall, it boasted garages, repair rooms, two fireproof, four-
ton elevators, a grill room with space for several hundred diners and discreet waiting areas for chauffeurs. At the building's
heart was a triple-height assembly hall, an exact replica of one
at the Château de Cheverny near Tours in France. Newspapers
estimated the cost at a staggering $1,700,000. This was quite a
statement for an organisation less than a decade old and whose
membership was capped at a select 1,000 (with a spot reserved
for the President of the United States). The statement was one of
power and intent: an embodiment of what automobile ownership
meant.

The very same year the ACA was founded – 1899 – the economist Thorstein Veblen published *The Theory of the Leisure Class* and birthed the phrase 'conspicuous consumption'. This is the idea that consumers buy some goods not because of their utility but because ownership of them confers status. Automobiles were perfect examples. Made on a small scale, they were enjoyed by the wealthy and powerful, more so for display than for the practical concerns of getting from A to B. Cars were relatively rare, especially outside major cities. They were often bespoke, providing ample opportunity for decorative flourishes and personalised gadgets that attracted considerable attention. As a 1904 issue of *The Motor* cosily put it, 'They are expensive, to be sure, but motoring and money have a way of whisking by together.'

Early advocacy groups and clubs reflected this. The various Peking–Paris organising committees attracted powerful individuals like the Marquis de Dion and Ludvig Alfred Nobel. So too did the popular representation of motorists. In Kenneth Grahame's *The Wind and the Willows*, published in 1908, the wealthy, spoilt Toad is a car-mad maniac who must be saved from himself by his down-to-earth friends Ratty, Mole and Badger.

Showrooms for automobiles and their accoutrements were located at exclusive addresses. Peugeot could be found on the Avenue de la Grande Armée, one of the spoke-like streets leading off the Arc de Triomphe in Paris. In London, Daimler had premises on Shaftesbury Avenue; Panhard and Dunlop on Regent Street; Fiat, Spyker and Mercedes on Long Acre near Covent Garden. At a time when the average yearly wage was £70, Gamages in Holborn – which billed itself as 'The Leading Motor Tailors' – sold motoring coats priced from £1 11s to £5 5s for a fleece-lined 'Scotia'. Automobile marketing materials were extravagant. Itala catalogues were decorated with gold leaf and silk ribbons, and bristled with fashionably bold, curvaceous fonts and illustrations. One brochure, from around 1908, featured a list of prestigious owners, including Queen Margherita of Italy, King Ferdinand of Bulgaria, Grand Duke Boris Vladimirovich (the tsar's first cousin), Hugh Fortescue Locke King (of Brooklands fame) and, of course, Prince Scipione Borghese.[1]

Just as horses and carriages required staff who needed to be paid, housed and liveried, so did motor cars. Chauffeurs, who were typically working- or lower-middle class men, became lightning rods for class anxieties since they had access to privileged knowledge of both automobiles and their owners. (The very word indicated trouble, deriving from the French verb *chauffer*, 'to heat', and harking back to steam engines, it recalled the phrase '*chauffer une femme*'.*) Popular culture accused them of putting on airs and transgressing social hierarchies. Owners without mechanical knowledge worried that these men were purposely damaging the vehicles in their care, because they received commissions from garages and dealerships. 'Every motor garage,' one *Daily Telegraph* article concluded, 'ought to have a large notice to chauffeurs quoting the effect of their accepting any gift without the knowledge of their master'.†2

During the eighteenth and nineteenth centuries, elites – particularly men – set a great deal of store by their horses, carriages and horsemanship. Being seen in certain city parks seated in the latest equipage with a beautifully matched pair was a mark of status. By the early 1900s, motor cars were being adopted into such ritualised sociability. A light-hearted etiquette guide for passengers – or 'the automobilist without an automobile' – was published in a Russian magazine in 1907. 'It is absolutely essential,' the author, Count Tishkevich, wrote, 'that the invitee generously praise the function of the motor and speed of the car . . . If the person inviting you asks, with poorly disguised indifference: "Do you think I drive well?" Answer boldly: "Yes, but you have one fault – you are too fearless."'

In some instances, the pageantry overwhelmed the driving function of automobiles entirely. Vehicles festooned with blooms would parade around village greens or racing enclosures and be judged like prize pumpkins. Dutch photographer and journalist Andrew

---

* 'Make love to a woman.'
† What the chauffeurs themselves made of such insinuations is difficult to know. Although, in a presumably unintentional act of friendly fire, the president of an American Chauffeurs Club insisted to the *New York Times* in 1905 that not *all* his peers were 'grafters, drunkards, moonlight excursionists, and incompetents'.

Pitcairn-Knowles made a study of such events in Belgium around 1900. His images show well-dressed motorists leaning out of vehicles to snatch up hoops or tap joisting posts, accepting trophies and rosettes and waving at obliging crowds. At a similar event held at Shanghai's Recreation Ground on Saturday, 10 June 1905, the winning car 'was simply smothered in white flowers, and made to resemble a monster swan'; another, 'an exceedingly ferocious dragon', made from bushels of corn and yellow flowers.[3]

Private individuals used motor cars as social tools. Leo Hendrik Baekeland, the wealthy American inventor, enjoyed impressing business associates by taking them out for a drive, and from 1914 for over a decade industrialists Henry Ford, Thomas Edison and tyre magnate Harvey Firestone went on regular motoring jaunts into the American countryside. Those invited to join them included the celebrated writer and naturalist John Burroughs and President Warren Harding (on this occasion they were tailed by droves of Secret Service agents). Dubbing themselves 'The Vagabonds', they camped, chopped firewood, slept in monogrammed tents, discussed politics and industry, and endlessly leaked anecdotes about their excursions to the press, helping to promote the automobile in the process. These were antecedents of an entire genre of literature – and later film – celebrating the 'road trip'.[4]

The idea that cars embodied elite status really came to an end in 1907. Already, on both sides of the Atlantic, increasingly affordable vehicles were being produced. Over the following century the automobile went from being an intrinsically luxurious object to an everyday one, a necessity. As the market became saturated, automotive firms adopted increasingly inventive techniques to sell more cars.

As early as the 1920s, General Motors (GM), under Alfred Sloan, marketed automobiles like clothing, superficially altering body styling and offering a rainbow of colour options. This encouraged people to keep up with what was considered fashionable and allowed near-identical technology to be sold to different customers; it ensured those same consumers came back as close to yearly as possible, well before their car needed replacing. The head of GM's styling department – known as the Art and Colour Section – Harley Earl, called this 'dynamic obsolescence'. By 1957 seventy-five different body styles were being produced for GM, with around 450 interior soft-trim combinations. Six years later this had proliferated to 140 body styles and 843 trim combinations. Nor were these tactics particularly hidden from consumers. One Buick advertisement listed all the features available on one model and accompanied this with the text 'You don't really need these, but how can you resist them?'

The mid-twentieth century saw a boom in consumption, especially in America. People were eager to spend money and admen were eager to help them spend it. This became something of an art form, in automobiles as in everything else. As Sloan put it, the 'laws of the Paris dress-makers have come to be a factor in the automobile industry – and woe to the company who ignores them.' While some marques continued to appeal exclusively to the very wealthy, more targeted specific consumers. Today, cars can be tweaked to appeal to the environmentally conscious (fuel efficiency), urban drivers (small, easy to park), parents (spacious boots, extra safety features), or ride-share drivers (hard-wearing interiors, cheap, impersonal).[5]

Ordinary drivers began to see their vehicles as extensions of their homes and themselves. Indeed, so closely do some identify with their automobiles that people who overestimate their own body size do the same with their vehicles, thinking they take up more space than they do. At the extreme, those who most closely perceive their cars as reflections of their self-identity (a phenomenon more common in men than women), are more prone to aggressive driving.[6]

Even the lives of those more emotionally detached from their automobiles were irreparably altered by them. The use of a vehicle became part of many people's jobs – as mechanics, valets, road workers and tyre specialists. Hauliers drove for a living, of course, but so too did taxi drivers, driving instructors and salesmen. And, as automobiles became more affordable, people began living farther from their workplaces and extended families. The car would carve out a niche in the lives of many millions in the form of the daily commute.

211

Automobiles became intertwined with national and political identity too. On 26 March 1923, Benito Mussolini, the recently elected Italian prime minister, raised a pickaxe and vigorously struck the ground forty-one times, ceremoniously breaking ground for the building of the Milan–Alpine Lake autostrada and kicking off two decades of feverish motorway building.

For Mussolini – and, a little later, for Adolf Hitler, who unveiled an ambitious plan to build 3,700 miles of four-lane autobahns in just five years – highways and cars were intrinsic to Fascist national identity. They symbolised potency and strength and fostered national excitement and pride, provided jobs and boosted the economy. Hitler was a great admirer of Henry Ford, believing mass production would reinvigorate Germany. He kept a framed photograph of the industrialist in his office as early as 1922 and championed several mass-produced goods, including a radio, the *Volksempfänger* or 'people's receiver', and, later, a car, the *Volkswagen*. Ferdinand Porsche, the Austrian engineer hired to design the latter, perfected the charismatic Volkswagen Beetle design between 1935 and 1939. Its purchase and use were heavily subsidised and promoted through a special *Kraft durch Freude* ('Strength through Joy') propaganda campaign, encouraging people to engage in healthful outdoor pursuits in the German countryside.[7]

Poor motoring infrastructure and lack of investment became shorthand for a sluggish state and a way of criticising regimes indirectly. One Russian driver, for example, jokingly reviewed a trip down an unfamiliar track just outside St Petersburg in the springtime. 'It was not bad at all: a wide strip of mud, cut through with deep ruts, stretched across the most wretched swamp. On the left – a sour grey sea. On the right – fragrant rubbish heaps. Ahead – nothing but mud. And above all this is a lovely grey sky. Delightful.'[8]

Fascists were hardly the only group that saw the automobile as a means of invigorating their country. The left-wing author of an article about the 1907 St Petersburg Automobile Exhibition wondered why, if nine-tenths 'have no chance of buying their own automobile in either the near or the distant future', so many thousands attended.

Why this interest in machines that serve only the needs of the rich and just cause problems for mere mortals? I believe it is because the crowd has a more-or-less clear sense that the automobile is intended primarily for them, for their use and benefit, and is only temporarily playing the role of an expensive toy in the hands of the privileged. No matter how rich the rich people are, the crowd is still richer than them.[9]

After the Russian Revolution in 1917, Vladimir Lenin returned to St Petersburg after nearly two decades in exile. At this critical moment for him and for his revolution, he arrived at Finland Station and chose to stand on top of a butch armoured automobile while addressing the crowd. 'The people need peace; the people need bread; the people need land,' he declaimed. 'And they give you war, hunger, no bread . . . We must fight for the socialist revolution, fight to the end, until the complete victory of the proletariat. Long live the worldwide socialist revolution!'

In 1930, that car – or at least a passable facsimile – was installed in Vosstaniya Square in Leningrad – replacing a much-mocked equestrian study of Alexander III – where it remained until 1994, by which time the city had reverted to its original name of St Petersburg. The symbolism, if a little on-the-nose, was apt. Like the Fascists Mussolini and Hitler, first Lenin and then Stalin would make the automobile a centrepiece of the fledgling USSR, in stark contrast to the plodding wreckage of the dead tsar's empire. In a fiery speech at the First All-Russia Congress of Working Cossacks in March 1920, Lenin told his audience that 'With this victory behind us, we must now do our utmost to consolidate it on another front, the bloodless front, the front of the war against the economic chaos to which we have been reduced by the war against the landowners and capitalists.'[10]

'Economic chaos' was an understatement. Even before the outbreak of the First World War, famine threatened annually. Agricultural production was inefficient and largely unmechanised: there were fewer than 500 tractors in the whole of Russia. Although the country had plenty of raw materials, mining and extraction methods were antiquated and industrial production was small-scale. The country needed to modernise rapidly. Automobiles, tractors and the factories in which they could be mass-produced were an intrinsic part of the Communists' vision for the future of their country. What they lacked was expertise, so Russia's Communist leaders turned to America.[11]

In 1907, Henry Ford was a carmaker like any other. Although he would later be famous for democratising the automobile, his early models were comparatively expensive. The average price for a Ford car between 1903 and 1908 was $1,600; Oldsmobiles, made by one of Ford's competitors, were selling for under $400.* Even so, the Model T was the car that would make Ford's name and shift public expectations about what an automobile was and who it was for. It was released in October 1908 and remained in production for nineteen years, during which time 15 million would be sold globally. It was basic – spawning a vibrant market for extras considered non-essential, such as rear-view mirrors, fan-belt guides and snowmobile conversion kits – and frequently cantankerous. Nevertheless, it inspired great affection. E. B. White, in a 1936 ode to the Model T for the *New Yorker* entitled 'Farewell, My Lovely!', wrote that this car 'was the miracle God had wrought . . . As a vehicle, it was hard-working, commonplace, heroic'.

Henry Ford's innovation was to marry the ethos of Frederick Winslow Taylor – whose ideas on industrial efficiency were highly influential in late nineteenth-century America – and production lines, already in use in meat-processing plants, and apply them to cars. This way, as he told an investor, his automobiles would 'come through the factory just alike; just as one pin is like another pin when it comes from a pin factory, or one match is like another match when it comes from a match factory.'

Production lines divorced workers from any claim to skill or overview of how automobiles were made. Instead, they repeated short series of choreographed movements, quickly and without

---

* At the turn of the century, those placing bets on American automobile manufacturers making it big would probably have picked Ransom E. Olds, maker of the Oldsmobile, rather than Ford. Oldsmobiles were produced in relatively large numbers – 600 in 1901, 5,000 by 1904 – and Olds himself became a millionaire.

error, from one end of their shift to the other, every single day. When asked what he thought of such repetitive labour, Ford admitted that it terrified him. 'I could not possibly do the same thing day in, day out, but to other minds, perhaps I might say to the majority of minds, repetitive operations hold no terrors.' His workers disagreed. Turnover at Ford's Highland Park factory reached 370 per cent in 1913, the year the moving assembly line was fully operational. Inducing workers to stay required the reduction of the working day, the headline-grabbing introduction of the five-dollar minimum wage and clamping down on attempts to unionise.

This was hardly the stuff of a workers' paradise, but Ford's methods *could* be relied upon to increase production and lower costs. Prices for a Model T fell from $859 in 1908 to $298 in 1923. Competitors were left scrambling to adopt his methods and match his prices, but it was too late. Model Ts were hatching like busy spiderlings from twenty-eight factories in America and fourteen distribution points or plants globally.

To remain competitive, manufacturers in Great Britain, Australia and across Europe adopted the principles of 'Fordism' in the 1920s and 1930s. An iconic example of this was Fiat's factory in Turin, designed by Giacomo Matté-Trucco and completed in 1923. Raw materials entered at ground level and then cars in production moved up five spiral storeys until they were released, completed, onto a steeply cambered rooftop test track. In the 1920s, the Soviet government approached Henry Ford and the man responsible for Ford's near-mythic factories, Albert Kahn.[*12]

Kahn was a Pickwickian character. His grey hair cascaded from a straight middle parting, a pair of round, wire-framed glasses sat atop a bulb of a nose. He was usually photographed in a tie, a white shirt with the sleeves rolled up to the elbow, a snug waistcoat and a shy half-smile. Born in 1869 in Germany, Kahn had

---

* Henry Ford himself was a rather prominent and popular figure in the USSR during the 1920s. His 1922 autobiography went through eight swift editions in Russian translation. 'Fordism', 'Fordisation' and 'Fordizatsia' were adopted into common parlance and used by industrial planners and orators to 'spellbind auditors'. There are even stories of banners bearing his name being held aloft by workers during parades.

emigrated to Detroit as a child and later trained as an architect. The profession suited him. By June 1942, *Time* magazine called the seventy-three-year-old the 'father of modern factory design' and 'the world's No. 1 industrial architect'. He was responsible for Ford's Highland Park and River Rouge plants, as well as those for Chevrolet, Oldsmobile, Cadillac and Chrysler.

Renowned as a fast, efficient worker with a sincere belief in the power of industry and the moving assembly line, Kahn had attracted the eye of the Bolsheviks. In 1930, a deal was struck: for $1,900,000,000, Kahn would construct four large automobile, truck and motorcycle factories and nine plants producing tractors and farm implements, as well as several hundred smaller-scale works for glass, asbestos, graphite, textiles, wood and shoes, essential for fulfilling the first and second Five Year Plans. A *New York Times* article reported Kahn as believing the contract represented 'a readiness of the Soviet Government to continue

Ford automobiles coming off an assembly line in Gorky (Nizhny Novgorod) in the 1930s under a banner of Stalin.

216

turning to America for assistance in the development of the country's industrial life and resources'. The Soviet slogan – 'to catch up with and surpass America' – put a different spin on the relationship.*[13]

The tension between Communist aims and the automobile, the *ne plus ultra* of individualistic consumption, was only too apparent to the Soviet leadership. Reconciling them meant two things. The first was that their automobile had to be utilitarian, cheap and simple to produce. The second was to prioritise headline-grabbing, breakneck production so that they were always seen to be outperforming the capitalist world and America in particular. The importance of the automotive industry to this tub-thumping nationalistic rhetoric was neatly summed up by Stalin himself: 'When we place the USSR at the wheel of an automobile and a *muzhik* [peasant] on a tractor, let the venerable capitalists boasting of their "civilisation" try to reach us.'[14]

The results were disastrous. The focus on speed at any cost and the pressure being applied from the top to meet quotas meant that quality plummeted, raw materials were squandered and the finished products were often of poor quality. Worse still, the money used to pay Kahn's firm and fulfil Stalin's other automotive ambitions was raised by exporting wheat to the United States, exports that continued even during periods of acute famine. In a directive written in August 1930, Stalin wrote:

Each day we are shipping one to one and a half million poods [16,000–24,000 tons] of grain [*sic*]. I think this is not enough. We must immediately raise the daily export quota to three–four million poods [48,000–65,000 tons] at a minimum. Otherwise we risk being left without our new metallurgical and

---

*When the Stalingrad tractor factory was completed on 17 June 1930, Stalin sent two telegrams. The first, to the factory, read, 'Greetings and congratulations on their victory to the workers and leaders of the first giant tractor plant in the USSR. The 50,000 tractors which you are to give our country every year are 50,000 shells shattering the old bourgeois world.' The second, to the Kahn employees, thanked 'our technical teachers, the American specialists and technicians, who have helped us in the construction of the plant'.

machine-building plants . . . In short, we must accelerate grain export at a mad tempo.[15]

Between 1932 and 1933, an estimated 5 million people would die of starvation across the Soviet Union, the vast majority in Ukraine. This period is now referred to as the Holodomor, from the Ukrainian words for 'hunger' (*holod*) and 'extermination' (*mor*). Throughout the crisis, grain exports continued and any mention of the famine was censored and denied.

No matter where people lived, or what ideology they subscribed to, over the course of the twentieth century cars acted as catalysts, effecting change on everything from social mobility to industrial outputs and national identity. Automobiles allowed people to travel more widely and to live further away from places of work. Time spent in vehicles spawned new genres of literature and carved out new niches in everyday lives. They were a supreme consumable good, and, like clothing, they signalled aspects of one's identity and self-image. Automobiles and the industries they spawned and then supercharged would also come to be critical to nation states. The miles of roads paved and the number of vehicles manufactured and sold became a proxy for economic vitality. As the French cultural critic Roland Barthes put it, when contemplating the new Citroën DS in 1957, cars are

almost the exact equivalent of the great Gothic cathedrals: I mean the supreme creation of an era, conceived with passion by unknown artists, and consumed in image if not in usage by a whole population which appropriates them as a purely magical object.[16]

18

Moscow – Kaunas

27 July–4 August

I believe that a car is as susceptible to training as a horse. I am sure that an engine responds well to good treatment, assiduous care, and is equally stubborn about unnecessary whipping!

Jean du Taillis, De Dion-Bouton team, formerly Spyker

The Italians entered Moscow like the spume on a cresting wave, carried along and aloft by well-wishers. From amidst the crowd, Barzini's impression of Russia's old Muscovy capital was of a scintillation of golden domes hovering over 'a white, diaphanous expanse of buildings'. As they drove through the city's outskirts, clusters of tall industrial chimneys put forth smoky blooms and workers poured out of factories in their hundreds to see the Itala pass by. Begrimed faces pressed against windows; squadrons of railway labourers came running to catch a glimpse of the automobile that had travelled overland all the way from Peking. Hands, hats and cheers were raised from trams and pavements, offices and schools. '*Viva l'Italia! Viva l'Italia!*'

As they approached the centre of the city, advancing along widening, asphalted boulevards, bells began to peal. (Moscow was famous as a city of churches and bells: residents liked to say the sweetness of their tones came from the high proportions of gold and silver used in the alloys.) The phalanx of vehicles, riders, motorcyclists and cyclists passed by St Basil's Cathedral and along the Kremlin's high walls, above which could be seen the

219

spires of three cathedrals and the tower of Ivan the Great. Finally, they turned right, heading north towards the newly opened Hotel Metropol. As they drove, they spotted guards armed with fixed bayonets on corners and jumpy Cossacks patrolling the streets.[*][1]

Although they had intended to remain in the city for just two days, they did not leave until the morning of Wednesday, 31 July, by which time they had been interviewed, feted and fed to within an inch of their lives. Notwithstanding – or perhaps in defiance of – the ongoing threat of revolution, the city put on a show for its guests as only Moscow knew how.

Borghese, often accompanied by Barzini, gave speeches and received medals and bouquets and applause in each of the city's smartest venues. Although reserved, he was courteous, with unimpeachable manners, and, thanks to his training in Italian politics, diplomatic. It helped too that the prince had social connections in the country: his mother-in-law, Maria Sergeevna Annenkova, was a Russian aristocrat.[†] He spoke a little Russian, as did Barzini, who had learnt a few words while reporting on the Russo-Japanese War. Both also spoke excellent French, a tongue in which many wealthy Russians were fluent. The press corps certainly found themselves forcibly charmed. 'My travel impressions?' the prince responded to one journalist. 'They are excellent. We were comforted by the welcome we found everywhere we went . . . this welcome makes us forget all the difficulties and remember only what was pleasant.'

---

[*] Moscow's December 1905 uprising, the denouement of that year's revolution, had left authorities wary. Kellogg Durland, the young, left-wing journalist who had travelled around Russia on horseback in 1906, observed that the city 'still cowered and quivered from the severe and bloody repression that followed the magnificent fight her mere handful of armed citizens maintained on the barricades for nine days'.

[†] Maria Annenkova's origins were mysterious. She had grown up at the Russian court under the protection of Nicholas I and claimed to be the last direct heir of Charles X, the French monarch from 1824 until 1830, and therefore a rightful heir to the Bourbon inheritance. That this claim was believed by the tsar is indicated by the fact that Maria was accompanied by a representative of the Russian court throughout her life, even after her move to Italy and marriage to Gaetano de Ferrari. It was apparently Maria who persuaded her husband both to buy the desolate Isola del Garda and to begin the ambitious project of building an imitation of the Doge's Palace in Venice surrounded by formal gardens, which can still be seen today.

As the racers travelled west, they found themselves increasingly surrounded by admirers who knew all about them and the Peking–Paris.

The motorists were treated to the spectacle of trotting races at the hippodrome, to watch 'the perennial struggle between Russian and American horses'. They spent an agreeable sunset on top of Sparrow Hill on the Moskva River's right bank. It was from this vantage point, just shy of a century earlier, that the emperor Napoleon was said to have watched the city burn.* Barzini, succumbing to a flight of fancy and Moscow's undeniable glamour, wrote that the 'dying sun tinged the superb city with blood, the gilded domes throwing off tongues of flame'.[2]

They dined at the city's brightest nightspots: the Hermitage and the Mauritania, in Petrovsky Park, where food was accompanied by 'cosmopolitan choruses', and the Yard, where musical

---

* Despite the intervening century, the French invasion of the area loomed large in the Muscovite imagination. The city arsenal housed the 875 cannons seized from the French, each with a little plaque bearing names like 'Le Valliant', 'La Ravissante', 'l'Eclair'. And on especially cold days, a Russian might be heard to say 'I feel as cold as a Frenchman!', a reference to the French Emperor's ignominious retreat through the snow.

performances began at midnight and ended with the dawn. They may also have visited the famous Slavianski Bazaar, on Nikolskaya Street, mere yards from their hotel and famous throughout Europe for its seafood. In the centre of the restaurant was a large tank covered with white water lilies and stocked with huge sturgeon and sterlet, brought in daily from the Volga. Individual fish were selected by diners and delivered to the kitchen by waiters who waded through the tank armed with small nets. 'We are in a most agreeable captivity,' Barzini telegraphed to the *Corriere* and *Daily Telegraph*, 'against which our fatigue will certainly not cause us to rebel.'[3]

On their final day, their automobile was publicly examined by Moscow society in the garage of the Hotel Metropol amidst great plumes of white smoke produced by the copious quantities of fresh oil being applied to the engine. Ettore Guizzardi had spent the three days since their arrival lavishing the Itala with attention: sluicing off the layers of mud and dust, tightening and replacing screws and polishing the paintwork. It was, mechanically speaking, in better condition than they had feared. It is likely that the footbrake, which had been useless for days, was replaced here, and a good deal of cosmetic repairs carried out.[*] Barzini was able to write, with a forgivable degree of obfuscation, that 'the Itala will reach Paris as it left Pekin.'

Paris, however, had unexpectedly receded further into the distance, thanks to an unwelcome addendum to their itinerary. Rather than bearing south-west from Moscow directly to Paris, the Italians would now turn north, taking in St Petersburg before turning around and setting their faces once more towards their goal. This detour would mean the addition of nearly 400 unnecessary miles, taking them through northern Poland, hugging the Baltic Sea, before dropping south to rejoin the scheduled itinerary at Berlin.

The reasons for this detour were unclear even in 1907 and, as it is wont to do, legend stepped in to fill the vacuum. Popular

---

[*] Guizzardi, decades later, remembered replacing the brake somewhere in Russia but not precisely where.

explanations involved the presence in the city of the prince's lover or his desire to attend a banquet given in his honour. In a report published the day the prince reached St Petersburg, on 1 August, *Le Matin* stated that 'The Itala is taking the Prince's chosen route.' Tantalising as these speculations are, they're difficult to square with the competitive seriousness that Borghese always demonstrated in his approach to the Peking–Paris. He was a single-minded man, who consistently pressed on with the journey despite the objections or finer feelings of those around him. (Even when, at Omsk, Barzini succumbed to exhaustion and fainted in the street, they still set off at dawn the next day.) Thus far, the prince's preferred itinerary had been one of early mornings, long days and infrequent stops.

Barzini, for his part, seemed equally resentful of the detour. 'We have only one regret,' he wrote. 'This diversion to St Petersburg represents a real interruption of the journey; we are deviating from a straight line.' The best explanation, and one that chimes with the legend of the prince's banquet, is supplied by the Russian editors of *Avtomobil'*. Their August issue implicates the St Petersburg Peking–Paris committee and particularly its chair, Mr Perelman. 'It took the committee a great deal of trouble, and a large number of telegrams, to convince the daring traveller to go through St Petersburg ... it would have been difficult to skip a visit to the committee whose efforts had ensured he never lacked for benzine or oil and whose intercessions opened doors for him everywhere.'[4]

Whatever the reasons, they set off characteristically early, sleepily packing themselves and their belongings into the car in the dark violet hours of 31 July. Even at four o'clock in the morning, a crowd had gathered – some in their own motor cars. The prince exchanged pleasantries with them as he donned his voluminous, ankle-length motoring coat, adjusting the elastic of its blousy sleeves around his wrists. Barzini, 'almost hidden in the depths of the backseat', was silent, appearing wan and frail – he had been ill the previous day, perhaps a reaction to the surfeit of rich food and wine after weeks of subsisting on black bread, tea and boiled eggs. Guizzardi, meanwhile, cheerfully whistled through his teeth as he secured the trio's baggage to the back of the car with ropes, an operation he had performed so many times it required no conscious effort. Finally, when satisfied, he swung himself up beside the prince: he would take over the driving once they were out of the city.

By six o'clock, the mist that had shrouded their farewells had burned away and they sped up. They had nearly 450 northwesterly miles to go before they reached St Petersburg and the mood in the car was one of intense focus. Barzini felt that they were all 'overcome by a desire, always growing stronger, to increase our pace'. The miles flew past unheeded. They were soon going at thirty miles per hour.[5]

A thousand miles behind them, the two De Dion-Bouton crews were struggling. Du Taillis held his tongue, and Collignon was far too loyal to say anything, but it was clear Cormier's insistence on taking the southern route through the Urals – rather than the more northerly one proposed by the Russian committee and taken by Borghese – had been a grave mistake.

The mechanic Bizac, a habitual pessimist who had for several days been muttering that things had been going too well, seemed almost relieved at the return of the usual dire driving conditions. Ink-black storm clouds, laden with rain, stalked them westwards and rain slackened the earth beneath their wheels. On one occasion, it took them a full hour to travel just under four miles as 'torrents of water rushed down the steeply sloping roads' and pooled under their feet in the open vehicles.

Between Birsk and Yelabuga, the ground was so sodden that Collignon felt that they were 'sliding, rather than driving . . . There is no road, just a swamp. In the summer, you take a boat to cross this country; in winter, you would use a sleigh.' On this occasion, at least, they were able to reach their intended destination. Just the day before, in a rugged region reminiscent of the Tyrol but with ground as 'greasy as the soils of Normandy', they had been forced to make camp two miles from the nearest village. Cormier's De Dion-Bouton, skidding violently down a steep slope after sundown, had come to rest with its rear wheel 'just on the edge of the void'. Shaken, all four men agreed it was too dangerous to continue. 'As we got rid of our tents in Omsk, we open our bonnets and sleep, in our cars, out in the Urals, surrounded by spectacular wilderness.'

This area was less populated, poorer, more rebellious, and suspicious of foreigners. Even when they did manage to stay in a post house, inn or outbuilding, rather than by the side of the road, they found themselves plagued by Siberia's pests at night. They had to throw out clothing and bedding that had become infested with lice. One night, after they had settled down in a barn, they were assailed by an army of rats. 'They came out of the stalls, the ruined hayloft or the very earth,' du Taillis reported with disgust. 'I tell you, no beast nor parasite in all creation let us sleep in peace in Siberia.'

With the exception of the ever-watchful officials, who clearly suspected them of being spies, no one was expecting them along the way. They struggled to secure fuel for their automobiles or lodging for themselves. Procuring food was difficult too. They relied on eggs and jam, which both travelled well but were hardly harmonious companions. One lunch consisted of 'some apples and local bread'. For several days they subsisted solely on tins of sardines, and they were pathetically grateful when they were able to find 'some tinned lobster and salmon as a change'.[6]

By the time they reached Birsk, their last significant stop before Kazan, they were thinner – especially du Taillis, his health still fragile after the four days suffering heatstroke and near starvation in the Gobi – and reaching the limits of their endurance. An air of suspicion – even hostility – followed them everywhere. Police met them on the outskirts of Birsk, and they proceeded into the town 'surrounded by a whole squadron of Cossacks, mounted on superb horses, revolvers in hand'. Although they would rest there for a few days, they did not think Birsk had much to recommend it. Du Taillis morosely commented on the lack of restaurants. Cormier, with his usual flair, described it as a 'small uninteresting town built on the Belaya River, which is heavily frequented by boats'.[7]

The French motorists set off again on 4 August, finally crossing the border between Asia and Europe. Although they stopped and

225

raised a toast to mark the moment, they were dispirited by their slow pace. Du Taillis's report contained an ostentatiously senti-mental address to their erstwhile colleagues, graciously mention-ing the prince before exclaiming, 'Ah! If only the Spyker had a driver less inclined to incredulity . . . If only Pons's perseverance had not been cut short . . . he would still have been with us to empty this glass of champagne, carefully set aside for this solemn toast.'

Neither Pons nor Godard, however, were content to fulfil the neat, subsidiary roles du Taillis and the organisers had allotted to them. The former, back in Paris and nursing his grievances, was insisting on giving the French public his side of the story. 'Here is the exact truth: I gave up the rally,' Pons wrote, 'because, on 18 June, my fellow travellers, failing to respect the laws of the most elementary humanity, [broke] the promise that each one had made not to abandon each other and to help each other.' Godard, meanwhile, was busy creating a ruckus of a different order.[8]

Despite his exuberance, none of his fellow racers, nor the journal-ists covering the race, knew or could dig up much about Godard. Du Taillis, who had shared many days alone with him in the Spyker, never managed to discover much about his life before the Peking–Paris starting line. He seemed unaware, for example, that Godard had served in the military, assuming instead that he was an old automobile hand.

There is no evidence, however, that Godard had *any* involve-ment with automobiles before he answered *Le Matin*'s Peking–Paris challenge. Accounts of the Peking–Paris written after the event have claimed all sorts of extraordinary histories for him. Allen Andrews's *The Mad Motorists*, for example, claims he was not only a professional driver 'known to the more observant students of form as potentially the most outstanding of them all', but also, on the side, a carnival per-former in Montparnasse, specialising in racing round a Wall of Death on a motorcycle.* If any of this is true, it passed contemporaries by.

* Andrews provides no sources for this, which is a shame: it would pre-date the current earliest confirmed Wall of Death by some years.

France's sporting magazines do not contain references to races he won or trips he undertook, either by automobile or motorcycle, in stark contrast to Collignon and Cormier, who were well known. Likewise, there is no evidence of chauffeuring experience, and some obscure references to reputed work as an automotive agent have left no trace. In fact, before entering the Peking–Paris, he seems to have made almost no impression on the world at all. He married Marie Milson, a woman ten years his senior, in 1897; the marriage certificate lists his profession as '*pépiniériste*', or nursery-man, a profession akin to that of his gardener father.

The clearest account of his life before the race comes from his military records. These – comprising a long official form in a variety of crabbed scripts – show he was in the army from 1897, when he signed up, until at least 2 July 1906. His military career had not been blemish-free. He was formally censured in November 1903, probably for disobeying orders; in March 1905, for 'abuse of trust'; in July 1906, for fraud, on which occasion he was also fined. Intriguingly, very faint notes made on his records in pencil suggest that he was still nominally in the army between 1907 and 1909, on sick leave.[9]

While his fellow participants may have been ignorant of these details, they were aware that Godard was penniless, thoughtless and unscrupulous. They knew he had persuaded the Dutch consul to pay for the passage of the Spyker. He had also let slip the stratagems he had used to secure his automobile and the host of spare parts, and crowed about promptly selling them to buy his steamer ticket. Du Taillis had personally had to advance Godard funds for food and fuel, and had heard him assuring the Dutch consul, who had also loaned him money, that Mr Spijker would pay him back. It had been clear for some time that Godard had no means of repaying his debts. He was boastful and rash and sometimes capable of cruelty, as when he had abandoned Pékin the dog in the desert.

For all these faults, he had also been an exuberant, resourceful and sometimes courageous teammate. It had been Godard who agreed – foolishly, as it turned out – to carry Pons's extra luggage when the Contal struggled in the early days of the race. He joked and sang and clowned when the vehicles stopped and was always ready with a tall tale if silence fell. His most significant moment of éclat had come in the Gobi on 19 June, when the Spyker had run out of fuel a hundred miles from the next fuel stop. Cormier and Collignon abandoned their stricken countrymen,* leaving them with no supplies. On the third day in the desert, as du Taillis was succumbing to thirst and heatstroke, Godard had walked off into the unknown in search of aid. Hours later, he returned riding double in a group of horsemen, like the hero of a romantic novel. No matter what else du Taillis thought of him, how much money he was owed or how many questionable character traits Godard revealed, the journalist owed his life to the driver.

For most of July, after the Spyker's magneto ignition had failed on the 6th, Godard's movements had been a mystery. Alone in his car after du Taillis joined the De Dion-Boutons, he was absent from published reports and his location from day to day went unrecorded. However, the day after meeting Godard in Tomsk, where he was trying to get the Dutch car back on the road, du Taillis was able to write to *Le Matin* announcing that the Spyker had been 'completely repaired in Tomsk' and that Godard would be 'leaving to take up the route again in Cheremkhovo'.[10]

Five days later, the Burgundian driver was only just setting off. Not only did this leave him 1,443 muddy Siberian miles behind the two other French cars and 2,650 miles behind the Itala, but there was some doubt about how he was going to obtain fuel. Those following the race, if they remembered him at all, probably assumed that Godard had been forced, without fanfare, to give up – which made the telegraph he sent *Le Matin* on 26 July announcing his rapid run from Cheremkhovo to Kansk all the more remarkable.

Other reports soon followed. Each was short on detail but bore news of astonishing feats of endurance, driving at – comparatively

---

* Notwithstanding the later concentration of blame for all the journey's ills on Prince Borghese, it is clear from the telegrams sent in the days after his arrival in Udde that du Taillis believed Cormier equally responsible. 'The Spyker ran out of petrol 150 kilometres [93 miles] from Udde. Prince Borghese was already ahead of us. Cormier carried on regardless. We have been in the desert for four days: four days of thirst, hunger and mortal fear.'

speaking – breakneck speeds. On 1 August, at 6.45 p.m., for example, he wrote from Petropavl, 'The Spyker arrived here in heavy rain that made the southern route impassable. [I] will try to reach another route, to the north, guided by a navigator.'

Near Omsk, he had been joined by Bruno Stephan, a young Spyker engineer, indicating renewed investment by the Spyker firm, probably as a result of burgeoning global publicity. If Godard had any more mechanical difficulties, they could now be fixed along the way. It is also possible Stephan had been given some financial discretion, since Godard's impecuniousness was seeping out into the European press. (The Dutch *Het Nieuws van den Dag* reported that 'Mr Spijker himself [told us] that Godard is an incredibly cunning man who has lied his way into receiving an automobile.') If Jacobus Spijker was going to capitalise on his initial investment – and be sure of recovering his car at the end of the race – he needed someone he could trust on the ground.

If this was the bet the Spyker's manufacturer had made, Godard seemed poised to make good on it. In six days, he covered ground that had taken the De Dion-Boutons fifteen. True, there was a slight difference in power – his automobile fielded 15 hp compared with the De Dion-Boutons' 10 hp. Even so, this was extraordinary. Two days later, on 3 August, he was in Chelyabinsk. 'Impossible to take the northern route. I am going on by the southern route.' Then came the kicker: 'I hope to rejoin the caravan at Kazan.'

Although secreted at the bottom of the third page of *Le Matin*, while Peking–Paris reports were usually front-page news, this revelation threatened to completely upend the race. A scant 300 miles now separated the two French De Dion-Boutons – pride of the powerful Marquis de Dion – and the Dutch car. And while the latter was flying, the former were getting slower and slower, like mechanical toys in desperate need of winding up. The 'grey twins' had straggled forty miles on 30 July, and seventy the day before. If Godard maintained his pace, he would pass them within hours. At his current rate, he might even plausibly catch up with – even overtake – the prince, still hundreds of miles north near St Petersburg.

From the perspective of *Le Matin* and the organisers, whose intention had been to promote the French automotive industry – and the De Dion-Bouton brand in particular – this threatened disaster. For weeks, an all-Italian automobile and team had hogged the glory and plaudits throughout the journey. Now it seemed the Dutch car and its fly-by-night driver might yet eclipse them all.[11]

St Petersburg, as the Italians soon learnt, was strikingly different from Moscow. Founded in 1703 by Tsar Peter the Great, it had been conceived and built as a modern, western European city. In place of the Byzantine-inspired domes of Moscow, there were large, classical palaces. Cafés and fashionable shops were strung along grand boulevards like bright glass beads. When Consuelo Spencer-Churchill, the American-born Duchess of Marlborough, visited in 1902, she was taken aback by its cosmopolitan air. The city had 'little of the Oriental splendour I had anticipated,' she groused. 'The wide, wind-swept avenues were lined with modern buildings in doubtful architectural taste.'

Tsar Nicholas II – who the duchess was seated next to at a grand dinner, finding him uncannily similar in appearance to his cousin, later King George V of England – disdained St Petersburg. Both he and his father had tried to Russify it, building churches and architectural flourishes and motifs that recalled Moscow's old town, but this attempt had only limited success. Although the American journalist Kellogg Durland admitted finding the city 'charming', he still described it as 'the rouge and enamel' worn by a 'sallow, ill-looking' Russia, 'beautifying herself, and coquetting successfully with many who see her'.[12]

The Itala had approached from the south on 1 August, passing through open country under persistent mizzle. The road was finally running smoothly beneath their wheels. In the four hours since their six o'clock departure that morning, they had been spirited eighty miles by their freshly polished automobile. For once, they were ahead of time, yet they remained subdued, insensible to the landscape around them even as it began to fill with smart villas, gardens and parks. The prince, driving in his long mackintosh, had not uttered a word for hours. All three were intent on getting to St Petersburg and then departing as soon as politely possible. If they hurried, they might arrive in Paris precisely two months after their departure from Peking.

Their Russian hosts, however, had other ideas. At Tsarskoye Selo, fifteen miles south of the city, a vivid convoy of thirty automobiles and one large, British-made bus, overflowing with people and bouquets, waved them down. They alighted, pasting pained smiles onto their faces. This did little to improve their appearance. Covered in filth, they resembled a colossal sculpture: 'Mud-men on a mud car,' according to one onlooker.

In truth, the crowd were not in much of a position to pass comment. The short journey from St Petersburg had been an education for all the foreign observers. Captain Windham, an English motoring enthusiast and aviation pioneer who had driven out from the city to meet the prince,* later wrote, with some fervour, that 'throughout my travels in all parts of the world I have never seen the equal [of that road] . . . enormous deep holes and ravines . . . mud in most places a foot thick. Cars skidded in all directions.'

This had evidently not deterred the enthusiasts, who presented a profusion of hands to be shaken and flowers to be held aloft. Bread and salt were solemnly given to them in a ritual welcome. The Italian military attaché pressed in to pump the motorists' hands. So too did the Italian consul, members of the Russian Auto Club, Ludvig Alfred Nobel and the indefatigable Mr Perelman. The motorists' evident desire to press on was blithely waved aside: they are early, they are informed, and this will not do. Besides, they must be properly celebrated. The entire group adjourned to the Tsarskoye Selo railway station for a long and early lunch, punctuated by the first in what seemed likely to be a long succession of congratulatory speeches.

The Italians, helpless in the face of this barrage of hospitality, fell in with their hosts' plans as gracefully as they could. Keen observers, however, might have noticed a strain in their manners and the frequency with which they shot glances at one another.

---

* The prince diplomatically told Windham how disappointed he was that there were no Britons competing, which pleased him no end. Windham himself was in St Petersburg to deliver the Anglo-Russian Entente as part of his duties as one of the King's Messengers, couriers employed to hand-deliver sensitive documents around the world.

Their bodies may have been stationary, but their minds craved a return to the onward flight of the Itala. Perhaps adding to their discomfiture were strange reports Barzini had begun receiving from colleagues at the *Corriere* and the *Telegraph*: the all but forgotten Spyker was back on the move and making up for lost time with a series of record-breaking dashes across the Siberian countryside arrowing towards Moscow. The Italians, far out of their way in St Petersburg, could only smile, quaff champagne and plot their route back south.[13]

19

THE TOLL

Automobiles and Safety

> The poetry of motion! The *real* way to travel! The *only* way to travel! Here to-day – in next week to-morrow! Villages skipped, towns and cities jumped – always somebody else's horizon! Oh bliss! Oh poop-poop!
>
> Mr Toad, *The Wind in the Willows*, Kenneth Grahame, 1908

> For over half a century the automobile has brought death, injury, and the most inestimable sorrow and deprivation to millions of people.
>
> Ralph Nader, *Unsafe at Any Speed*, 1965

In 1905, two years before the Peking–Paris, Georges Méliès, the world's first narrative film-maker, released the first smash hit. *Le Raid Paris–Monte Carlo en Deux Heures* imagined the Belgian monarch and car nut, King Leopold II,* as the archetypal reckless patrician driver on a farcically speedy journey from one holiday destination to another. During his two-hour drive, the luxuriantly fur-coated king flattens a policeman, who needs to be reinflated by curious onlookers, crosses the Alps, knocks over fruit stands

---

* King Leopold was said to be the first European monarch to take up automobiling and would remain an enthusiast until his death in 1909, often visiting the Salon de l'Automobile in Brussels. He personally owned the Congo after seizing it in 1885, developed and exploited the rubber industry there, and was responsible for atrocities against his colonial subjects.

and postmen, tussles with Dijon city officials and generally leaves a trail of explosions, death and destruction in his wake, only to be greeted in the millionaires' playground with rapturous applause.

It was a spoof, of course, but real stories of aristocratic drivers mowing down other road users were not unusual. William Kissam Vanderbilt II, one of the world's wealthiest men and an early De Dion owner, was mobbed and nearly lynched in Italy in 1906 after running down a five-year-old boy while out motoring near Pisa. Vanderbilt, it transpired, certainly had not helped matters by drawing his revolver on villagers who came to the boy's aid. During the Peking–Paris, the only injuries reported were to the motorists themselves, but judging from the number of horse-drawn carts they overturned, especially in Russia, it would not be surprising to discover there were many more.[*1]

Vehicular injuries, accidents and deaths were hardly unknown before the automobile, but its invention made the process brutally efficient. In the early days, people were simply not used to vehicles that could travel so much faster than horse-drawn ones. The first recorded pedestrian fatality by an automobile was forty-four-year-old Bridget Driscoll of Croydon, who was killed on 17 August 1896 in Crystal Palace Park in London while out walking with her daughter and her friend. The vehicle responsible had a top speed of thirteen miles per hour, but was probably moving considerably slower. By 1907 the danger was evident: *The Economist* conducted a survey and found that while there was only one accident per seven horse-drawn vehicles in London, each motor bus caused slightly more than one accident.

Today, 1.53 people die for every 10,000 vehicles on the road in America; in 1913, the rate was 33.38. In the four years following Armistice Day, more Americans were killed in automobile

---

[*] The Itala caused havoc just outside Perm one morning by upsetting an entire convoy of dairy carts. 'In a moment, all the *telegas* were overturned, milk was flowing everywhere, and the peasants, encouraged by their wives, were throwing themselves in our direction. What was to be done? What can you do when you are on a 40 hp motor car, and threatened by a crowd of *moujiks* anxious and able to fall upon you? . . . [R]egretfully, but firmly, we put down the speed lever, and our machine took to its heels, and was soon far out of reach of the peasants' sticks.'

234

accidents than had died in battle in France during the First World War. During the 1920s, when there was a rapid shift to automobile traffic and very few laws or customs governing behaviour, pedestrian deaths were especially high. But even a generation later, the situation was little better. In 1951 the American Air Force found that they were losing more men – either through death or injury – to automobile accidents than to combat in Korea.[2]

The dangers automobiles posed to pedestrians (particularly children), other road users and drivers were clear from the moment they were invented.

The death toll was obvious to many from the earliest years of automobile use. Children, who habitually played in the streets around their homes, were over-represented among early casualties and fatalities. The use of walking harnesses or reins for children became widespread as safety measures in urban areas. A twenty-five-foot obelisk was erected in Baltimore in 1922 to commemorate the 130 children killed in the city the year before; in 1925, more than 7,000 children were killed by cars and trucks in America at a time when only 17 per cent of the population owned one. A 1920 *Literary Digest* article found one person was being killed every thirty-one minutes. In fact, it continued, automobiles were responsible for

> three times as many deaths as result from all the accidents combined on all the steam and electric railways, on all river, lake, and coastwise boats, in all the coalmines, in all foundries and blast-furnaces, in all the factories and machine-shops, and in every other more or less dangerous industry of whatever sort in the country.[3]

Speed, spectators and the heightened visibility of motor sports added to the carnage. Émile Levassor, an automotive pioneer, died after swerving to avoid a dog while taking part in the 1896 Paris–Marseilles race. Marcel Renault perished in 1903 following a crash during the Paris–Madrid. Twenty onlookers were bowled over like ninepins – and two fatally injured – during a twenty-four-hour contest at Brighton Beach in New York in 1907, when a driver lost control and broke through a fence. Earlier at the same event, John Halfpenny, a mechanic, was carted off the course when 'flames suddenly shot out of the gasoline box' on which he'd been working and 'his arms were burned to the elbow'.

Fifty-six people were killed during the thirty-year run of the Italian Mille Miglia endurance race. It was finally banned in 1957 after two fatal crashes, the second of which, involving a four-litre Ferrari 335 S in the village of Guidizzolo, killed the driver, navigator and nine spectators, five of whom were children. Two years earlier, at Le Mans, a high-speed collision caused a Mercedes 300 SLR to explode, killing the driver instantly and hurling large pieces of burning debris onto the stands. Eighty-three were killed and 180 injured.*[4]

Even on ordinary roads, without high speeds and with the

---

* The incident had an instant impact on motor sport, not least because British Pathé captured chilling footage of the disaster and the immediate aftermath. While several teams – including Mercedes and Ferrari – withdrew out of respect to the victims, Jaguar continued and went on to win by five laps. Several countries, including France and West Germany, immediately banned motor sports until safety reviews could be conducted and regulations amended.

greater legal restrictions and cultural familiarity born of more than a century of experience, automobiles exert a grim toll. Today, around 1.3 million people die each year in road traffic crashes. Over half of these are other road users: pedestrians, cyclists and motorcyclists. In America, vehicle crashes have been the leading cause of death for children every year for the six decades leading to 2020.* Elsewhere, the toll is even greater: around 90 per cent of all road traffic deaths occur in low- and middle-income countries, with rates being highest in Africa.[5]

From the beginning, automobile manufacturers have been reluctant to acknowledge the dangers posed by their products, choosing to focus instead on the carelessness of a few drivers or, more often, other road users. This was even the case in instances where it was obvious deaths could have been prevented by altering the non-essential design elements of cars. For example, hood ornaments (the small models and logos mounted onto car bonnets) and tail fins (introduced in the late 1940s to give cars a speedier appearance and reaching their apotheosis in the 1960s) were especially dangerous for other road users. Vehicles did not need to be travelling very fast for these sharp appendages to cause serious – often fatal – damage. The situation was not much better in the car either: rigid steering assemblies ran through bodies like lances, while instrument panels acted on the head and face, according to one expert, like 'a steel beam or an anvil'.[6]

Despite efforts by legislators and pressure groups to make cars and roads safer, change has usually been slow and hard-fought, in part because of the political influence car manufacturers enjoy. In the 1960s, car production consumed 21 per cent of all steel, 49 per cent of lead, 61 per cent of the rubber, 32 per cent of the zinc and 58 per cent of all upholstery leather sold in America. As a result, the motor lobby were not shy about exercising their political influence. 'Need I comment,' Richard Cross of the Automobile Manufacturers Association asked Congress in 1964 during a hearing on excise tax cuts, 'on the essential role of the automobile in the growth of the American economy?' Congress got the message.[7]

---

* Since then, firearms have been the leading cause of children's deaths.

This is not to say that automotive companies had it all their own way. In 1965, a young journalist by the name of Ralph Nader published *Unsafe at Any Speed*. Despite its publisher's fears that no one would buy it, Nader's book would jostle Truman Capote's *In Cold Blood* at the top of the bestseller lists for months and go on to inspire several pieces of legislation, including the 1966 National Traffic and Motor Vehicle Safety Act. It argued that automobiles were costing Americans billions of dollars and innumerable lives and that automobile makers were aware of both problems and potential solutions but were determined to put profits above the health and well-being of the nation and their customers. It was a radical accusation against an industry that had come to define American industrial might.

From the beginning of the automotive era, motorists and manufacturers fiercely defended and advanced their rights. It is well known – and has been for decades – that increased speeds lead to increased fatality rates. The risk of death for pedestrians struck at forty miles per hour is four and a half times greater than those struck at thirty. This danger also applies to those in cars: the likelihood of fatality for passengers and drivers struck side-on at forty miles per hour is 85 per cent. At the turn of the twentieth century, speed limits were kept low: the median state-designated limit in America in 1906 was ten miles per hour. But these were difficult to enforce and were soon overturned. By the 1920s, when most lawmakers were car owners themselves, speed limits had increased significantly. Over the course of the century, as cars became faster, drivers fought to use the increased speed, even though they were aware of the risks this would pose.

Britain's 1903 Motor Car Act, which increased the speed limit to fourteen miles per hour, was bitterly opposed by those who wanted restrictions removed completely and resented the stipulation for cars to carry identifying registration marks. Early motoring magazines often published maps depicting the location of police 'speed traps', so that motorists could evade them. After lobbying, all speed constraints in Britain were removed in 1930, leading to record numbers of fatalities. Limits were reimposed four years later, despite objections from the AA and RAC.[8]

Other proposals for safer cars and driving have faced similar stiff opposition. Seat belts, for example, the ancestor of which was invented in 1885, would not become compulsory to wear in many countries for a full century. Manufacturers did not want to install them, querying their effectiveness and pointing out that they raised costs, while drivers complained they were uncomfortable, inconvenient and infringed their rights. In America, a Gallup poll in 1984 found 65 per cent opposed mandatory belt laws, even though 86 per cent believed they saved lives. One Michigan lawmaker who proposed a seat-belt bill in 1982 allegedly received hate mail comparing him to Hitler. The following year, a tussle over Reagan's roll-back of a law requiring new cars to be equipped with either air bags or seat belts made it all the way to the Supreme Court.

Nor is America the only country where such requirements have received pushback. In the United Kingdom, the RAC submitted a paper in 1973 that called compulsory seat belts a 'drastic measure . . . which [will] in no way affect the safety of any other road users'. Police argued that enforcing such a law would damage their relationship with the public. A working group of bureaucrats worried about children, pregnant women, 'dwarfs and persons who are deformed, grossly obese, exceptionally tall' and 'some old people . . . [who] would deprive themselves of the pleasure they get from a car outing rather than put up with the nuisance involved in what they regard as an unnecessary, newfangled idea'.[9]

Measures against drink-driving or the use of mobile phones while driving and requiring child seats or stricter speeding regulations meet identical reactions. In Bangladesh in July 2018, young people protested against the country's lax road safety laws after two students were hit by a speeding bus. The government's response was to accuse the protestors of conspiring with their opposition and turn a blind eye when masked groups began attacking them. In many countries where laws are passed but not strictly enforced they tend to be ignored.[10]

Societies – particularly ones with large numbers of drivers – have wrestled with the issue of assigning responsibility for automotive safety. Even when fatalities are involved and the evidence is overwhelming, juries often acquit those accused of dangerous driving, probably because they identify more with the driver than they do with the victim. A study in Britain in 2006 found that fewer than half of drivers convicted of killing a cyclist were given any jail time at all. A *New York Times* article entitled 'Is it OK to Kill Cyclists?' found that, all too often, no attempt was made to prosecute drivers involved in fatalities. In one incident, where a teenage driver simply ran over a cyclist from behind near Seattle, the attending police found the driver was not drunk and issued him a $42 ticket. As they saw it, he had not been driving recklessly: there was no blame to be assigned.[11]

Driverless vehicles have recently been touted as panaceas: without fallible, distractable people behind the wheel – we are told – drivers, pedestrians and cyclists alike will enjoy a world without traffic jams or road deaths. Tesla's CEO, who called self-driving cars a 'solved problem' in 2015, claimed two years later that 'Getting in a car will be just like getting in an elevator. You just tell it where you want to go and it takes you there with extreme levels of safety.' So far, this has not been the case, as a series of well-publicised accidents using Tesla's autopilot and Uber's self-driving modes has illustrated. To date, liability has largely rested with drivers, who are meant to remain fully alert and ready to take over even when 'driverless' modes are in operation.[12]

If safety in the case of direct, physical accidents has proved difficult to manage, more insidious has been the issue of pollution. As early as 1950, experts demonstrated an incontrovertible link between automobile exhaust fumes and Los Angeles's terrible smog. This not only caused measurable and expensive damage to infrastructure, buildings and other cars, but also appeared to have health implications. In the decades since, the evidence has become overwhelming.

Nader took aim at the industry for this too, relating with some glee an anecdote that perfectly illustrated the industry's chutzpah. It involved Harry A. Williams, managing director of the Automobile Manufacturers Association, and his speech at the first American National Conference on Air Pollution in 1958:

It must have been impossible for our elders to imagine life in this land without the polluted air in which they lived – before people were liberated from the congested cities by the motor vehicle and the pavements it demanded and eventually got.

There were reeking livery stables in every neighborhood.

Cow-barns were the customary auxiliaries to dairies.

There were malodorous privies in every backyard . . .

Yet this world, in which such filth-originated diseases as typhoid were common and regular visitors, disappeared in the smoky haze of the very automobiles which chugged erratically on the scene only one lifetime ago.

Williams claimed his industry could not be expected 'to concern itself with how the consumer uses or misuses the product long after its sale to the public'. Nevertheless, it had already 'eliminated pollution of air and water' from factories and foundries and had spent millions studying air pollution. 'A million dollars a year – spent on one problem by one industry – is still a substantial outlay.' Martin Agronsky, a veteran NBC commentator, responded with admirable sangfroid: 'With all due respect to a twenty-billion-dollar industry, I am not impressed.'[13]

His scepticism, as it turned out, was warranted. Vehicles with combustion engines emit numerous pollutants: nitrogen oxides, sulphur oxides, carbon monoxide; particulate matter (essentially everything in the air that is not a gas), heavy metals including cadmium, lead and mercury. Emphysema, bronchitis, lung cancer, high blood pressure, Alzheimer's and heart disease and attacks can all be triggered or exacerbated by breathing in this cocktail.

Air pollution particles, once inhaled, get into the bloodstream and are carried into virtually every cell in our bodies. Iron-rich nanoparticles caused by traffic have been found in the mitochondria – our cells' energy producers – inside heart tissue. Pollution has been found in the placentas of women who have just given birth; high levels have been linked to low birth weight, premature birth and increased miscarriage rates.[14]

Dirty air causes around 7 million early deaths annually worldwide: more than AIDS, diabetes and traffic accidents combined.

In Europe alone, air pollution is responsible for the death of fifteen times more people than car crashes.[15]

While this is truly a global problem – it is estimated that just 5 per cent of us breathe truly clean air – Asia, and South Asia in particular, is its epicentre. Of the world's ten most polluted cities, six are in India, three in Pakistan and one in China. India's five-year National Clean Air Programme, launched in 2019, alleges vehicles are responsible for over a quarter of the country's air pollution. Controlling their emissions, upgrading India's motor fleet and trying to neutralise road dust are pivotal to the programme. The problem is also acute in China, the world's biggest automobile market since 2009. In 2013, after several 'airpocalypse' smogs in Beijing, the Chinese government announced a war on pollution. The following year, it pledged that more than 6 million cars in Beijing no longer meeting exhaust emission standards would be removed. Five years later, well before the Covid pandemic, Chinese consumers were still spending 4 billion yuan annually (£480 million) on anti-pollution masks.[16]

At international climate talks held in Glasgow in November 2021, six auto manufacturers – including GM, Ford and Mercedes-Benz – and thirty countries announced they would phase out the manufacture and sale of petrol-powered vehicles globally by 2040. (China, the USA and Japan did not sign up to the pledge, which in any case was not legally binding.) Electric vehicles – praised for their cleanliness and lack of stinking fumes back in 1907 – are fast becoming the most widely accepted solution. In terms of emissions, the cars themselves are certainly less polluting, although of course the impact of the vehicle depends on how the electricity is generated. There is also the vexed question of raw materials used in batteries, such as lithium and cobalt. Sourcing, refining and processing them has been linked with human rights abuses and environmental damage. Around 70 per cent of the world's cobalt supply, for example, comes from the Democratic Republic of Congo, where mines are often small and unregulated and there is widespread use of child labour.[17]

Automotive companies have issued assurances about tightening their supply chains, ensuring batteries are recycled and developing ones that do not use problematic raw materials. As sceptics have been quick to point out, however, manufacturers do not have the best track record when it comes to safety.

In April 2017, a Detroit judge issued a $2.8 billion criminal penalty to Volkswagen (VW). Five years earlier, researchers at the University of West Virginia discovered that the firm had rigged some diesel cars with software allowing them to cheat American emissions tests. When not being run under test conditions, cars fitted with 'defeat devices' were emitting pollutants up to forty times over the legal limit. VW later admitted to making 11 million such vehicles. The resultant series of penalties and lawsuits are estimated to have cost the firm around $30 billion.* The fines, while steep, did little real damage: in the first quarter of 2021, thanks to strong performance in China, Volkswagen sales rose 13 per cent to $75 billion.

Automobile manufacturers' promises of freedom and excitement have always rung truer than those about safety and cleanliness. Around 1900 this mattered less, since danger was an accepted part of locomotion. Almost every day, front pages were emblazoned with the details of accidents involving various forms of transport. Trains skipped tracks, horses shied and bolted, airships fell from the sky, ships sank and automobiles crashed. Over the course of the following decades, however, safety has become a higher priority. We expect to be able to travel in comfort and without fear of injury or death. And now that the public knows more about the damage that can be done to us and our environment by chemical emissions, we expect polluting companies to be held to account by our legislatures. Failure to do so increasingly carries political penalties. All of which means the continuing risks posed by automobiles – even tempered by a century of safety measures – seem disproportionately high. It is impossible to know exactly what will come next and how exactly transport will evolve. What *is* certain is that the cost of largely ignoring safety and cleanliness concerns back in 1907 has proved incalculable.

---

* VW are not an outlier: Toyota, GM and Takata have all been fined in the past decade for failing to protect consumers or comply with regulations such as the Clean Air Act.

20

Kaunas – Liège

4–8 August

Since I've been in this car, we've stopped living: we haven't eaten,
we haven't slept, because we're always hurrying somewhere and
can't afford to waste time on these dreadful roads. I've had it up
to here!

Bruno Stephan, Spyker team

The two De Dion-Boutons straggled into Kazan on Wednesday,
7 August, woebegone after days of too much mud and too few
miles. They were now a full fortnight behind Prince Borghese and
his Itala. Any lingering hopes of catching up had evaporated into
the Uralian mists.

True, the prince had headed hundreds of miles out of his way to
St Petersburg, but the crews of the De Dions were exhausted, dis-
heartened and progressing slower than ever. It had taken them forty-
eight hours to reach Kazan from Yelabuga, a small town on the
banks of the Kama: 150 miles in exchange for two full days' driving. Rain clouds stalked them. Collignon had lost his hat and now
drove bare-headed through the drizzle, hair the texture of seaweed
dripping sullenly into his beard. Du Taillis – always inclined to be
an eccentric dresser – had acquired a voluminous set of rubber
overalls, which he called his 'drivers' umbrella'. These he paired
with a French colonial officer's hat – a small, straight-sided cap
with a curving peak – which, had there been any sunshine, would
have shaded his eyes behind their round, wire-framed spectacles.

245

The sun, however, was nowhere to be seen. 'This wretched rain that has been dogging us not only makes the journey more arduous,' Cormier wrote, 'but it also tires us. It seems to have made it its mission to rob us of the fortitude we so desperately need.'

They resolved to spend a few days in Kazan, the largest city they had seen since Omsk. Photographs from the era show graceful, cobbled boulevards lined with two- and three-storey shops, mosques, cathedrals and hotels. Men in long coats and a wide array of headgear stride about purposefully or gather in sociable knots at street corners, their voices no doubt raised above the clatter of wheels and hooves as carters guided their charges around the city.

It certainly cheered the French motorists. Cormier dutifully recorded the population – 130,000 – before, buoyed by its brisk, confident air, describing it with an uncharacteristic eye for detail. 'Just near the river is the Muslim quarter, with wooden houses. In the centre, however, the buildings are made of brick or stone. We really feel like we are in a major European city.' Electric trams, a hotel with 'conventional beds (bed base and blanket), very few insects . . . These are all reasons to make us smile.'[1]

Before sunrise the next morning, the four Frenchmen were awoken by the snorting roar of a four-cylinder engine in the street below. There was a brief silence, and then they heard a round of muffled curses and what sounded like a whispered disagreement. Groggily, du Taillis opened his shutters and looked outside: the errant Spyker, her irrepressible driver and Bruno Stephan, the young Dutch engineer, had arrived.

The scenes that followed were experienced very differently by the individuals involved. Du Taillis thought Godard, Stephan and the Spyker seemed like exhausted heroes:

[They were] in a pitiful state: mudguards and lanterns were half broken; the whole chassis and the once green dome which housed the luggage had disappeared under the black splashes from the muddy tracks; the tyres . . . were worn down all round, right down to the last tread. Godard had lost a lot of weight and had hollow cheeks, a pale complexion and dark circles around his eyes, which witnessed clearly enough enormous fatigue, stoically borne.

Stephan, the twenty-year-old Spyker employee, looked ill and frail too. 'I'm not used to this way of life,' he apparently wailed to Cormier. 'If I'd known the Peking–Paris would be like this, I would have stayed in Holland.'[2]

Godard and Stephan outlined their movements over the previous weeks to du Taillis, timings that were relayed to *Le Matin* and published with little fanfare near the bottom of a longer report on 10 August:

Thursday, 25 July – Depart Cheremkhovo; arrive Nizhneudinsk.

Friday, 26 July – Arrive Kansk.

Saturday, 27 July – Arrive Krasnoyarsk in the early afternoon and continue on to Achinsk, arriving at 11 p.m.

Sunday, 28 July – Pass through Mariinsk and Tomsk and camp sixty miles further on.

Monday, 29 July, and Tuesday, 30 July – Arrive Omsk at 8 a.m. on Tuesday, 30 July, after a full 24-hour sprint. Joined by Bruno Stephan.

Wednesday, 31 July – Depart Omsk at five in the morning and travel 160 miles before the springs break, then limp on to Petropavl to get them fixed.

Thursday, 1 August, and Friday, 2 August – Drive non-stop to Chelyabinsk, arriving at 7 a.m. on 2 August.

Saturday, 3 August – Arrive in Ufa after taking a different, more southerly route.

Tuesday, 6 August – In Birsk.

Wednesday, 7 August – Depart for Kazan.

To du Taillis, Godard's efforts over those days made him the true hero of the entire affair. 'Such a performance ranks in the category of outstanding,' du Taillis stoutly declaimed in his 1908 book. 'It also makes one reflect on the lightning speed he could achieve, that would have allowed the Spyker to get ahead of all its competitors, even the Itala, if Godard's rather hot-headed valour had been complemented by the knowledge and experience of a professional mechanic.' His belief was that, had Stephan or someone like him been there from the start, Godard would have won the Peking–Paris.[3]

In the decades since, many ardent enthusiasts of the Peking–Paris story have agreed. T.R. Nicholson's *Adventurer's Road* (1957) admiringly described Godard as 'utterly exhausted and near the end of his tether' in Kazan, after 'his extraordinary race against time across Siberia'. Seven years later, Allen Andrews, author of *The Mad Motorists*, devoted an entire chapter – 'Godard Rides Again' – to the driver's supposed exploits, which Andrews artfully embroidered. Just past Omsk, for example, Andrews has him rescue a baby that had been thrown from a cart. 'Here we are in

Charles Godard courted attention and controversy from the beginning.
Decades later, he was portrayed as a hero; a counterpoint to Prince Borghese.

the middle of the Peking–Paris race, and we get stuck with a baby,' his Godard exclaims. 'I bet it's one of Borghese's.' By the time he reaches Kazan, his hands are 'blood-raw' from the strain of completing 'in a fortnight what Borghese had done in three weeks, and the De Dions had covered in nearly five weeks'.[4]

Not everyone, however, was convinced by Godard's driving genius. In the chilled morning air of the courtyard, Cormier was by turns truculent, sullen and disbelieving. Stephan, who had never met the De Dion crew before, found that he and Godard were treated, not 'as long-lost comrades but as suspects'. 'Godard and I were interrogated separately in the hotel,' he later wrote to a friend in the motoring industry, 'without being granted the opportunity to discuss in between.'

The suspicions of the De Dion-Bouton crews were understandable. Godard told them he had travelled 170 miles on 25 July between Cheremkhovo and Nizhneudinsk (actually, it is nearer 230 miles), and 500 miles in a single push to reach Omsk on 30 July. His countrymen, traversing the same terrain, had often been hard-pressed to travel sixty miles in a day.

Rumours swirled back in Europe that Cormier suspected the Spyker crew of foul play and was lobbying *Le Matin* and the organising committee to have them expelled from the race. Precisely what he believed the Spyker crew had done to make up the distance he never made clear. He also – perhaps because of his own lack of charisma – failed to sufficiently convince anyone else of Godard's perfidy. (In any case, the organisers were extremely reluctant to cause a public scandal.) By the time he wrote his book about the Peking–Paris months later, Cormier had tacitly accepted Godard's version of events.[*]

Still, there can be no doubt that the story Godard was asking his fellow competitors to accept was inconsistent and vague. He provided dates and the names of major towns he had passed through, but little else. No indication of the weather or driving conditions or, in most instances, even the precise route that might have helped explain his astonishing speed. On the other hand, his movements were largely corroborated by the telegrams he had sent to *Le Matin* along the way.[†] These would have been difficult to fake, given telegraph operators' records and the fact that the messages were all written in French and in a similar style.[5]

---

[*] It is possible that by this time Cormier had recovered from his initial sour grapes, or maybe pressure had been applied by sponsors and organisers who wanted to tamp down any hint of scandal.

[†] There were some inconsistencies even here, however. For example, Godard told du Taillis he had arrived at Chelyabinsk early in the morning of 2 August and left later that day. However, the telegram he wrote to *Le Matin* from that town was sent on 3 August.

However, no one seems to have probed – then or since – the two most suspicious aspects of Godard's narrative. The first was his detour through Ufa: this was the 'southern route' from Chelyabinsk he had mentioned in his telegram to *Le Matin* on 3 August. The second was the strange silence about the Spyker's whereabouts between 3 August, when they arrived in Ufa, and 6 August, when they reached Birsk, just sixty miles to the north.

Had Godard and Stephan continued at the same pace they claimed to have managed since Cheremkhovo, they would have easily outstripped the two De Dion-Boutons, which reached Birsk on 2 August. They might even – on reaching better roads near Moscow – have given the prince a run for his money. It is possible they spent three days resting or doing essential repair work in Birsk, but if so, the story they told in Kazan on the morning of the 7th about their utter exhaustion after travelling without pause for days was clearly a lie.

A clue to this second mystery may lie in the first. That the Spyker went from Chelyabinsk to Ufa is suspicious: by road, it entailed a significant southerly detour. However, Ufa *was* the next stop on from Chelyabinsk on the Trans-Siberian Railway line.

While Godard was busy explaining himself to the other French drivers, Prince Borghese, Guizzardi and Barzini were dashing south from St Petersburg towards the German border. Their time in the city had been short. Rumours circulated that Borghese had only visited the city because he had a mistress there. If this was true, his heart was clearly a good deal fonder in her absence than her presence: he remained in St Petersburg scarcely a dozen hours, most of which were spent in rounds of very public engagements. The Russian Automobile Club and the Motor Car Society held a banquet in his honour; he was presented with medals struck to commemorate the trip.* The Itala's co-driver, the mechanic

---

* These medals evidently meant a lot to the prince. They were carefully kept and they remain in the possession of a branch of his family. They are heavy and beautifully detailed. One has gold worked into a scrolling ribbon-like bow above a round face. On one side, a small Itala is sculpted in gold on a white enamel base, surrounded by the words 'Peking – St Petersbourg – Paris – 1907', while the obverse carries the glorious

Guizzardi, stood proudly by: he neither attended the banquet nor received any recognition of his role.

At four o'clock the next morning, Friday, 2 August, the Italians set off down silent, mist-veiled streets. The city's emptiness and stillness made it somehow seem to loom larger around them. The rattling growl of the engine seemed louder too. They passed underneath the colossal Narva Triumphal Arch, erected in 1814 to celebrate the Russian victory over Napoleon, with its distinctive, pistachio-green verdigris patina, and headed south through slumbering suburbs.

Their route took them through Gatchina – here they caught a glimpse of the golden facade of its palace, home of Tsar Nicholas II's mother, the dowager empress – Luga, Pskov, Dvinsk (now Daugavpils, in Latvia), Kovno (now Kaunas, Lithuania), Wierzbowo (Poland), Stargard (Poland). Close to the border with Germany, the bonds of the Russian Empire seemed to weaken. Customs, manners, clothing, accents and even languages shifted rapidly.

Back near Dvinsk they encountered a group of 'True Believers', a marginalised Russian sect cleaving to the teachings of the seventeenth-century Orthodox church. At Kovno they met a company of Polish journalists, who professed their admiration for Italians because of their struggles for freedom, and dashingly recounted an altercation with the Russian authorities over their 'Warsaw Automobile Society' banner. (The carrying of flags was forbidden in the region and the Russian gendarmes could not read Polish.) '[We] are among people who seem to have been pushed back to the gates of the Empire,' Barzini wrote. 'Near the boundaries are the hated races; you might think they have settled here so as to be ready for flight.'[6]

To the west of St Petersburg, the major artery roads improved dramatically and Borghese and Guizzardi took full advantage: 510 miles spooled beneath their wheels in the day and a half after leaving St Petersburg. They were travelling so fast the newspapers

monogram, in red, white and pale blue enamel, of the Russian Automobile Club. The second is designed around a decorative gunmetal spoke motif, also worked with fine blue and red enamel, decorated with a goldwork crown.

could no longer keep up. By the time they reported his arrival in one city, he had sped past three more: 'Journey from Moscow!', 'Prince Borghese Arrives at St Petersburg!', 'Brilliant Reception!', 'The Motor Car Race!', 'Prince Borghese Arrives in Germany!', 'Peking–Paris Race Ending!' Only *Le Matin* seemed keen to slow events down, determinedly trying to stoke interest in the French automobilists, thousands of miles in the rear.

Despite this, Borghese was chafing at the impediments besetting him. In each town they passed through, fresh impositions had been placed on his time by organisers, sponsors, local dignitaries and hosts of well-meaning well-wishers. Banquets heaped upon banquets, speeches upon speeches. A lifetime of training had lacquered the prince in the layers of politesse worthy of a consummate gentleman, but the stresses of the journey were beginning to wear them thin.

As an aristocrat from one of the most prominent families in Italy, he was used to managing his own affairs to his own tastes. He also possessed a level of focus – bordering on selfishness – often seen in sportsmen and explorers. It was these qualities, coupled with the smoothing power of his wealth, that had ensured their success to date. Borghese had taken a bet on driving a powerful, heavy car when received wisdom suggested that something light and manoeuvrable was the better choice. He was much better prepared than the other drivers, and he had the unquestioning loyalty of Guizzardi, who not only kept the Itala functioning with daily care and on-the-spot repairs but was also able to drive – a huge advantage that went largely unremarked at the time. Through his own force of will and steadiness of purpose, Borghese had managed to keep the morale of his team up despite accidents, poor weather and a gruelling schedule with fewer rest stops than the others allowed themselves. All he wanted now was to reach his destination and regain control. It was this overpowering emotion that had sent him scurrying from St Petersburg. At another town, when the local governor's wife sent a messenger entreating him to bring the Itala to a charity fair to help her raise money for the Red Cross, he simply declined, retiring to the peace of his hotel room instead.[7]

The principal emotion the Italians experienced when crossing the border into Germany on 4 August was relief. 'There seems to be an intelligent fury in the machine,' Barzini wrote, tactfully implying that even the Itala seemed eager to reach its destination now. The border itself was marked by a little bridge with a chain across the middle and insignias at each end, the double-headed eagle of Russia facing off against the single-headed eagle of Germany. Without delay, officials examined and stamped their papers. The chain was lowered and the Itala nosed forward between two imperial guards in crisp salute.

If Borghese expected that, once out of the Russian Empire, he would have more command over his movements, he was soon set right. The journey from the border to Berlin, scheduled to be completed in three days, was accomplished in two. 'Prince Borghese ... decided today to double his stage and arrived unexpectedly in Berlin at six o'clock in the evening. It goes without saying,' Henri des Houx, Berlin correspondent for *Le Matin*, reported waspishly, 'that [we] had been warned, in good time, about the prince's whim.'[8]

The Italians were quietly installed – 'almost incognito' according to des Houx – in the Hotel Bristol on Unter den Linden, the wide, grand boulevard in the city's historic centre. Here they were firmly buttonholed by the irked organisers:

They will rest until tomorrow morning when a new series of trials will begin for Prince Borghese, which he has declared to me to be much more of a challenge for him than those he has just endured. But he promised me that he would gird himself with all his intrepid courage and, contrary to his normal habits, would submit himself obediently and point by point to all the various items on the programmes intended to ensure him everywhere, especially in Paris, a reception worthy of his Herculean work. So tomorrow in Berlin, everything will be arranged as if he had not arrived the previous day.[9]

Count Adalbert von Francken-Sierstorpff, a co-founder of the Imperial Automobile Club, who would later become a member of the International Olympic Committee and campaign for consistent traffic signage throughout Europe, led the toast at the celebratory luncheon the following day.

'When you commenced your tour,' he told the prince, 'doubt prevailed in Germany as to whether it would succeed. What you were attempting was thought to be impossible . . . Your enterprise has added a new page to the glory of motoring . . . [showing it to be] one of the most effective means by which nations are brought together and learn to understand and value one another. More than this, you have been a pioneer of civilisation.'

On this triumphal note, it seemed the prince's faux pas was forgiven. The assembled dignitaries and journalists raised their glasses to Borghese, smiling as they put squabbles aside to bask in the moment.[10]

21

ON THE OFFENSIVE

Automobiles in the First World War

No other man-made device since the shields and lances of ancient
knights fulfils a man's ego like an automobile.

Attributed to Lord William Rootes, 1958

Earl Frederick Roberts, a slight but richly moustached general, veteran of the Indian Rebellion, Afghan and Boer wars and
commander-in-chief of the British army, rose to his feet in Britain's
House of Lords on 10 July 1906 to give one of the most impas-
sioned speeches of his career. 'I sometimes despair,' he cried, 'of
the country ever becoming alive to the danger of the unprepar-
edness of our present position until too late to prevent some fatal
catastrophe.' The danger he referred to was global war and his
warning proved prescient. Eight years later, Austria-Hungary
declared war on Serbia, dragging most European nations and their
colonies into a conflict that would last four years and kill around
8.5 million troops. This would also be the first campaign in history
to see the widespread use of automotive power – and the invention
of the tank – which remain mainstays of modern warfare.[1]

Warning tremors had been shaking loose old certainties
for decades. The European powers' empires appeared to be
approaching their zenith, accumulating more wealth than ever
before, but trouble was brewing. Rebellions broke out in colo-
nised nations, such as that against the French in Morocco in the
summer of 1907. Anti-imperial views – like those of economist

John A. Hobson – were discussed more openly and by more proponents.* Squabbles broke out between imperial powers over influence in Asia. Britain and France were alarmed by the expansionist ambitions of Russia and Germany. Small, warning international crises flared up and were then snuffed out again, like matches struck in night-time drizzle. An indiscreet announcement that the American fleet would transfer from the Atlantic to the Pacific, for example, led to a global panic that there would soon be war with Japan.

These anxieties found expression in exhaustive rounds of diplomacy by heads of state. Kaiser Wilhelm II, the rather erratic German emperor, visited Denmark in July 1907.† (The visit was a damp squib. It rained so much that when Wilhelm shook hands with a commanding officer on parade, the latter's sodden white glove adhered to the emperor's hand as he withdrew it.) The emperor was visited in turn by the tsar, who was courting Germany, much to the dismay of the Francophile faction of his own court. Kaiser Wilhelm II used the meeting with the tsar as an opportunity to display an 'unusually large assemblage of German warships'.[2]

Relationships between individual rulers played out against a subtler, more ominous backdrop, characterised by mistrust on all sides. The Triple Entente, an association between France, Russia and Britain that built on previous agreements, including the Entente Cordiale of 1904, would snap into place at the end of August 1907. It was hoped it would provide a counterweight to the Triple Alliance between Germany, Austria-Hungary and Italy. Russia's role in the Triple Entente was considered essential to off-set German expansionist aims.‡ It was an irony lost on no one

---

* Hobson's book *Imperialism* (1902) argued that the drive for expanding a country's empire had a great deal less to do with 'civilising' nations – a common justification – than with capitalism. His ideas strongly influenced Lenin and Trotsky

† The kaiser was, incidentally, a keen motorist. He ordered roads 'as smooth as a parquet floor' to be built through his hunting estate at Schorfheide in the summer of 1907 to better accommodate his passion.

‡ Hereafter the principal combatants in the First World War will be collectively referred to as the Central Powers (Germany, Austria-Hungary and Turkey) and the Allies (France, Britain, Russia, Japan, the United States (from 1917) and, despite being a signatory of the Triple Alliance, Italy).

that the Hague Peace Conference was taking place amidst all this manoeuvring.[*3]

The hollowness of any talk of peace was evident to the Peking–Paris motorists, especially in the uneasy borderlands between the Russian and German empires. The Russian army carried out large-scale manoeuvres near St Petersburg the week that Prince Borghese passed through. The final approach to the strategically important town of Kovno was a 'fortified slope' bristling with military paraphernalia, and the roads approaching the border were suspiciously well maintained. (Barzini noted drily that excellent roads were 'at least one practical use of war'.[4])

In Germany the Italians encountered a battery of artillery, a patrol of Hussars and an infantry regiment on the march in a single day. The French drivers happened across German 'batteries, squadrons, infantry or artillery regiments taking up positions, or squadrons of Uhlans[†] doing an exercise to capture artillery'.[5]

The First World War would both affect the development of the automobile and be profoundly altered by it. Early outings of traction engines in the Boer (1899–1902) and Russo-Japanese (1904–05) wars were widely regarded as failures, since they struggled in anything but the smoothest terrain. It was perhaps understandable that in 1901 the *New York Times* declared the idea of motorised vehicles in war 'more fanciful than practical'. Nevertheless, by 1903 a 'novel military automobile', intended to transport troops but also act as a 'travelling blacksmith shop, carpenter shop, and saddlery for field use', had been constructed in New York and sent to Washington to be inspected by the War Department. An article devoted to the topic appeared in the September 1907 issue of *Avtomobil'*. It noted horse-drawn transport was currently an 'inescapable aspect of modern armies' but stated that

---

* The second Hague Peace Conference had lofty aims. It ran from June to October 1907 and was to build on resolutions secured at the first, held in 1899: placing limitations on armaments, establishing an International Court of Appeal, adapting the Geneva Convention for maritime and airborne combat – military dirigible balloons were being experimented with and were greatly feared – and generally seeking to regulate and curb warfare and the build-up of armaments.

† Light cavalry units that traditionally bore lances with spear-like points at both ends.

'mechanical transport has been shown to be particularly beneficial not only in the matter of constantly changing troop locations, but in staging points between troops and railway lines.'

Both Germany and the United Kingdom would establish volunteer motor corps in the early 1900s, soon superseded by more official units that chiefly used automobiles to supply troops and move heavy equipment, especially guns, to the front and away from it. Earl Roberts proclaimed that 'in future wars motor cars will play a very important part.' His subordinate, Lieutenant Colonel Mark Mayhew, acknowledged that while motor cars 'cannot leap fences and make short cuts across fields', as horses could, this was compensated for by the motor car's ability to cover great distances without tiring.[6]

Even supporters saw motor cars as supplementary to horses and mules, rather than a full replacement, until well into the First World War. A learned person using the name Pegasus wrote to *The Economist* in 1907 to assure the editor that 'for the purposes of war and sport – closely allied, because in its origin sport is

The significance of automobiles in the First World War was clear by November 1918; they were indispensable.

258

mimic warfare – no substitute will be found for the blood horse'. This scepticism was due both to early failures and to their association with wealthy dilettantes. An article in a British military journal from 1901 typifies the latter attitude: 'The number of fatal accidents which attended the recent Paris-to-Madrid Motor Race ought surely to convince the public that motor racing has very little to recommend it. Whilst Englishmen will ever continue to admire the professional or amateur jockey . . . there can be very little enthusiasm excited for the man who wins a motor race simply because he is rich enough to own a machine of the very latest pattern.'[7]

Nevertheless, the public success of the Peking–Paris vehicles over what had previously been considered impassable terrain made their potential military value explicit. While in St Petersburg, Prince Borghese mentioned to journalists 'the practical value of the motor-car to commerce and to military equipment'. An editorial in the *South China Morning Post* proclaimed the race had 'furnished valuable information as to the capabilities of motor cars in transport and field service in warfare'. A Russian magazine article, despite the headline 'Peking–Paris', did not mention the race once, except to state that cars were 'increasingly conquering the world', before going on to talk about military applications: 'We are far from proposing that twentieth-century automobiles might replace the war chariots of the ancients on the battlefield, but in skilled hands they could undoubtedly provide powerful support to the army . . . covered in light armour, they could help to protect communication lines from guerrillas between outposts and railway stations.'[8]

When the First World War broke out on 28 July 1914, most European armies possessed some motorised vehicles but remained reliant on combinations of rail and animal power.* Over the first year, the use of automobiles would be a ramshackle, hand-to-mouth affair. The French, in their initial retreat towards Paris before the

---

* Widespread reliance on animals would continue throughout the war. In all, some 16 million animals would be drafted, primarily horses and mules, but also oxen, camels, donkeys, elephants, pigeons, cats, dogs, pigs, mice, reindeer, canaries and at least one fox cub, as the mascot of an RAF squadron.

Battle of the Marne, commandeered every Parisian Renault taxi they could find to help move soldiers faster. Similarly the British, in addition to the vehicles purchased in the build-up to war – including sixteen Leyland lorries, two five-seater cars with limousine bodies and seven two-seater Vulcan motors – requisitioned large numbers of commercial vehicles and London omnibuses.* Several of the latter were transported to the front on such short notice they arrived still bearing advertisements for West End shows, including 'England Expects' and 'One Damn Thing after Another'.

It wasn't long before the use of automobiles became more deliberate. They were nimbler and less easy to target and sabotage than railways, as the Central Powers discovered to their cost early on.† They were also more robust than horses on the battlefield. Horses were more vulnerable to machine-gun fire and shelling and required a great deal of care to treat and recuperate. Vets noticed that horses, like people, suffered from shell shock and stress. After especially dirty runs it might take a dozen man-hours to clean a single horse and its harnesses. Horses also required large quantities of food, which was difficult and costly to grow in wartime and then transport to the front where it was needed.

Automobiles soon proved their mettle in a variety of military roles: Ford Model Ts, for example, were converted into ambulances. Rolls-Royce Silver Ghosts were requisitioned and refitted with light armour and, occasionally, swivelling gun turrets. Itala began producing ambulances, vans and rugged, jeep-like vehicles for the Italian army.[9]

Tanks evolved from early experiments with specialised automotive vehicles in warfare, such as a self-propelled armoured steam traction engine built in 1900 by John Fowler & Co. for the Boer War. Although various armoured and armed vehicles had

---

* The RAC also helped organise the British Motor Service Volunteer Corps, which enlisted civilians and their cars to go to the front. There is a story – one hopes apocryphal – that a chauffeur, employed by a parochial lord, was donated to the war effort, along with his lordship's Rolls, without his consent.

† Germany's Schlieffen Plan depended on the lightning mobilisation of troops and supplies along railway lines in the west before Russia could mobilise in the east. Belgium, well aware of this and that German troops would need to pass through Belgium on their way to Paris, sabotaged their own railway network to thwart the German advance.

been proposed and designed before 1914, the war accelerated the need for a vehicle fully suited to battle conditions. This meant substituting wheels for harder-wearing tracks – already in use on some agricultural vehicles – that could cope with uneven, broken-up terrain. The first tank – nicknamed Little Willie* – was created in July 1915 in Britain as a result of work by the Landships Committee under Winston Churchill, First Lord of the Admiralty, to break the deadlock of trench warfare. At their best, these early tanks were able to cross trench systems, punching through enemy lines and helping infantry to advance.[10]

America, which would not join the war until April 1917, had nevertheless already exported tens of thousands of horses and automobiles to the Allies. The *New York Times* announced with some relish in April 1916 that an 'enormous export business' had built up thanks to the war. In total, some $360 million worth of 'horses, mules and motor cars' had been shipped to Europe, including around 3,000 motors to France and 13,000 to Britain between July 1915 and January 1916. John Burroughs, out on a Vagabond excursion in rural Virginia with Henry Ford, recalled seeing army trucks 'rolling eastwards toward the seat of war, some loaded with soldiers, some with camp equipments . . . On other highways the weapons and materials of war were converging toward the great seaports in the same way. The silent, grim, processions – how impressive they were!'[11]

The exigencies of war – hotly followed by the influenza pandemic, which infected a third of the global population and killed somewhere between 25 and 50 million – consolidated several automotive trends. By 1907, American motor-car production was beginning to challenge that of France, but by 1918 it would dwarf Europe in its entirety, with little hope of the battered old world clawing back lost ground. Six million motor cars were registered in America in

---

* Little Willie was essentially made by mounting an armoured car body onto a tractor chassis. An updated version, inevitably named Big Willie, soon followed. (The name was said to refer to Kaiser Wilhelm.) The French were simultaneously working on creating tanks and would independently produce a similar design, known as the Schneider, in early 1916, which was swiftly followed by the Renault F. T., a lighter and more successful design than the bulkier and very slow British ones.

1918 alone, one million more than the previous year. This would also affect who cars were for and how they were made. European brands like Renault still created a variety of models at different price points with a high profit margin; in America, Ford's mass production of one car at the lowest possible price was creating an automotive juggernaut.[12]

The combined assault of the war and a devastating pandemic meant more women became involved in the workforce. Women were needed as everything from farmers to munitions workers, and from factory workers to nurses. For many countries, this represented a significant cultural shift. Even if it was presented at the time as a temporary extension of their 'natural' sphere in supporting male relatives, by the time war ended traditional gender roles had loosened. Women expected more freedom – something immediately apparent in the fashions for shorter skirts and hair that appeared. Many combatant nations extended the right to vote to women either during the war or shortly afterwards.

In this context too, automobiles, which unlike teams of horses could be easily managed by one person and did not require a great deal of physical strength to operate, met the moment. Western societies could no longer afford to be po-faced about women drivers. Many prominent racers immediately volunteered their skills as ambulance drivers, mechanics and chauffeurs for military and medical personnel. These included Muriel Thompson, from Aberdeen, who had chauffeured Emmeline Pankhurst; Violette Morris, an accomplished sportswoman who would later become a Nazi spy; Gwenda Hawkes, who drove ambulances on both the Russian and Rumanian fronts for the Allies and was awarded a Cross of St George; and Christabel Ellis, who became a commandant in the transport section of the Women's Legion.

Other female drivers also put their skills to good use. In Paris, Marie Curie abandoned her lab at the outbreak of the war, fashioned a mobile X-ray unit for use at the front using a Red Cross car and funds donated by the Union of Women of France, and set to work training 150 other women to operate similar vehicles.[13]

Another reason vehicles came into their own in the wake of the war was that the equine populations of combatant nations were seriously depleted. Britain, for example, lost 484,000: one for every two men killed. And while there is no reliable estimate for the total number of horses killed in combat, we do know that 7,000 died during a single day in 1916 at the Battle of Verdun. The automobile was ready to take up the strain.

War had also put a thumb on the scale when it came to the fuel question. While electricity, alcohol and steam were all of interest – especially in oil-strapped European countries – before the First World War, by 1918 it would be increasingly Sisyphean to persuade anyone these were worth investing in. Oil-derived fuels had proved both too convenient and too versatile when the principal concern was moving supplies and troops as quickly as possible. A British trade body admitted in 1921 that a 'feature of the recent war was an extraordinary development of the uses of petroleum in munitions factories, motor transport services, aviation, tanks, and warships of all classes'. More was being produced too: global output doubled between 1907 and 1919, largely thanks to America, which had supplied the Allied armies. 'Truly,' Lord Curzon, a member of the British War Cabinet, remarked in November 1918, 'posterity will say that the Allies floated to victory on a wave of oil.'[14]

Back in 1907, participants and observers of the Peking–Paris believed the automobile would prove an agent of peace. Even as the race reinforced the idea of the car's military utility and the coverage stoked nationalist and imperial sentiments, the knitting together of East and West was a continual theme. A Chinese diplomat in Berlin gave a speech – 'in perfect French,' one observer enthused – assuring his audience that 'this will certainly result in closer relations between foreign countries and our own. China loves progress and welcomes anything that can bring it closer to the nations of Europe.' Mr Hoelping, an official delegate of the Imperial Automobile Club, proudly told participants, many of whom would later fight against his country, 'We Germans have been able to judge how many international prejudices have disappeared thanks to motoring . . . Let us toast the union of France and Germany on motoring, with the hope that this union will bear fruit, not only in automotive matters, but in other areas also.'[15]

22

Liège – Paris

8–10 August

> This is the winning post. Our long race is about to end.
>
> Luigi Barzini, Itala team

On the evening of 9 August in Meaux, a small town thirty miles east of Paris usually singularly concerned with the production of brie, all was pandemonium. 'There is endless cheering,' a foreign correspondent staying there wrote. 'The building is overrun: I am telegraphing in the middle of a stampede.'

The public rooms of La Belle Sirène hotel, where the Italians were lodging, buzzed as loudly as the telegraph office. They brimmed with journalists covering the race, some of whom had driven with the Italians from Berlin. Motor cars and bicycles crowded the hotel forecourt, its garden and the street outside. Although everyone else seemed to be revelling in the melee, Barzini was subdued. He had reached Meaux – in his customary, cramped seat amongst the luggage at the back of the Itala – at five thirty in the afternoon on Friday, 9 August. Capitalising on the short day's driving, he wrote and dispatched a long report to the *Corriere della Sera* and *Daily Telegraph*.[1]

This was the penultimate day of their odyssey: on Saturday they would reach Paris and glory. But if the atmosphere should have felt jubilant, for Barzini it was more unsettling, as it often is for travellers at the conclusion of a long journey. Together with the prince, he had tried to calculate the miles covered but could not – 8,000?

10,000? What with the detours and doubling back, they could not say precisely and would never go back to measure it.* It felt as if Paris were a powerful magnet that had been drawing them towards it over two continents and as many months: only now that they were so close had Barzini discovered his own belly was filled with jittering iron filings. He went to bed but sleep eluded him. Several times during the night he got up to look out of the window at the glowing sky over the City of Lights. Checking that it was still there, 'as though I wished to persuade myself and conquer an unreasonable sense of doubt'.[2]

As he lay down again, his imagination pulled him back in time, backwards along 'the road already travelled, which the rapidity of thought brings near us again in violent foreshortening'. Uneven slabs of stone beneath Peking's Deshengmen Gate. Rock-strewn passes, sweat coursing down between heaving shoulder blades. Golden dragons embroidered on the silks of officials' robes at Kalgan. The headlong unison of the wild equine herds on the outskirts of the Gobi Desert. Stuttering thumps as the Itala passed along the railway line at Lake Baikal. Days of mud; days of birch trees; days of boiled eggs and tea; days of grasses bending away in successive silver ripples as they passed.[3]

When they crossed from Germany into Belgium and on towards France, public attention had focused on the Itala like the scorching beam from a small boy's magnifying glass used to harry an ant. Even the bombardment of Casablanca and a ghoulish murder case from Monaco† could not budge the Peking–Paris from the coveted positions above the fold. Less formal networks were

---

* Someone retracing the route today using the modern roads would clock just under 8,000 miles. The original drivers did far more, because of the vagaries of the tracks followed, the difficulties of the terrain, reversals around Lake Baikal, getting lost and so on. As none of them had milometers, and for large portions of the journey the maps they had were not reliable, their own estimates vary wildly.

† Dubbed 'The Trunk Murder' by the press, the case involved the killing of Emma Levin, a wealthy Swedish widow, by Vere and Marie Goold. Vere, an Irish sportsman and former Wimbledon tennis player, and his wife, the proprietor of a dressmaker's shop, stabbed and dismembered Mrs Levin, stuffed her body in a trunk and checked it in at the cloakroom of Marseilles railway station. Porters became suspicious when blood began leaking out of the trunk.

attuned to them too. Pied Piper-like, Prince Borghese seemed to draw forth the population in every town and village he passed through. Crowds lined the roads they drove down and burst like confetti over town squares and village greens.

They had left Berlin two days before, on Wednesday, 7 August – the same day the use of war balloons was being discussed with great acrimony at the Hague Peace Conference – and had passed through red-brick Brandenburg before continuing to Magdeburg, with its sky-piercing cathedral perched on a rocky outcrop over-looking the Elbe.* It was market day, and the city was filled with brightly dressed, voluble men, and women in white headdresses. The weather was heavy and mercurial. A storm broke as they passed through Brunswick; by the time they reached Hanover, forty miles west, the sun scorched a yellow hole in the sky. In the spa town of Herford they found themselves surrounded by inva-lids in wicker bath chairs,† there to take the waters. One, bracing himself with his arms, half raised himself and cheered '*Evviva!*'[4]

The next day the advance guard of the Peking–Paris, swollen to half a dozen since Berlin by additional Italas bearing mem-bers of the press and representatives of the Imperial Automobile Club, set off from Bielefeld, where the three Italians had spent the night. Above the persistent hum and growl of their engines could be heard the snapping of various flags and banners. The red, white and black of the German Empire; the tricolours of France and Italy, the latter bearing the arms of Savoy: a white cross on a blood-red shield edged with azure. Here and there signs reading 'Peking–Paris' (and, occasionally, 'Peking–Matin', the new slogan the newspaper was trying to deploy). Every now and again one of the convoy would pull out, putting on a spurt of speed to drive alongside Borghese, the challenging machine's occupants waving

---

* These rocks, officially the Domfelsen, jut out into the river, where they can be seen when the water is low. Since this heralded unusually dry conditions associated with drought and a failing harvest, they were known locally as Hungerfelsen – literally 'Hunger Rocks'.

† Bath chairs – an early form of wheelchair – were invented in Bath in the mid-eighteenth century. They were used by the elderly and infirm to travel around the town. They were pushed from behind and the occupant could steer themselves using a rod that connected to a wheel at the front.

their arms and whooping like children at a fair. At eleven o'clock they crested a hill at Schlebusch and a luminous riverine valley opened up beneath their wheels, a city set in its centre like a jewel. Racing down, they passed into Cologne's dense thicket of Gothic spires and belfries and crossed a bridge spanning the Rhine made from boats tied together in a line.

With every flurry of copy sent by the journalists to their parent papers, victors' laurels settled more securely on the brows of the motorists. Borghese became a modern Hercules in a voluminous driving coat and a floppy waterproof hat commanding a latter-day chariot. '[Slender] in appearance despite his great, slightly stooped height . . . one quickly notices that all [the prince's] muscles, stripped of superfluous stoutness, are as hard as steel; and the energetic and calm expression on his face explains how the effort and success were achieved.'

Guizzardi was ennobled as the faithful retainer and mechanic; Barzini, the gifted bard. One correspondent noted 'his witty face, sharp as a blade'. Another, surrendering himself utterly to hyperbole, commended Barzini's 'Adamantine fibre, under the mantle of an almost diaphanous grace, an imperturbable serenity even during the harshest incidents of his nomadic life, ready with wit and a zinger even beneath the thunder of the cannon'.[5]

Just after – and probably as the result of – a long lunch at Cologne punctuated by effusive toasts, a motor sent by the Imperial Automobile Club to pilot the Italians through Germany slammed into the side of a house. The convoy stopped long enough to ensure there had been no fatalities and then set off once more. In another village, a schoolmaster assembled his pupils to watch the race as an incentive to study geography. Later, an old woman mistook them for some motorists who had run over her hen a few days previously, yelling and shaking her fist at them from her window.

On they drove. Flashing through landscapes of soft hills and wide valleys. The vegetation rushing by began to remind them of home. Hot herbaceous smells; the nostril-singeing stench of hot oil and exhaust fumes. Road-baked dust billowed out from every wheel. On past Aix-la-Chapelle (Aachen), to the Belgian border. Here a border guard, neatly puncturing the glossy word picture drawn by the pack of journalists, refused to believe the begrimed personage behind the wheel of the leading Itala was a prince. 'You – a prince! Nonsense. You are a Belgian chauffeur and are breaking the law by going too fast. You know the regulations – ten kilometres an hour!'

They spent Thursday night at Liège, a city filled with relics of its wealthy medieval past. It is threaded like a bead onto the River Meuse: the ancient border between the Holy Roman Empire and France. They used the river as they had done the telegraph wires in Mongolia, following its twists for over a hundred miles bearing south-south-west. Huy, Namur, Heer-Agimont flashed by. By mid-morning they were at the Belgium–France border near Givet.

Here, at least, the customs officials were primed and the formalities kept to a minimum. Some glasses of champagne were solemnly poured; a short series of toasts given. Borghese, cheered by their pace and the proximity to their destination, had returned to his habitually urbane self. He smiled and thanked everyone profusely, before backing away to the Itala. A sharp, upward turn of the crank, and then the prince swung himself once more back into the driver's seat – here, at the well-publicised end of the journey, Guizzardi largely remained a passenger. At long last, they were driving on French soil.[6]

Meanwhile, the French motorists were over 2,000 miles away from their homeland in the bosom of the Russian Empire, still grappling with a race in which they appeared to have become a footnote before it was even over. The surprise reappearance of Charles Godard and his brightly striped Spyker threatened to obscure the efforts of Cormier, Collignon and their De Dion-Boutons. The organisers – no doubt under pressure from the Marquis de Dion – were left in something of a quandary.

Le Matin dealt with the difficulty by downplaying or even expunging mentions of the Spyker team from reports and valiantly talking up the achievements of the French-made cars wherever possible. The Peking–Paris committee's head, Georges Bourcier Saint-Chaffray, wrote: 'Alongside the Itala, the two little

Dion-Boutons are advancing side by side, without ever having to be parted, without any need for forced stops or breakdowns.'

On the ground, the men were still chewing over Godard and Stephan's account of speeding across 2,700 muddy miles in a fortnight. On Thursday, 8 August, hours after the Spyker crew had unexpectedly arrived in Kazan, du Taillis had dispatched a terse telegraph to *Le Matin*: 'Godard told me about his route; he sent a tyre to *Le Matin* as undeniable proof of the distance covered ... The Dion team refuses to believe that the 15 hp Spyker actually covered the distance from Cheremkhovo to Kazan. But it is proven that the Spyker did leave from Cheremkhovo. The rest of the trip will be checked.' A few hours of wrangling later, they agreed to drive on in convoy, but the atmosphere remained ugly, any lingering sense of camaraderie extinguished.[7]

Leaving Kazan, they took a wrong turn, irritably heading back to the city after losing four hours and a break in the weather into the bargain. On 10 August they achieved a grand total of twenty-two miles on roads made mud-slick by unceasing rain. Cormier, filled with impotent fury at Godard, his own place so far behind the prince and the weather, was short with his teammates and anyone else within his orbit. A correspondent in Moscow noted how peremptorily Cormier treated him, getting his own back in print by snidely noting that while Cormier was 'luxuriating in a bathtub', Collignon was faithfully attending to the vehicles. Du Taillis, who had never warmed to the driver, reported with satisfaction that he 'cursed the rain with dreadful oaths'.[8]

Cormier's mood would not have been improved had he known the truth about the Spyker's movements over the previous fortnight. Unfortunately for him, the mystery would persist for over a century. Godard, although never shy about talking to anyone about his achievements, particularly the press, offered no additional gloss and resolutely stuck to his story, with all its inconsistencies, until his death in 1919. The young Dutch mechanic Stephan, who lived into the 1960s and was the only man who could corroborate Godard's version of events, also chose to remain silent. At least publicly.

In the autumn of 1963, fifty-six years after the race had ended, Stephan received a letter from a Dutch Spyker enthusiast, Mr C. Poel. Poel had in turn been commissioned by Allen Andrews, a British author, to fact-check his forthcoming book about the Peking–Paris race. 'Unknowingly,' the eighty-eight-year-old Stephan eventually replied to Poel, 'you have . . . compelled me to face a dilemma in my old age.'*

In his book, Andrews made no bones about his partisanship. Godard, he states in the preface, was 'the most brilliant automobilist of them all in terms of sheer driving nerve and endurance . . . a historian should have no favourites, but all the world loves a rascal.' Many of those who know about the race today agree with Andrews's assessment. Indeed, it was in this account that many of the most enduring and eccentric anecdotes about the Peking–Paris were introduced.

The denouement of the book is Godard's heroic run from Cheremkhovo to Kazan. This passage includes record-breaking twenty-four-hour dashes around Omsk, bloodshot eyes, bizarre incidents involving a runaway cart, a cast-off baby and a mysterious, alcoholic countess. Elsewhere, Andrews had Godard fixing a hole in the Spyker's rear axle with 'a plug of raw bacon, forced in with a wooden spigot and held by an iron bracket'. Where such details came from is anyone's guess. Andrews provides no references for them.[9]

The typewritten letter Stephan sent to Poel in 1963 is a study of a man wrestling with his conscience. 'Sixty-five years [*sic*] I have remained as silent as the grave on this. A great deal of thought has gone into my decision whether or not it would be better to continue in my silence.' He hesitatingly agreed to tell Poel the truth, but appealed to his correspondent's patriotism and well-known love of the Spyker brand and begged him to keep the truth secret. 'I am of the opinion that it would serve no purpose to publicly reveal

---

* The only mention of this extraordinary letter that I was able to find was an article from 1998 in the Dutch newspaper *de Volkskrant*. It does not seem to have become more generally known. All the Peking–Paris enthusiasts – including the Dutch ones – that I spoke to during my research were unaware of it and of the crucial evidence it provides regarding one of the most sensational episodes of the Peking–Paris story.

the historical facts at this time. The story as currently written by Andrews contributes to the well-deserved renown of the Spyker car. Why should that reputation be damaged in this day and age?'[10]

Even though he did not want the truth to become public, Stephan, nearly eighty years old, was ready to come clean. 'One would need,' Stephan wrote, 'to have a very generous definition of "poetic license" to justify describing the inaccurate accounts in Mr Andrews's Spyker chapters as historical facts . . . The entire epic about us racing against the clock from Irkutsk to Kazan is completely made up by Godard. Not a single word of it is true.' In fact, Stephan told Poel, while they had 'encountered many difficulties along the way . . . the kind of journeys described by Mr Andrews never took place.'

In fact, he confessed, the vast distances covered in that fortnight had been achieved – under the cover of night – by train and by boat. Godard, with an eye towards allaying suspicions, had sent postcards and written telegrams to *Le Matin* at their various stops. He also, according to Stephan, bullied the twenty-year-old Dutchman into silence. Godard 'claimed that he was the one calling the shots and that I was to act according to his will. I was just a mechanic who, except for the mechanical part, played no role whatsoever in the journey. As a Spyker employee, I had to consider the fact that it would be in the best interest of the firm to catch up with the others.'

Stephan admitted to lying to the other competitors when they reached Kazan to back Godard up. The only other person he ever told was his boss Jacobus Spijker when they met near the end of the race. 'Of course, at that moment, he agreed that we should hush up the true story, which is exactly what happened. I kept the secret with me and did not even share the truth with my family.'[11]

Poel evidently agreed to Stephan's condition of continuing secrecy. *The Mad Motorists*, Andrews's book, was published the following year, reviving interest in the Peking–Paris story, greatly burnishing the reputation of Godard and his Spyker and tarnishing that of Prince Borghese.

Even before Andrews's book, Godard's charm and the organisers' desire to declare the event a success meant compelling evidence that Godard was little better than a confidence trickster was ignored. His lies to secure his vehicle, spare parts and passage to China were known to participants and organisers, and discussed in some newspapers, but were largely expunged from the books that were published in the following years. The most famous account – that by Barzini – barely mentioned Godard at all.

In Meaux, the morning of Saturday, 10 August, was spent in fluttering, frustrated anticipation of a two o'clock departure. Although they had only thirty miles to travel and were due at the offices of *Le Matin* on the Boulevard Poissonnière between four and five o'clock, progress through the throng was expected – even required – to be stately. Tens of thousands of ticket holders for festivities at the Tuileries Gardens in Paris would anticipate a decent show, and, since these had now sold out,* timetables had been printed in various Parisian newspapers so that even those without tickets could witness the spectacle from the Place de la Concorde. Aware of the prince's distressingly free-spirited approach to his own schedule, representatives from the Peking–Paris committee and *Le Matin* tightened the leash, repeatedly impressing upon him the importance of doing things their way. There would be no repeat of the Berlin fiasco.

Borghese, however, seemed tractable enough. He allowed himself to be interviewed and photographed by dozens of newspapermen. Tactfully soothing ruffled Gallic pride, he told several French journalists that there was almost as much French blood in his veins as there was Italian, various relatives having married into La Rochefoucauld and Bonaparte families. (He was perhaps a little more truthful when he confided to the *New York Times* correspondent that he was 'naturally pleased as an Italian to win the race on an Italian machine'.) He joked that Barzini, who after two months of struggling through long days of driving followed by long nights of writing and sending telegrams, was delighted that

---

* An initial run of 14,000 tickets had been distributed and, when these were almost immediately snapped up, a further 16,000 printed.

he could finally get some sleep. Barzini, who had barely closed his eyes the night before, smiled wanly but did not contradict him.[12]

As the morning progressed a motley battalion of automobiles – large, small, steam-, electric- and petrol-powered – streamed into Meaux past crumbling keeps and towers, the remnants of Roman-era defensive walls. Some bore the names of newspapers. Others contained members of various automotive trade bodies and sponsors. 'Horns sounded, sirens shrieked, and motors and men were mingled in inextricable confusion in the courtyard of the hotel.'

The Itala, being given its final toilette by Guizzardi in a shed near the hotel, was proving a draw. The prince gave a good-natured lecture on its finer points to an audience consisting of his brother, Prince Rodolfo Borghese, the Count and Countess of Montbel, and 'the hunters, bakers boys, and ploughmen of Brie'. Guizzardi, standing proudly by, interjected at intervals in a strange mixture of Italian, French and a smattering of Chinese phrases. A formal lunch was laid on, hosted by Henri Fournier. This legendary French racing driver had set a land-speed record of seventy-six miles per hour in 1902. (A performance somewhat less startling than the fact that his average speed on the first leg of the Paris–Vienna race, held earlier that year on public roads, had been seventy-one miles per hour.)[13]

By quarter past two the Itala, its rump bulbous with roped-on luggage, was leading a procession of high-spirited motorists through the throngs of Meaux. From his seat by the fuel tanks, Barzini could see the backs of Borghese's and Guizzardi's heads, swivelling left and right as they turned to wave and smile. Borghese had shaved with a fresh razor, but otherwise looked much the same as he had for the previous two months: his pate concealed by a broad-brimmed waterproof hat, his body enrobed in a full-length driving coat. Cheers crescendoed around them. Handkerchiefs and hats were raised aloft. '*Bravo, mon gars!*' someone yelled. '*Evviva!*' They moved forward by inches. Cyclists plunged around the Itala like a school of porpoises at the bow of a steamer. Hands reached in from all sides to grasp theirs. Women in gowns and large hats like decorative *petit fours* proffered flowers with gloved fingers.

Twenty minutes later they had travelled seven miles and were past Couilly. The sky was clouding over but they continued undeterred, reaching Chessy at two forty-five and Lagny at three. Borghese began smiling in earnest, his reserve pierced at last by the sheer weight of this welcome, the enthusiasm of the throng and the satisfaction of reaching this ultimate goal. '*Bravo!*' '*Vive le Prince!*' Later, pink-faced from a heady combination of champagne and the fulsome praise meted out to him in a cascade of speeches, he would modestly turn aside from the acclaim. At that moment, however, he was in the full flush of his success, and it showed. He raised his right hand from the wheel, waving and smiling all around.

At Bry, a well-fed man standing by the side of the road with an enormous bouquet either missed his aim or mistook his man, because his flowers landed in the motor behind the Itala, right in the lap of Henri des Houx, *Le Matin*'s dignified Berlin correspondent. Near Joinville, the tree-lined roads were edged with spectators as if the automobiles were prize-winning cyclists during a stage of the Tour de France. '*Vive le Prince!*' '*Vive le Prince!*' '*Bravo!*' '*Vive le Prince!*' Through Bois de Vincennes, where omnibuses and trams filled with people came to a standstill around them, their occupants clapping their hands and waving through the windows. The wind picked up and rain began to patter around them at Saint-Mandé, and then started to pour down, accompanied by crashes of thunder. No one seemed to care. A woman with a baby in her arms rushed forward, clapping her child's hands together in silent applause. A brass band in a large motor bus materialised at the front of the three bemused motorists and began playing the triumphal march from Verdi's *Aida*. Rain splashed up from the gleaming surface of their instruments and ran in rivulets down their straining, red cheeks.

By the time they reached the centre of Paris, the cars were proceeding at walking pace, brushing the legs of the spectators, nosing through the crowd. The wide streets were black with people, the air above their heads 'a whirl of hands, hats, and handkerchiefs'. Members of the Republican Guard, on horseback, helplessly tried to direct the flow. Near the Place de la République, the

Itala became overwhelmed entirely: 'literally covered in a moving mass of men and children, like a hunted animal covered by dogs'.

Pulling up at last outside the editorial offices of *Le Matin* on the corner of Boulevard Poissonnière and Rue du Faubourg Poissonnière, they saw the facade bedecked with flags, an exact echo of the barracks of the French Legation in Peking two months before to the day. A bevy of photographers appeared as if by magic, and to the full-throated roar was added the hum and whir of cameras. Cinematographers bent behind their apparatuses and yelled out to Borghese and his companions to look down their lenses, disembodied hands slowly revolving the handles. Borghese guided the Itala up over the pavement, took the car out of gear, engaged the brake and turned off the engine. For a moment, all three men were still and utterly silent while the crowd around them surged and buffeted. Barzini turned towards Borghese, who seemed suddenly stunned by the enormity of what they have accomplished,

In Paris, vast crowds braved torrential rain to welcome the winners of the Peking–Paris.

both his hands still resting on the wheel. It is done. They have driven from Peking to Paris.[14]

The earlier storm, short though it had been, had played merry havoc with the preparations in the Tuileries Gardens. Paper lanterns had been ripped from the trees. Boxes of sparklers soaked through so that when lit they could only sputter sadly. Worst of all, the enormous one-hundred-metre square canvas screen had pulled free of its supporting poles and blown away like a sail, only to land in a nearby pond. This had been particularly galling since it was involved in the apex of the evening's festivities. Thirty-five thousand guests would soon be descending on the park, all of whom had been faithfully promised that at nine o'clock precisely images from the Peking–Paris race would be projected onto the canvas, accompanied by commentary from the well-known author and journalist Hugues Le Roux, broadcast through an amplifying phonograph.

The staff of *Le Matin* – from junior journalists to clerks and newspaper boys – as well as their wives, sisters and mothers, were deployed to set things right. The canvas had to be retrieved, dried and remounted, new fireworks found, bottles of champagne unpacked, the phonograph replaced and floodlights set back in their places – all before the guests began arriving at eight. They swarmed over the grass and pathways. Representatives from the organising committee reminded anyone who would listen that if men could travel all the way from Peking, then surely Parisians could ensure an appropriate welcome for them, no matter what the weather.

A mile to the north-east, in the midst of a select banquet, Borghese got to his feet and looked around at a table of expectant faces comprising the great and the good of European society. It had been a tumultuous afternoon. The culmination of two months of days that began before dawn and usually ended after dark. Days of hands numb from the vibrations of the steering wheel. Days of mud and dust and rain and the roar and stink of engines and motor oil. All of which, of course, only made arriving in Paris all the sweeter. Half a million people had choked the streets of the French capital, in defiance of the summer squall, to welcome him and his companions. And it felt as if every one of them had personally congratulated him, wrung his hand, slapped his back.

There had been a succession of speeches, toast and gifts. A cup cast in bronze featuring a steering wheel surmounted by a lion on a pedestal of Carrara marble was much admired. For his efforts, Barzini received an inkwell shaped like a globe wrapped in telegraph wires, a pen crossing from one pole to the other. 'We launched the Peking–Paris challenge,' Henry de Jouvenel, editor-in-chief of *Le Matin*, intoned, 'not despite it being utopian, but because it was utopian.' Another speech was given from abroad, using a telephone receiver held up to a gramophone, so that the speaker's voice boomed around the room. The *Corriere della Sera* would later proclaim this marvel to have been 'truly *up to date*'.

Finally, it was the turn of the man of the hour. Borghese, raising his coupe, thanked the people of Paris for their welcome and the previous speakers for their compliments. 'But,' he said, 'you have exaggerated, gentlemen. We are not heroes; we are simply patient men.' He paused and smiled as the assembled guests pooh-poohed his modesty. 'Yes,' he continued, 'it is true, our only virtue was patience. Well, perhaps we had one other: perseverance.'[15]

23

AFTER THE RACE

Contal

After walking out of the Gobi Desert in late June 1907, Auguste Pons returned to Paris by train, enraged by his treatment at the hands of his fellow motorists but not in the slightest bit chastened. He continued to believe, fervently, that small, light vehicles were the future. Determined to prove the point to the world, he immediately applied for the New York–Paris race.

Sponsored once again by *Le Matin*, in conjunction with the *New York Times*, this sought to capitalise on the global success of the Peking–Paris and was due to set off from New York less than six months later, on 28 January 1908, but was later delayed until 12 February. The route would take the drivers westwards across America from New York to San Francisco, where the automobiles would be loaded onto a ship heading north to Seattle. From there, they would drive to Alaska, cross the Bering Straits, drive across Japan and then into Russia at Vladivostok. The route would then rejoin that of the Peking–Paris at Irkutsk.

For this race, Pons forswore the Contal Mototri in favour of a French-made, single-cylinder Sizaire-Naudin fitted with Michelin tyres. 'My machine,' a proud Sizaire-Naudin agent told the *New York Times*, 'will be the only light automobile in the race. It will have only twelve horsepower.' Pons, however, was perhaps not quite the competitor he believed himself to be. On the first day, the French team took two wrong turns – one of which led them

The 1908 New York–Paris race was attended by lots of publicity and huge crowds, like this one in Berlin.

straight back to New York – and then, having decided to make up some lost time in the dark, suffered a broken differential four miles from Poughkeepsie.* Pons had to return the seventy miles to the city by train to pick up a replacement. Two days later, at Red Hook, more mechanical difficulties forced them to stop again.

By 18 February, six days into the race, Pons would once again be forced to withdraw. He announced to the *New York Times* that an irreplaceable part of the automobile had been cast with a flaw in the steel – an admission that probably did not endear him to his sponsors – and was only talked out of sending to Paris for a replacement when it was pointed out that the delay would mean the straits between Alaska and Russia would have thawed and would be impassable. He offered to start afresh in an entirely new

---

* The differential is an essential part of the system that applies torque or force to the drive wheels. It is called a differential because it allows the wheels to turn at different speeds, something which is necessary to allow a vehicle to turn.

American car, and finally, in desperation, went back to take another look at the Sizaire-Naudin, vowing that he would drive twenty hours a day to make up the lost time if he had to. 'Mr Pons,' the *New York Times* concluded, 'is a most unfortunate racer.'

More unfortunate still were the family he had left behind and to whom he decided he would not return. His young daughter, Alice Joséphine Pons, who later styled herself Lily Pons, would go on to become a famous American opera singer, with a career spanning from the 1920s until the 1970s. She performed at the Metropolitan Opera nearly 300 times and appeared in several hit films, including *That Girl from Paris* (1936) and *I Dream too Much* (1935), in which she starred opposite Henry Fonda. She was exceptionally savvy at marketing herself and frequently gave interviews; she rarely mentioned her father.[1]

Itala

The hero's welcome extended to Borghese and Barzini in Paris knew no bounds. Restaurants began serving Bombe Borghese with Barzini wafers. They were lauded as latter-day Marco Polos, the Guglielmo Marconis of exploration, Phileas Fogg and Passepartout made flesh. For Italians, it was a moment of national triumph, made all the sweeter because it involved beating the French. 'The race announced by *Le Matin*,' one Italian journalist had gloated right before Borghese entered the French capital, 'will be won by an Italian car and an Italian driver.'

Rather than basking in the glory in Paris – and despite the pleas of the organisers – the three Italians set off almost immediately for home. They spent a night in the Alpine village of Courmayeur on Thursday, 15 August. Over the following days, tens of thousands of spectators – including workers released from factories closed early in the Itala's honour – roared their approval in Milan, Turin and Rome.

Although towards the end of the Peking–Paris Borghese expressed a desire to travel across America by motor car, and was at one time rumoured to be considering taking part in the 1908 New York–Paris race, he did not perform a motoring encore. Instead, he returned to the world he had inhabited before, trying to

281

rebuild his family's fortunes and splitting his time between Rome and his wife's home on the Isola del Garda. He served as a member of the Italian parliament, with a particular interest in agricultural reform. He also fought in Libya in 1911 and later in the First World War, first as a captain and then as a lieutenant colonel.

His wife, Anna Maria de Ferrari, drowned in Lake Garda in November 1924, leaving the island to her husband and their daughters. It remains the family home. Two years later he married Teodora Martini, a woman roughly thirty years his junior; his family disapproved of the match. He died in Florence, on 15 March 1927, aged fifty-six.[2]

Luigi Barzini would also return to the life he had known before the race, albeit with his professional prestige greatly enhanced. His book about the race, *La Metà del Mondo Vista da un Automobile* (sold under the title *Peking to Paris* in English), was an instant bestseller and was translated into eleven languages.

He was sent to Libya by the *Corriere della Sera* as a war correspondent and later spent time on the French and Italian fronts during the First World War. He founded *Il Corriere d'America*, an Italian-language newspaper in New York, but returned to Italy during the 1930s and 1940s and worked at various publications during the Fascist period. After the Liberation of Italy in 1945, Barzini would be condemned as a collaborator, forbidden to continue working as a journalist and stripped of his pension. He died in poverty in Milan in 1947.

He had four children. One son, Ettore (born in 1911 and so perhaps named in honour of his former Peking–Paris companion Ettore Guizzardi), was arrested by the Fascist regime for his Communist beliefs. He was passed to the Nazis and died at Mauthausen concentration camp in 1945. Another, Luigi, became a journalist like his father. His 1964 portrait of his countrymen, *The Italians*, is considered a classic.

Ettore Guizzardi outlived his companions by decades, finally passing away at the age of eighty-two in 1963. He had remained in the prince's employ after the race, marrying at a simple service held on the Isola del Garda, photographs of which were lovingly added to the family photograph album. He volunteered for both the First and the

Second World Wars as a chauffeur and later became an officer in the motor corps. He continued to drive right up until his death, apparently favouring a decrepit, thoroughly Italian, Fiat 500 'Topolino'.

De Dion-Bouton

If there had ever been any doubt about the partisanship of *Le Matin* for the two automobiles supplied by the Marquis de Dion, it was laid to rest after the Itala's victory. The newspaper doggedly reported the automobiles' progress through August, doing everything in their power to conjure a sense of French victory. A week after Borghese's arrival at their offices, they published a piece entitled 'On the Road to Victory' by the indefatigable Bourcier Saint-Chaffray. 'Is it not most admirable that these two cars' – the existence of the Spyker was largely ignored – 'always arrive at each stage together, that . . . for sixty-six days, neither of them have had the slightest failure? Isn't it more difficult to prove the reliability and solidity of two cars than the reliability and solidity of just one?'

By the time they reached Warsaw and Berlin, on 20 and 22 of August respectively, mention of any other vehicles or indeed the De Dion crews had disappeared almost entirely and the automobiles themselves were being lauded by ever more purple prose. The cars were returning with 'an air of indestructibility that the shiny steel and new brass of their youth could not have conveyed, like leather-skinned old soldiers'. They arrived in Paris on 30 August at five in the afternoon, nearly three weeks after Prince Borghese, to a rapturous celebration from the organisers. 'Belated but jubilant all the same,' the *Telegraph* reported. 'If not quite as triumphant as [Borghese's], their entry was a popular success. Crowds watched them parade up and down the Boulevards with drums and trumpets playing, and French, Dutch and Chinese flags flying.'[3]

The Peking–Paris seemed to quench whatever lust for adventure Georges Cormier possessed. He settled down to concentrate on the business of selling automobiles. He retained a close relationship with the Marquis de Dion, became the president of various trade bodies, including the *Chambre Syndicale du Commerce et de la Réparation de l'Automobile*. He returned to China in 1908 to set up an automobile concession and founded an automotive agency in Shanghai in 1911. He fought in the First World War – first with the 14th Artillery Regiment automotive service and later seconded to the De Dion factories – was awarded the *Légion d'honneur* and would later die, a much-respected industry expert, on 14 August 1955. A vocational school in Coulommiers – just to the west of Paris and specialising in automotive mechanics – is named after him.

Of Victor Collignon, Jean Bizac and Jean du Taillis – who joined the de Dion-Bouton team after the Spyker's magneto gave out – little trace remains. Collignon was involved with advising the Marquis de Dion on improvements to his automobiles before the New York–Paris race, but afterwards disappeared. He may well be one of the five Victor Collignons who died during the First World War. Jean du Taillis also fought in the war. His poor eyesight excluded him from battle duties and instead he dealt with logistics, concerned with the movement of heavy goods to and from the front. He was later transferred to the 14th Artillery Regiment – the same regiment as Cormier. He was ill-suited to warfare, however, and absconded without leave several times, before finally being released in October 1918. He returned to North Africa, where he continued to write, specialising in illustrated travel guides for motorists. He was fined 100 francs in 1921 for driving without a licence. He died in Algeria on 9 October 1932, aged fifty-nine.[4]

Spyker

Godard and his Spyker remained with the two De Dion-Boutons as they continued westwards, but it was clear to all that the driver was in disgrace. Not because he had cheated – this remained only a rumour – but because of the lies and debts he had incurred getting the Spyker to China in the first place. He was excluded from official receptions and banquets and finally, quietly, removed from the Spyker altogether at Berlin* and replaced, possibly by the

---

* Some reports suggest that he was actually arrested for unpaid debts either in Berlin or at the French border as he made his own way back after being removed from the Spyker. If this was the case, it happened quietly enough not to have made much of a splash and any legal difficulties soon evaporated.

anxious Mr Jacobus Spijker himself. 'Godard,' *Le Matin* reported firmly, 'was forced to leave the caravan for health reasons.' *Le Matin*, it seemed, preferred to draw a heavy veil over the whole affair.

If they hoped their discretion would be repaid by Godard's dignified retreat into obscurity, they were mistaken. A few short months later he announced that he would take part in the New York–Paris race, not on a Spyker, unsurprisingly, but a Motobloc, from a small French carmaker based in Bordeaux. Louis Vuitton once again provided the luggage.

Godard initially proved a hit during this chaotic, ill-thought-through race. He began sporadically adopting the title of 'Baron' and was in close contact with the *New York Times*, which enjoyed spoofing his French accent in print. ('Ah,' he apparently told a man in Buffalo over whom he accidentally spilled some water, 'I demand pardon for ze water: he make ze gentleman one grand stiff.') However, by mid-March he was falling behind, hindered by the thick snows and bad roads of rural America. He evidently decided to reuse the method that had worked so well for him the year before and loaded his stricken vehicle onto a train in secret in Carroll, Iowa, bound for San Francisco. On this occasion, however, he was spotted by locals, photographed and disqualified.[5]

Godard would never again emerge from obscurity. Records show that he travelled: addresses include London in 1909, Athens in 1911, Germany in 1912 and Russia – the Grand Hotel in St Petersburg, no less – in 1913. The following year, during general mobilisation in France, he was put into the Light Infantry and seconded to work in a car factory in 1917. He died on 15 August 1919, aged forty-two, at the Beaujon Hospital in Paris, from '*paralysie générale*': a description often associated with untreated syphilis. On admission he gave his profession as 'chauffeur'.

EPILOGUE

A wise traveller never despises his own country.

Carlo Goldoni, *Pamela*, 1749

In 2018 Anton Gonnissen, a fifty-eight-year-old Belgian architect from the Flemish city of Ghent, found himself worrying about an unfinished story in the history of motor racing. This was that of Pons, Foucault and the Contal Mototri: the only team of the original five that failed to complete the Peking–Paris. True, in Gonnissen's opinion the tiny Contal had been unsuited to the rigours of the journey and Pons naive to the point of recklessness to think otherwise, but it still bothered him. The two men had been forced to walk out of the Gobi Desert and the race – an ignominious end for such intrepid pioneers – and no one even knew where their vehicle had ended up. He figured that this story deserved a proper ending and believed he was the man to finish it.

It helped that Gonnissen already had a Peking–Paris rally under his belt: a revival event had sprung up in 1997 and had run sporadically ever since. In 2013 he and Inge, his wife and co-driver, had travelled the 8,700-mile course from Beijing in a handsome 1950 Bentley Mark VI Special. The experience had given him an appreciation for the terrain and the strain it put on antique automobiles, but it had also ignited a love of historic endurance rallies and of the brotherhood that formed on the road. He knew just how difficult it would be to repeat the experience in what was essentially an underpowered, three-wheeled motorcycle. There was also the fear that, as with Pons, he would fall behind, the goodwill of his fellow motorists would evaporate and history would repeat itself all over again.

Over the following months he tracked down an original Contal – one of only a handful that survive from that era – and a modern replica from Australia that he bought and had shipped to Ghent. Since neither were up to scratch, and his quest was rapidly developing into a full-blown obsession, he decided that the only sensible

In 2019, sat in a Contal, Anton Gonnissen and Herman Gelen vowed to finish what Pons and Foucault had started.

course of action would be to build a new one of his very own from scratch. He based the design for the chassis on the original model, with two wheels at the front and one at the back, and added a 700 cc engine capable of speeds up to fifty miles per hour. For extra verisimilitude, he bought the rights to the old company name, so that this new motor would truly be a Contal Mototri.

With the vehicle in hand, all Gonnissen needed was a co-driver. This was easier said than done: the design of the vehicle was such that whoever agreed would be perched in a small seat slung between the front wheels with nothing between themselves and the oblivion of oncoming traffic except some fresh air and Gonnissen's skill as a driver. Not for nothing were such seats referred to around 1900 as '*le tue belle-mère*': 'the mother-in-law killer'. Inge gamely agreed to get in for a short test drive down to their local town but, on returning, she told her husband flatly that there was no way she would be riding in it one mile more. In her stead, and with only six months to go before the trip began, Anton enlisted Herman Gelan, a colleague and running buddy from his architecture practice.

The pair set off from Beijing on Sunday, 6 May 2019, in the company of 119 other teams. As in 1907, their vehicle was the smallest and least powerful; among their competitors was a 1930 Ford Model A, a 1959 Morris Minor and a 1975 6750 cc Rolls-Royce Silver Shadow. Within days, they had passed through the Gobi landscape that had ensnared Pons and Foucault 112 years previously. Like their predecessors, Gonnissen and Gelan frequently had to get out and push their Contal – all 240 kilos of it – up and over hills, through sand and mud. Over the next month there were days when they despaired, when it rained and their hands grew numb and they shivered like stray dogs. Conscious of the faith Gelan was putting in him, Gonnissen drove always conservatively, careful to stay out of trouble and control his speed.

Together, they passed through Mongolia, into Russia, dipped into Kazakhstan and then back into Russia. They passed through Ufa, Nizhny Novgorod and Moscow, into Poland and Germany, and then, after thirty-four days, found themselves approaching the finish in the Place Vendôme. As they crossed the line, a spare, vaguely familiar figure stood in the crowd, his grey hair brushed back from patrician features. Later that evening he introduced himself as Prince Paolo Costantino Borghese, grand-nephew of Scipione.

'Congratulations,' he told them with a smile. 'You must have balls of steel.'[1]

ACKNOWLEDGEMENTS

This book happened – with apologies to Ernest Hemingway – first gradually and then suddenly. Years were spent in the initial research stages before a pitch was written and a deal agreed; more were spent in the writing, the majority of which was accomplished during the two years of coronavirus lockdowns; by the time we were out the other end, I was a new mother and time and deadlines began galloping away from me. All of which is to say that a huge debt of gratitude is owed to the many, many people who helped at every stage of this project's life, from inception, through the research and before, during and after publication.

My profound thanks go out to: Imogen Pelham, my agent, without whose steadfast enthusiasm, encouragement and wisdom this book would never have been written. Georgina Laycock, Lauren Howard and Kate Craigie, my editors at John Murray, both for their faith in my vision for what this book could and should be, and for their sound, practical advice. When chapters needed knocking into shape, they knocked. To the extended teams at John Murray and Hachette, Tim, Nathan, and everyone involved in taking this from manuscript to book and into the hands of readers.

Margaret Morrison, Emily Plank, Joris Canoy, Laura Shanahan, Dominique Hoffman and Tui McLean all provided invaluable translations of French, Italian, Russian and Dutch primary sources. Murièle Ochoa Gadaut trawled through French archives to dig out official documents that illuminated the lives of three of the participants. A lot of people generously shared their expertise, whether it was showing me round collections, sharing family lore, treasures and photographs and reading draft chapters. These include but are not limited to Alberta Cavazza, Giovanna Ruffini, Prince Paolo Costantino Borghese, John de Bry, Tomas de Vargas Machuca, Tony Jardine, Gerard Brown, Daniel Ward, Anton Gonnissen, David Ayre, Davide Lorenzone, Andrew Lewis, Rupert Grey, Luke Hayman and Brita Asbrink.

291

ACKNOWLEDGEMENTS

The staff at the London Library, the Special Collections at Queen's University, Belfast, the Museo dell' Automobile in Turin, Brooklands and the Victoria and Albert Museums and the archivists and custodians at Michelin, Pirelli, Spyker, Louis Vuitton and the RAC Club were all exceedingly generous with their time, experience and deep knowledge of their subjects. My thanks also to Lana, for her patience during the Russian lessons.

The Society of Authors gave me a generous grant allowing me to dedicate myself to this project full time. Paul French, Tim Cross, Antony Beevor, Rob Gifford, Tom Standage, Simon Akam, Rory Maclean and others provided encouragement and invaluable advice about tackling this large and ambitious story. The community of non-fiction writers at the London Library honestly shared their struggles with their own work and listened to me talk about mine. And finally to Olivier, who reads every draft first and gives great notes.

PICTURE CREDITS

Alamy Stock Photo: xiv/Photo, 44/Chronicle of World History, 52 and 200/Sueddeutsche Zeitung Photo, 64/The Print Collector, 122/Heritage Image Partnership Ltd, 140/Chronicle, 178/VTR, 189/photo-fox. BoJack/Shutterstock.com: 288. DEA/A.Dagli Orti/ De Agostini via GettyImages: 94. Hulton Archive/Getty Images: xxiv. ©Hulton-Deutsch Collection/CORBIS via Getty Images: 20, 157, 163, 258. Keystone/Hulton Archive/Getty Images: 235. National Library of France (BnF): 248, 276. Public Domain: 7. Sovfoto/Universal Images Group via Getty Images: 216. Touring Club Italiano/Marka/Universal Images Group via Getty Images: 76, 108, 131, 221. Ullstein bild via Getty Images: 280.

Maps and Illustrations: © Nathan Burton.

NOTES

PREFACE

[1] Comyns Beaumont, a British journalist and author, was prescient on this point, although it should be noted that he also argued that Babylon was built by giants and, moreover, was really the original name for York, and that ancient Egyptians had in fact come from Wessex and Wales.

[2] **bold and gigantic** 'The Paris–Peking Motor-Car Race', *Sydney Morning Herald*, 8 February 1907; **the most stupendous** 'From Pekin to Paris Finished Without Mishap', *Evening Bulletin* (Honolulu), 3 September 1907; **between 500,000 and 600,000** 'Peking–Matin Triumph: All Paris Acclaims Prince Borghese', *Le Matin*, 11 August 1907, trans. Margaret Morrison; **The winner was** Our Own Correspondent [Paris] and Our Own Correspondent [Berlin], 'Pekin to Paris: Great Preparations in the French Capital', *Daily Telegraph*, 7 August 1907.

[3] **Greatest event in the automotive** Lakhvari, 'Significance of the Peking–Paris Automobile Contest', *Avtomobil'*, 1 August 1907, pp. 1814–15, trans. Dominique Hoffman; **worldwide economic and social** Lakhvari, 'Significance'; **The *Sunday Times* dared** 'Pekin to Paris: Prince Borghese Arrives', *Sunday Times*, 11 August 1907.

[4] **poetry of motion** 'Peking to Paris Race', *South China Morning Post*, 27 July 1907; **universal attempts** '*Evviva*', *Sunday Times*, 11 August 1907.

[5] **Luckless enthusiasm** 'The Triumph of the Horse', *The Economist*, pp. 1495–96.

PROLOGUE

[1] **It is three minutes** Description of the start stitched together from sources including Victor Collignon, *Mes Souvenirs de Route: 16,000 Kilometres sur De Dion-Bouton (Pneus Dunlop)*, 1908, p. 8; B. L. Putnam Weale, Photographs Belonging to Sir Robert Hart, Sent to Him by B. Lenox Simpson (pen name: Putnam Weale), 1907, Sir Robert Hart Collection, Queen's University Belfast, MS 15.1.86.69a (ii–xi); Luigi Barzini, *Peking to Paris: Across Two Continents in an Itala*, trans. L. P. de Castelvecchio (Harmondsworth: Penguin, 1986), pp. 23–26; Jean du Taillis, *Pékin–Paris: Automobile en Quatre-Vingts Jours*, (Paris: Félix Juven, 1907), pp. 49–54, trans. Margaret Morrison.

CHAPTER 1. HITTING THE ROAD

[1] **By the time everything was ready** Francis Trevithick, *Life of Richard Trevithick, With an Account of His Inventions* (London: E. & F. N. Spon, 1872), pp. 106–08.

[2] **A few days after** Trevithick, *Life*, p. 117.

[3] **1888 | 1895 | 1901** Clay McShane, *Down the Asphalt Path: The Automobile and the American City* (New York: Columbia University Press, 1994), p. 104.

[4] **Socialistic feeling** 'Motorists Don't Make Socialists, They Say', *New York Times*, 4 March 1906; **advertisements often** see for example 'Motor Cars and Motor Boats: Motoring', *Daily Telegraph*, 6 July 1907; **No SHOCKS** 'Motoring', *Daily Telegraph*, 12 January 1907.

[5] **marketing opportunities** For a detailed analysis of Michelin and their marketing, see Stephen L. Harp, *Marketing Michelin: Advertising and Cultural Identity in Twentieth-Century France* (Baltimore: Johns Hopkins University Press, 2001); **[footnote]** 'Peking–Parijs: Godard en de Spijker-Wagen', *Het Nieuws van den Dag*, 29 July 1907, trans. Joris Canoy; 'End of the Pekin–Paris Run', *Horseless Age*, 28 August 1907, p. 283; du Taillis, *Pékin–Paris*, p. 262; **1906 total sales** Harp, *Marketing Michelin*, p. 18.

[6] **One wore old clothes** Hiram Percy Maxim, quoted in Virginia Scharff, *Taking the Wheel: Women and the Coming of the Motor Age* (Albuquerque: University of New Mexico Press, 1992), p. 13; **half metaphysics** E. B. White, 'Farewell, My Lovely!', *New Yorker*, 16 May 1936; **crank handles** Dorothy Levitt, *The Woman and the Car: A Chatty Little Handbook for All Women Who Motor or Who Want to Motor* (London: John Lane, Bodley Head, 1909), pp. 41–42; **the most humiliating** quoted in Scharff: *Taking the Wheel*, pp. 58–59.

[7] **Many automotive firms** See discussions in 'The *Belle Epoque* and the First World War: Industry, Sport, Utility and Leisure, 1903–1918', in Hugh Dauncey, *French Cycling: A Social and Cultural History* (Liverpool: Liverpool University Press, 2012), pp. 75–101; Michael Taylor, 'The Bicycle Boom and the Bicycle Bloc: Cycling and Politics in the 1890s', *Indiana Magazine of History*, 104.3 (2008), pp. 213–40; David Rubinstein, 'Cycling in the 1890s', *Victorian Studies*, 21.1 (1977), pp. 47–71.

[8] **Even allowing for** Geoff Tibballs, *Motor Racing's Strangest Races* (London: Portico, 2016), pp. 13–15.

[9] **[footnote]** The Festival of the Horse, *Le Matin*, 17 June 1907, trans. Laura Shanahan.

[10] **Passe Partout** 'The World Girdlers', *The Autocar*, 29 March 1902, p. 305; 'The Motor World – Week by Week', *The Tatler*, 8 October 1902, p. 55.

[11] **Paris–Madrid** 'La Course Sanglante: La Première Étape de Paris–Madrid', *Le Matin*, 25 May 1903; 'Six Persons Killed in Automobile Race', *New York Times*, 25 May 1903; 'Paris–Madrid', *Automotor Journal*, 30 May 1903.

[12] **A 2021 study** Fédération Internationale de l'Automobile (FIA), *A Report on the Global Contribution of Motor Sport to Economy and Community Development* (EY-Parthenon/FIA, July 2021).

CHAPTER 2. PARIS – PEKING: 30 JANUARY–9 JUNE

[1] **carefully selected small** 'Paris–Pékin Automobile: A Prodigious Challenge', *Le Matin*, 31 January 1907, trans. Laura Shanahan.

[2] **Mr Contal, accepting | worthy of Jules Verne** 'Paris–Pékin Automobile', *Le Matin*, 1 February 1907, trans. Laura Shanahan.

[3] **boisterous, roly-poly** Griffith Borgeson, 'The Automotive World of Albert de Dion', *Automobile Quarterly* 15.3 (1977), pp. 266–85.

[4] **By 15 February 1907** 'Peking–Paris Challenge: The Entrants', *Le Matin*, 15 February 1907, trans. Laura Shanahan; 'Peking–Paris Challenge', *Le Matin*, 17 February 1907, trans. Laura Shanahan; 'Peking–Paris Motor Race: Official Departure Date Set', *Le Matin*, 28 March 1907, trans. Laura Shanahan; 'The Pekin to Paris Drive', *The Autocar*, 2 March 1907, p. 297; **[footnote]** 'Entrants', *Le Matin*; **[footnote]** 'Peking–Matin: Prince Borghese has Left Moscow', *Le Matin*, 1 August 1907, trans. Margaret Morrison.

[5] **hardy annuals** 'From Paris to Pekin', *The Autocar*, 9 February 1907; **self-promotion** E. Kuzmin, 'Letter to the Editor: Paris–Peking', *Avtomobil'*, 1 April 1907, pp. 1675–77, trans. Dominique Hoffman; **a colossal joke** 'Notes and Comments', *South China Morning Post*, 11 April 1907.

[6] **unachievable mission** 'Peking to Paris: The Contestants Meet and Make Several Important Decisions', *Le Matin*, 22 February 1907, trans. Laura Shanahan; **a momentous decision** Barzini, *Peking to Paris*, p. 17.

[7] **By the time** Barzini, *Peking to Paris*, pp. 17–18; **a defiant article** Georges Bourcier Saint-Chaffray et al., 'The Peking–Paris Challenge: Competitor Cars Left Marseille Yesterday', *Le Matin*, 15 April 1907, trans. Margaret Morrison; **[footnote]** Gerville-Reache and Georges Bourcier Saint-Chaffray, 'Alger–Toulon Automobile', *Le Matin*, 24 November 1904; 'La Course Alger–Toulon Automobile', *Le Matin*, 15 May 1905.

[8] **aboard the *Océanien*** Du Taillis, *Pékin–Paris*, p. 18; **feverish expectancy** Paul Gauguin, *Noa Noa: The Tahitian Journal*, trans. O. F. Theis (New York: Dover, 1985), p. 1.

[9] For biographical information on Jean du Taillis, Charles Godard and Georges Cormier, I am indebted to the work of researcher Murièle Gadaut, who dug out birth and death certificates and military service records for all three.

[10] **blonde-bearded, gangly** Barzini, *Peking to Paris*, pp. 18–19; **although prone to** du Taillis, *Pékin–Paris*, p. 20; **The purser** du Taillis, *Pékin–Paris*, p 18 **less endearing traits** du Taillis, *Pékin–Paris*, pp. 21–27, 44–45.

[11] **On debarking** Du Taillis, *Pékin–Paris*, pp. 18, 27–28; **[footnote]** 'Da Berlino: L'Arrivo dei Ritardatari della Pechino–Parigi', *Corriere della Sera*, 24 August 1907; **work something out** du Taillis, *Pékin–Paris*, p. 18.

[12] **His career had taken off** Luigi Barzini, *Vita Vagabonda: Ricordi Di Un Giornalista* (Milan: Rizzoli, 1948), p. 207.

[13] **Masterpieces such as** Eduardo Ximenes, 'Corriere di Parigi', *L'Illustrazione Italiana*, 25 August 1907, p. 200; **fifteen kilograms of luggage** Barzini, *Peking to Paris*, p. 16. Prince Scipione Borghese's personal effects from the trip are in a private collection in Turin.

[14] **interested in politics** 'Peking–Paris Motor Race: Prince Scipione Borghese', *Le Matin*, 2 April 1907, trans Laura Shanahan; **since he was fifteen** Barzini, *Peking to Paris*, p. 11.

[15] **The Groundhog** Du Taillis, *Pékin–Paris*, p. 17; **Bizac had never** Georges Cormier, *Le Raid Pékin–Paris en 1907* (Paris: Édition 'Publi-Inter', 1954), p. 15, trans. Emily Plank; *lay it on thick* du Taillis, *Pékin–Paris*, p. 19.

[16] **the proud proprietor** 'Le Raid Pékin–Paris', *La Vie au Grand Air*, 7 September 1907, p. 173; **Du Taillis joked** du Taillis, *Pékin–Paris*, p. 43; **successfully competing** Georges Prade, 'La Course Bordeaux–Périgueux–Bordeaux: Toujours Plus Outre', *La Vie au Grand Air*, 17 June 1900, p. 522; 'Résultats Techniques de la Semaine', *La Vie au Grand Air*, 8 December 1905, p. 1026; **Cormier's account** Georges Cormier, 'Mon Tour d'Europe', *La Vie au Grand Air*, 13 December 1902, pp. 842–43, trans. Laura Shanahan.

[17] **cans of preserved meat** Collignon, *Souvenirs*, p. 8; **reconnoitre the route | dear friends** Collignon, *Souvenirs*, pp. 3–6, 5; **a good mood in motion** 'The Glorious Three: Triumphant Entry into Paris', *Le Matin*, 31 August 1907, trans. Margaret Morrison; **The engine is his passion** Jean du Taillis, 'Will we See the End of the Accursed Forest?', *Le Matin*, 22 July 1907, trans. Margaret Morrison; **bamboo rods** Barzini, *Peking to Paris*, p. 13.

[18] This description of Pons is taken from photographs, his own actions and words about the race in the piece written for *La Vie au Grand Air* and the recollections of his fellow travellers.

[19] **showing off its adaptations** F. Savorgnan di Brazza, 'Da Pechino a Parigi in Triciclo', *L'Illustrazione Italiana*, 16 June 1907, p. 583; 'Taking Their Road with Them: The Ingenious Equipment for the Peking to Paris Motor Race', *Illustrated London News*, 27 April 1907; **There is no debating** du Taillis, *Pékin–Paris*, p. 52.

[20] **tremendously impressive** Ellen N. La Motte, *Peking Dust* (New York: The Century, 1919), p. 16; **Peking is composed** Sarah Pike Conger, *Letters from China, with Particular Reference to the Empress Dowager and the Women of China* (Chicago: A.C. McClurg, 1909), p. 2; **seat of imperial power** Jung Chang, *Empress Dowager Cixi: The Concubine Who Launched Modern China* (London: Vintage, 2014), p. 6.

[21] **A few are wide** La Motte, *Peking Dust*, p. 17; **Half-wild dogs** Chang, *Empress Dowager*, p. 5; **After dark** Harry de Windt, *From Pekin to Calais by Land* (London: Chapman & Hall, 1889), p. 97; **a heavy perfume**, Paul Claudel, quoted in Xuelei Huang, 'Deodorizing China: Odour, Ordure, and Colonial (Dis)Order in Shanghai, 1840s–1940s', *Modern Asian Studies*, 50.3 (2016), p. 1093; **[footnote]** du Taillis, *Pékin–Paris*, p. 33; **dense, impalpable, penetrating** du Taillis, *Pékin–Paris*, pp. 29–30.

[22] **A proud stronghold** Barzini, *Peking to Paris*, p. 6; **The very presence** Barzini, *Peking to Paris*, p. 22.

[23] **flattered themselves** Du Taillis, *Pékin–Paris*, pp. 40–41; Barzini, *Peking to Paris*, p. 7; **the scarcity of** 'Motoring', *North-China Herald*, 8 January 1897, p. 18; **Precisely four years later** 'Motor-Cars', *North-China Herald*, 9 January 1901, pp. 45–46; **By June 1902** 'Paterfamilias', 'Letters to the Editor: The Motor Car Again', *North-China Herald*, 18 June 1902; **By the summer of 1907** 'Street Traffic and Motors', *South China Morning Post*, 8 February 1907; 'The Motor Industry in Hongkong', *Hong Kong Telegraph*, 27 April 1907; 'An Innovation at Shanghai: Motor Parade', *South China Morning Post*, 13 June 1905; **When a motoring reader** 'Motorist' and 'Another Motorist', 'Letters to the Editor: The Motor Siren', *North-China Herald*, 16 August 1907.

[24] **A smart 1901** Chang, *Empress Dowager*, p. 404; **General Yuan** Pamela Crossley, 'In the Hornets' Nest', *London Review of Books*, 17 April 2014; **[footnote]** 'Car Fit for an Empress Dowager', Bowers Blog, 2017; John Rogers, 'Rare Art from China's 19th Century Woman Ruler Come to US', AP News, 11 November 2017.

[25] **profanations of the west** Barzini, *Peking to Paris*, pp. 7, 9; **more sympathetic** du Taillis, *Pékin–Paris*, p. 46.

[26] **called a meeting** Du Taillis, *Pékin–Paris*, pp. 43–44.

CHAPTER 3. HART AND HUMILIATION

[1] **chilblains** *The I.G. in Peking: Letters of Robert Hart, Chinese Maritime Customs 1868–1907*, I, ed. John King Fairbank, Katherine Frost Bruner and Elizabeth MacLeod Matheson (Cambridge, MA: Belknap Press, 1975), p. 1398; **woodland city** T. Hodgson Liddell, *China: Its Marvel and Mystery* (London: George Allen & Sons, 1909), pp. 125–26.

[2] Biographical information on Sir Robert Hart taken from Edward B. Drew, 'Sir Robert Hart and His Life Work in China', *Journal of Race Development*, 4.1 (1913), pp. 1–33; Henk Vynckier and Chihyun Chang, '"Imperium in Imperio": Robert Hart, the Chinese Maritime Customs Service, and Its (Self-) Representations', *Biography*, 37.1 (2014), pp. 69–92; Hart, *I.G. in Peking*; *Entering China's Service: Robert Hart's Journals, 1854–1863*, ed. Katherine Frost Bruner, John King Fairbank and Richard J. Smith (Cambridge, MA: Harvard University Press, 1986); *Robert Hart and China's Early Modernization: His Journals, 1863–1866*, ed. Richard J. Smith, John King Fairbank and Katherine Frost Bruner (Cambridge, MA: Harvard University Press, 1991); **[footnote]** Hart, *I.G. in Peking*, pp. 1158, 1170.

[3] **Taiping Rebellion** Rania Huntington, 'Chaos, Memory, and Genre: Anecdotal Recollections of the Taiping Rebellion', *Chinese Literature: Essays, Articles, Reviews*, 27 (2005), p. 62; Philip A. Kuhn, 'Origins of the Taiping Vision: Cross-Cultural Dimensions of a Chinese Rebellion', *Comparative Studies in Society and History*, 19.3 (1977), pp. 350–51.

[4] **trading access was prized** Chang, *Empress Dowager*, pp. 23–25; **Britain demanded** Chang, *Empress Dowager*, p. 28; **A few decades later** Isabella L. Bird, *Chinese Pictures: Notes on Photographs Made in China* (London: Cassell, 1900), p. 126; **They deface**, Earl Li, quoted in Chang, *Empress Dowager*, p. 81.

[5] **unspeakable bullying** Chang, *Empress Dowager*, p. 30; **She, in turn** 'The Empress Dowager', *North-China Herald*, 2 May 1900; **[footnote]** Bird, *Chinese Pictures*, p. 100.

[6] **For the first time**, A. H. Smith, quoted in Chang, *Empress Dowager*, pp. 316–17; **Food prices** 'Kuangyuanhsien – North Szechuan', *North-China Herald*, 2 May 1900; 'From Chinese Sources', *North-China Herald*, 18 July 1900; 'The Boxers at Wênchow', *North-China Herald*, 18 July 1900; 'With Seymour's Column: A Diary of the Attempt to Relieve Peking', *North-China Herald*, 18 July 1900; **Little by little** Conger, *Letter from China*, pp. 90, 94, 98–107.

[7] **The Boxer Protocol** Frank H. King, 'The Boxer Indemnity: "Nothing but Bad"', *Modern Asian Studies*, 40.3 (2006), pp. 663, 669–70; **very dark indeed** Hart, quoted in King, 'Boxer Indemnity', p. 663.

[8] **As a result of the world war** Mao Zedong, 'To the Glory of the Hans: Toward a New Golden Age', *Hsiang-Chiang p'inq-Lun*, July 1919.

[9] **In Tajikistan** 'How Vladimir Putin's Embrace of China Weakens Russia', *The Economist*, 25 July 2019.

[10] **While spending** Dominic Ziegler, 'China Wants to Put Itself Back at the Centre of the World', *The Economist*, 6 February 2020; **the overland route** State Council Information Office: The People's Republic of China, 'What Are Six Economic Corridors Under Belt and Road Initiative?', 2020.

CHAPTER 4. PEKING – KALGAN: 10–14 JUNE

[1] **the vehicles passed** Cormier, *Le Raid*, p. 17; Putnam Weale, Photographs; **After a few hundred yards** du Taillis, *Pékin–Paris*, p. 53; Cormier, *Le Raid*, p. 17.

[2] **Straw-hatted Peking policemen** Barzini, *Peking to Paris*, p. 28; **wicker watering cans** du Taillis, *Pékin–Paris*, p. 32; **A manager from** 'Peking–Paris Motor Race: Where they Will Go', *Le Matin*, 3 May 1907, trans. Laura Shanahan; **a convoy** Barzini, *Peking to Paris*, pp. 9–10; **made up of steps** Collignon, *Souvenirs*, p. 5.

[3] **all were heavily laden** Luigi Barzini, 'Pekin to Paris: The Great Motor Race', *Daily Telegraph*, 10 June 1907.

[4] **ancient dust fell** Du Taillis, *Pékin–Paris*, p. 53; **Our departure** Jean du Taillis, 'The Spectacular Challenge: Cormier, Collignon, Godard and Pons have Completed the Peking–Nankou Stage', *Le Matin*, 13 June 1907, trans. Laura Shanahan.

[5] **The perfect swimmer** Du Taillis, *Pékin–Paris*, p. 55.

[6] **On horseback** John Grant Birch, *Travels in North and Central China* (London: Hurst & Blackett, 1902), p. 17.

[7] **three or four hundred yards** Cormier, *Le Raid*, pp. 17–18; **draw straws | Only just out of Peking** du Taillis, *Pékin–Paris*, p. 56.

[8] **To-day the participators** Barzini, 'The Great Motor Race', *Daily Telegraph*, 10 June 1907; **There are no formalities** 'Entrants', *Le Matin*.

[9] **The pace was glacial** Du Taillis, *Pékin–Paris*, p. 56; **This would technically** Luigi Barzini, 'Paris to Pekin: Italian Car at Nankau [*sic*]', *Daily Telegraph*, 12 June 1907.

[10] **I have seldom** De Windt, *Pekin to Calais*, p. 118.

[11] **bounded and swayed** Barzini, *Peking to Paris*, pp. 28–29; **The Itala's bodywork** Barzini, *Peking to Paris*, p. 23.

[12] **the days before Christ** Cormier, *Le Raid*, p. 18; **almost European** Barzini, 'Great Motor Race'; **like everything else in China** du Taillis, *Pékin–Paris*, pp. 62–63; **The Frenchmen deployed** du Taillis, *Pékin–Paris*, p. 63; **after trying and failing** Barzini, *Peking to Paris*, pp. 30–31.

[13] **grandiose in its harshness** Du Taillis, *Pékin–Paris*, p. 67; **[footnote – Nankou] | [footnote – merely something]** Cormier, *Le Raid*, p. 20, 19; **travelled on horseback** de Windt, *Pekin to Calais*, p. 128; **150 men and several mules** Jean du Taillis, 'The Spectacular Challenge: Cormier, Collignon, Godard and Pons have Completed the Peking–Nankou Stage', *Le Matin*, 13 June 1907, trans. Laura Shanahan; **removed yet more** Cormier, *Le Raid*, p. 19; **faithfully promised** Barzini, 'Great Motor Race'; **set between walls** Barzini, *Peking to Paris*, pp. 60–61; **as if at the prow** du Taillis, *Pékin–Paris*, p. 93; **gigantic willow** Barzini, *Peking to Paris*, pp. 57–58.

[14] **become road workers** Cormier, *Le Raid*, p. 20; **it took Borghese** Jean du Taillis et al., 'The Spectacular Challenge: Through the Mountains of Nankou', *Le Matin*, 14 June 1907, trans. Laura Shanahan; **We are not** Jean du Taillis et al., 'Peking–Paris: Along the Precipices', *Le Matin*, 16 June 1907, trans. Laura Shanahan.

[15] **Nothing was heard** Barzini, 'Great Motor Race', p. 47; **Barzini quickly came to admire | feet touched the ground** Barzini, *Peking to Paris*, pp. 37–38; **He admired | On 12 June** du Taillis, *Pékin–Paris*, pp. 75, 73–74; **It was all very well** du Taillis, *Pékin–Paris*, p. 87.

[16] **think too much** Du Taillis, *Pékin–Paris*, p. 76.

CHAPTER 5. AGE OF INVENTION

[1] **pitiable screams** 'Suspected Murder at Salt-Hill', *The Times*, 3 January 1845; 'The Salt-Hill Murder', *The Times*, 25 March 1845; 'The Murder at Salt-Hill', *The Times*, 16 January 1845; **multiple ancestors** information on the invention and early history of telegraphy can be found in: Tapan K. Sarkar et al., *History of Wireless* (Hoboken, NJ: John Wiley, 2006), pp. 54, 58–64; John Moyle, *Timelines of Telegraphy* (Penzance: Cable and Wireless Porthcurno and Collections Trust, 2008); Hugh Barty-King, *Stupendous Engine: The First Fifty Years of Telegraphy* (Penzance: Cable and Wireless Porthcurno and Collections Trust, 2005).

[2] **He grumbled** H. D. Traill, *Number Twenty: Fables and Fantasies* (London: Henry Holt, 1892), p. 2; **distance and time | annihiliated time and space** Iwan Rhys Morus, '"The Nervous System of Britain": Space, Time and the Electric Telegraph in the Victorian Age', *British Journal for the History of Science*, 33.4 (2000), pp. 463, 456; **Commentators likened the telegraph** see for example J. Henniker Heaton, 'An Imperial Telegraph System', *Nineteenth Century*, 45 (1899), pp. 906–14; George Ethelbert Walsh, 'Cable Cutting in War', *North American Review*, 167.503 (1898), pp. 498–502; for a discussion of the telegraph's cultural impact, see Morus, 'Nervous System'; **a discovery which operates** Lord Salisbury, quoted in Iwan Rhys Morus, 'The Electric Ariel: Telegraphy and Commercial Culture in Early Victorian England', *Victorian Studies*, 39.3 (1996), p. 339.

[3] **The triumph of speed** 'To-Day', *Daily Telegraph*, 25 June 1907; **the lullaby** du Taillis, *Pékin–Paris*, p. 141.

[4] **Guglielmo Marconi** For more on Marconi, his work and its significance, see Sarkar et al., *History*; Marc Raboy, *Marconi: The Man Who Networked the World* (Oxford: Oxford University Press, 2016); G. A. Isted, 'Guglielmo Marconi and the History of Radio – Part I', *GEC Review*, 7.1 (1991), pp. 45–56; and G. A. Isted, 'Guglielmo Marconi and the History of Radio – Part II', *GEC Review*, 7.2 (1991), pp. 110–22; **wireless greenfingers** Isted, 'Marconi I', p. 54; **stunned the world** Our Correspondent [St John's], 'Wireless Telegraphy across the Atlantic', *The Times*, 17 December 1901; 'Mr. Marconi on Wireless Telegraphy', *The Times*, 21 February 1902; **In the history** Our Correspondent [St John's], 'Marconi in Newfoundland', *The Times*, 3 January 1902.

[5] **Rats rats rats rats** Quoted in Thomas McMullan, 'The World's First Hack: The Telegraph and the Invention of Privacy', *The Guardian*, 15 July 2015; **Don't be alarmed | Thanks old man** poems quoted in Isted, 'Marconi II', p. 112.

[6] **Evelyn Nesbit** It's difficult to explain how fully the story of Evelyn Nesbit captured imaginations. For more information, see Simon Baatz, *The Girl on the Velvet Swing: Sex, Murder, and Madness at the Dawn of the Twentieth Century* (New York: Mulholland, 2018); Paula Uruburu, *American Eve: Evelyn Nesbit, Stanford White, the Birth of the 'It' Girl, and the Crime of the Century* (New York: Riverhead Books, 2008); **Return from the North Pole** 'Retour de Pole Nord', *Le Matin*, 22 February 1907, trans. Laura Shanahan; **improbable expeditions** 'Airship to the Pole: A Jules Verne Project', *Daily Telegraph*, 25 May 1907; **mastery over distance** 'Paris–Pékin Automobile: A Prodigious Challenge', *Le Matin*, 31 January 1907, trans. Laura Shanahan.

[7] **an intoxicating journey** Luigi Barzini, 'Pekin to Paris: Nearing the Great Desert', *Daily Telegraph*, 10 June 1907; **an enormous plateau** Collignon,

Souvenirs, p. 12; **escort of cavalry** Barzini, 'Nearing the Great Desert'; **Readers of the *Corriere*** 'La Pechino–Parigi: La Strada Percosa e Quella da Percorrere', *Corriere della Sera*, 21 June 1907.

[8] **regional newspapers** See for example 'Motoring in the Desert', *Lancashire Daily Post*, 21 June 1907; 'Late News Condensed from the Coast Files', *Hawaiian Gazette*, 21 June 1907; 'The Peking–Paris Motor Race', *Sydney Morning Herald*, 3 June 1907; **On 30 June** 'Urga Lama Has an Auto: Shows it to Peking-to-Paris Racers, Who Find that it Won't Run', *New York Times*, 30 June 1907.

[9] **smoking opium** Barzini, *Peking to Paris*, p. 65; **Pong-Kong** Jean du Taillis, 'Peking–Paris: 240 Kilometres in 13 Hours, through the Red Dust of the Desert', *Le Matin*, 19 June 1907, trans. Laura Shanahan; Barzini, *Peking to Paris*, p. 94.

[10] **A newspaper correspondent** Barzini, *Peking to Paris*, p. 75; **Barzini told a colleague** Ximenes, 'Corriere'; **At one office** Barzini, *Peking to Paris*, p. 225.

[11] **The first electric | broken and bent** Christopher Otter, 'Cleansing and Clarifying: Technology and Perception in Nineteenth-Century London', *Journal of British Studies*, 43.1 (2004), pp. 58, 60.

[12] **between 18 and 21 June | Unless I am very** Leo Baekeland Diary Volume 1, 1907–1908, pp. 45–46, 96, Leo Baekeland Papers, Archives Center, National Museum of American History, Smithsonian Institution, Washington DC, pp. 45–46, 96; **[footnote]** F. A. Freeth, 'James Swinburne. 1858–1958', *Biographical Memoirs of Fellows of the Royal Society*, 5 (1960), p. 257.

CHAPTER 6. KALGAN – URGA: 14–24 JUNE

[1] **as if waiting** Barzini, *Peking to Paris*, p. 67; **350,000 chests | Street tumblers** de Windt, *Pekin to Calais*, pp. 150, 152–55; **building of the Trans-Siberian Railway** du Taillis, *Pékin–Paris*, p. 97; **buttercup yellows** Cormier, *Le Raid*, p. 25.

[2] **The Italians reached** Barzini, *Peking to Paris*, p. 69; **two days later** Jean du Taillis et al., 'Peking–Paris: At the Doorway to the Desert', *Le Matin*, 17 June 1907, trans. Laura Shanahan; **Although much of the Gobi** de Windt, *Pekin to Calais*, pp. 188–89; **Fourteen days** de Windt, *Pekin to Calais*, pp. 189–90.

[3] **Previous travellers** De Windt, *Pekin to Calais*, pp. 154–77; **convoy of nineteen camels** Barzini, *Peking to Paris*, pp. 9–19; 'Pekin–Paris: The Tremendous

Challenge', *Le Matin*, 22 June 1907, trans. Laura Shanahan; **chipping 'taels'** du Taillis, *Pékin–Paris*, pp. 99–100.

[4] **Mongolians in caps** Du Taillis, *Pékin–Paris*, p. 100.

[5] **Two tracks led** Barzini, *Peking to Paris*, p. 73; **around twenty-five days** de Windt, *Pekin to Calais*, p. 176.

[6] **short 12 June bulletin** Luigi Barzini, 'Paris to Pekin: Italian Car at Nankau [*sic*]', *Daily Telegraph*, 12 June 1907; **all doing well** du Taillis, 'Completed the Peking–Nankou Stage'.

[7] **Cormier groused** Cormier, *Le Raid*, p. 26; **the princely driver** du Taillis, *Pékin–Paris*, p. 99; **camel bones** Cormier, *Le Raid*, p. 34; Collignon, *Souvenirs*, p. 15; **the rain means mud** du Taillis et al., 'Doorway to the Desert'; **slowness of their progress** du Taillis, *Pékin–Paris*, p. 107, 111; **drivers of the larger cars** Jean du Taillis et al., 'Peking–Paris: Their Climb to the Heavens', *Le Matin*, 18 June 1907, trans. Laura Shanahan.

[8] **a deliciously soft** Du Taillis, *Pékin–Paris*, p. 117; **a novel soup** Barzini, *Peking to Paris*, p. 82; **Darkness seems to** Barzini, *Peking to Paris*, p. 83; **The next morning** Auguste Pons, 'Lache en Plein Desert!', *La Vie au Grand Air*, 24 August 1907, p. 129; **Scratching** du Taillis, *Pékin–Paris*, p. 119; **Unhappily preoccupied** du Taillis, *Pékin–Paris*, p. 118; **In countries like this** du Taillis, *Pékin–Paris*, p. 120.

[9] **the tracks did not converge** Cormier, *Le Raid*, pp. 27–28; du Taillis, *Pékin–Paris*, pp. 120–21; **After half an hour** du Taillis, *Pékin–Paris*, p. 121.

[10] **decided to place our trust** Cormier, *Le Raid*, pp. 28–29; **Du Taillis wrote** du Taillis, *Pékin–Paris*, pp. 128–29.

[11] **a different story** Barzini, *Peking to Paris*, p. 96; **lagging behind** du Taillis, 'Red Dust'; **Barzini telegraphed** Barzini, 'Nearing the Great Desert'.

[12] **Just before their departure** Du Taillis, *Pékin–Paris*, p. 130.

[13] **Rather than carrying** Cormier, *Le Raid*, p. 30; du Taillis, *Pékin–Paris*, p. 110; **Pékin, a small Pekinese dog**, du Taillis, *Pékin–Paris*, p. 130.

[14] **Fifty miles into** Du Taillis, *Pékin–Paris*, p. 133; **[footnote]** Georges Cormier, 'Pékin–Paris', *L'Auto*, 16 July 1907, p. 3.

[15] **Safe in the lead vehicle** Barzini, *Peking to Paris*, pp. 100–03.

[16] **Stunned by the sun | a little humiliating** Pons, 'Lache en Plein Desert'. The last line, which reads as if added by an editor, absolves the other racers of any guilt. This sentiment is very much at odds with the balance of the article.

[17] **I must once again** Cormier, *Le Raid*, p. 33; **hospitable, selfless** Cormier,

Le Raid, p. 34; ***Sprechen Sie Deutsch?*** Barzini, *Peking to Paris*, pp. 87–89; **Mongolians? Honest barbarians?** Barzini, *Peking to Paris*, p. 95.

[18] Account of du Taillis and Godard's time stranded in the Gobi taken from: du Taillis, *Pékin–Paris*, pp. 131–40; Jean du Taillis, 'The Tremendous Peking–Paris Challenge: Lost in the Desert', *Le Matin*, 24 June 1907, trans. Laura Shanahan.

[19] **an imposing edifice** De Windt, *Pekin to Calais*, p. 269; **a compact mass** de Windt, *Pekin to Calais*, p. 267.

[20] **without blending | trenches, ditches** Barzini, *Peking to Paris*, p. 122; **The reigning *Kootookta*** de Windt, *Pekin to Calais*, p. 270; **an intelligent, energetic** Barzini, *Peking to Paris*, p. 123; **indifferent to their rulers** Thomas W. Knox, *Overland through Asia: Pictures of Siberian, Chinese, and Tartar Life* (London: Trübner, 1871), p. 351; **[footnote]** de Windt, *Pekin to Calais*, p. 291.

[21] **Russian, French, and Italian** Barzini, *Peking to Paris*, p. 127; Luigi Barzini, 'Pekin to Paris: Chinese Governor in a Motor Car', *Daily Telegraph*, 24 June 1907.

[22] **The Itala had roared** Barzini, *Peking to Paris*, p. 122; **The two De Dion-Boutons** Cormier, *Le Raid*, pp. 34–35; **a mad dash | Prince Borghese had set off** Jean du Taillis, 'Champions of the Desert: 617 Kilometres in 23 Hours', *Le Matin*, 25 June 1907, trans. Laura Shanahan; **in a sorry state** Cormier, *Le Raid*, p. 37; **[footnote]** Special Correspondent [St Petersburg] and Auguste Pons, 'Peking–Matin: Borghese Arrives in St. Petersburg', *Le Matin*, 2 August 1907, trans. Margaret Morrison; **One downpour** Barzini, *Peking to Paris*, p. 134.

CHAPTER 7. THE DIVIDED EMPIRE

[1] **If I fail** Ivan Vasilev, quoted in Orlando Figes, *A People's Tragedy: The Russian Revolution 1891–1924*, 2nd edn (London: Bodley Head, 2014), p. 173.

[2] **A Bolshevik** Figes, *People's Tragedy*, pp. 177–78; **People have died** Maxim Gorky, quoted in Figes, *People's Tragedy*, p. 179.

[3] **The country spanned** John E. Bowlt, *Moscow and St. Petersburg in Russia's Silver Age: 1900–1920* (London: Thames & Hudson, 2008), p. 33; **disastrous handling** Figes, *People's Tragedy*, pp. 157–62; **The country's population** Bowlt, *Moscow and St. Petersburg*, p. 33; **nearly 80 per cent** Figes, *People's Tragedy*, p. 105; **little or no land** Hugh O'Beirne, 'Report by Mr. O'Beirne Respecting Agricultural Distress in Russia' (St Petersburg: British Foreign Office, 30 January 1907), pp. 109–110v; **overcrowded and dirty** Figes,

People's Tragedy, pp. 111–12; **A 1904 survey** Douglas Smith, *Former People: The Last Days of the Russian Aristocracy* (London: Pan, 2012), p. 21; **By 1913** Bowlt, *Moscow and St. Petersburg*, p. 34.

[4] **There are no less than** H. A. Munro-Butler-Johnstone, *A Trip up the Volga to the Fair of Nijni-Novgorod* (Oxford: James Parker, 1875), pp. 1–2; **The empire's 5 million | In Poland** Figes, *People's Tragedy*, pp. 80–81; **Sergei Witte** Smith, *Former People*, pp. 36–37; Simon Sebag Montefiore, *The Romanovs: 1613–1918* (London: Weidenfeld & Nicolson, 2016), pp. 469–71.

[5] **Its buildings | [footnote]** A. I. Gertsen, quoted in Orlando Figes, *Natasha's Dance: A Cultural History of Russia* (London: Allen Lane, 2002), pp. 8–9.

[6] **During the reign | By the end of** Daniel Beer, *The House of the Dead: Siberian Exile under the Tsars*, 2nd edn (London: Penguin, 2017), pp. 18–19; **master of hospitality** du Taillis, *Pékin–Paris*, p. 206; descriptions of post houses by Western writers are plentiful and rarely flattering – see for example Knox, *Overland through Asia*, pp. 269–70; de Windt, *Pekin to Calais*, pp. 474, 493, 501; Cormier, *Le Raid*, p. 53; *podorojne* de Windt, *Pekin to Calais*, p. 493; **One winter traveller** Knox, *Overland through Asia*, p. 512.

[7] **Between 1801 and 1917 | Many were guilty** Beer, *House of the Dead*, pp. 4, 22, 24, 27; **some 45,000** Figes, *People's Tragedy*, p. 202; **they were issued** George Kennan, *Siberia and the Exile System*, I (London: James R. Osgood, McIlvaine, 1891), p. 290; **The journey from Moscow** Beer, *House of the Dead*, p. 39; **ugliest road** Anton Chekhov, quoted in William Atkins, 'On the Island of the Black River', *Granta*, 21 November 2019.

[8] **punishments varied** Kennan, *Siberia*, I, pp. 315–16; **impossible to conceive** John Dundas Cochrane, quoted in Beer, *House of the Dead*, p. 86.

[9] **What am I going to do** Quoted in Smith, *Former People*, p. 23.

[10] **Khodynka Tragedy** Sebag Montefiore, *Romanovs*, pp. 498–99; **reactionary father** Figes, *People's Tragedy*, p. 20; **the last spectacular**, Alexander Mikhailovich, quoted in S.V. Mironenko and Andrei Maylunas, *A Lifelong Passion: Nicholas and Alexandra, Their Own Story* (London: Doubleday, 1997), p. 227.

[11] **'Hubby' and 'Wifey'** Orlando Figes, *People's Tragedy*, p. 25; **You are mistaken** Tsarina Alexandra, quoted in Figes, *People's Tragedy*, p. 26. There is no shortage of books about this period of Russia's history and many spend some time discussing Tsar Nicholas II's faults, in part because so many of his friends, family and ministers were only too happy to record their (often scathing) views for posterity. For further discussion of Nicholas II's character, I recommend the first chapter of Orlando Figes's *A People's Tragedy*, a masterful and very weighty

account of the period 1891–1924; Robert Service's biography of Nicholas, *The Last of the Tsars*; and Simon Sebag Montefiore's *The Romanovs*. To get a better idea of the milieu from which the Tsar sprang: *Former People* by Douglas Smith. For something colourful and digestible: *October* by China Miéville.

[12] Descriptions of Rasputin taken from Sebag Montefiore, *Romanovs*, pp. 535–37, and Figes, *People's Tragedy*, pp. 29–33; for a detailed insight into the relationship between the Romanovs and Rasputin, see Service, *Last of the Tsars*.

[13] **[footnote]** Figes, *People's Tragedy*, p. 197; **The public were disgusted** 'Why the Czar's Army Was Beaten', *New York Times*, 15 February 1907; **By October 1905** 'The Siberian Revolt', *North-China Herald*, 14 June 1907; Claud Russell, 'Summary of Events in Russia for the Fortnight Ended June 6, 1907' (St Petersburg: British Foreign Office, 5 June 1907), pp. 386–386v; **In the countryside** Figes, *People's Tragedy*, p. 182; **Russia was on fire** Grand Duke Mikhailovich, quoted in Smith, *Former People*, p. 54.

[14] **already begun** A. G. Bulygin, quoted in Figes, *People's Tragedy*, p. 186; **declare martial law** Beer, *House of the Dead*, pp. 364–65; **continual chaos** '1,242 Assassinated, 1,080 Executed', *New York Times*, 10 April 1907; **much-needed loans** A. Raffalovich, 'Financial Progress of the Russian Government', *New York Times*, 6 January 1907.

[15] **In January 1907** Ernest Scott, 'Report by Mr. Scott on Events in Russia for the Fortnight Ended January 2, 1907' (St Petersburg: British Foreign Office, 2 January 1907), pp. 285–286v; **famine threatened** O'Beirne, 'Agricultural Distress'; **Warsaw's chief of secret police** 'Warsaw Police Chief Slain', *New York Times*, 1 February 1907; **local officials** Ernest Scott, 'Report on the Events in Russia for the Fortnight Ended February 13, 1907' (St Petersburg: British Foreign Office, 13 February 1907), pp. 323–25; **no improvement** Claud Russell, 'Summary of Events in Russia for the Fortnight Ended April 24, 1907' (St Petersburg: British Foreign Office, 25 April 1907), pp. 366–366v; **By late May** Stephen Bonsal, 'Czar Again Faces Revolt: Open Revolution is Already in Progress in a Dozen Provinces', *New York Times*, 28 May 1907; **mysterious murders** Our Own Correspondent, 'The State of Russia', *North-China Herald*, 26 July 1907; **On Sunday, 16 June** 'Dissolution of the Russian Duma', *Daily Telegraph*, 17 June 1907.

CHAPTER 8. URGA – IRKUTSK: 23 JUNE–1 JULY

[1] **scale on his map** Du Taillis, *Pékin–Paris*, p. 164; **The roads beyond Kiakhta** Barzini, *Peking to Paris*, pp. 153–54.

[2] **The early part of the route** Jean du Taillis, 'Peking–Paris Challenge:

The Mountains of Russia', *Le Matin*, 29 June 1907, trans. Laura Shanahan; **Fifteen minutes later** Barzini, *Peking to Paris*, pp. 135–36.

[3] **Belts of copse-wood** De Windt, *Pekin to Calais*, p. 308; **traversing a true labyrinth** Barzini, *Peking to Paris*, p. 148; **like sailors** Luigi Barzini, 'Pekin to Paris: Strange Adventures of a Motor Car', *Daily Telegraph*, 27 June 1907; **the heavy Itala** Barzini, 'Strange Adventures'; **The Spyker once** Collignon, *Souvenirs*, p. 16; **delighted to learn** Cormier, *Le Raid*, p. 43; **When Borghese, Barzini and Guizzardi** Barzini, 'Strange Adventures'; **Two days later** Jean du Taillis, 'Le Matin's Tremendous Challenge: The Cars Go for a Swim', *Le Matin*, 30 June 1907, trans. Laura Shanahan; **remove the sensitive magneto** Guizzardi, quoted in Valerio Moretti, 'Il Terzo Uomo della Pechino Parigi', *Auto Italiana*, 18 July 1963, p. 17, translation mine. Descriptions of the Iro crossing can be found in du Taillis, 'Cars Go for a Swim'; Barzini, *Peking to Paris*, pp. 152–53; du Taillis, *Pékin–Paris*, pp. 166, 171.

[4] **Here is Kiakhta!** Du Taillis, 'Cars Go for a Swim'; **Europe had come to meet us** Barzini, *Peking to Paris*, p. 155; **[footnote]** du Taillis, *Pékin–Paris*, pp. 175–76; du Taillis, 'Cars go for a Swim'; Cormier, *Le Raid*, p. 159.

[5] **a village** Barzini, *Peking to Paris*, p. 161; **Many of the homes** de Windt, *Pekin to Calais*, pp. 323, 332; **gold candelabras** Cormier, *Le Raid*, p. 44; **The altar alone** de Windt, *Pekin to Calais*, p. 332.

[6] **national beverage** Munro-Butler-Johnstone, *Trip up the Volga*, p. 84; **raw material** Knox, *Overland through Asia*, p. 299; **You have no idea** Barzini, *Peking to Paris*, pp. 161–62; **sour in 1856** Knox, *Overland through Asia*, pp. 306, 318; **They estimate** Jean du Taillis, 'Peking–Paris Challenge: On the Banks of Lake Baikal', *Le Matin*, 2 July 1907, trans. Laura Shanahan.

[7] **triumphal wooden arches** Barzini, *Peking to Paris*, p. 175; Robert Service, *The Last of the Tsars: Nicholas II and the Russian Revolution* (London: Macmillan, 2017), p. 5; **They passed through** Barzini, *Peking to Paris*, p. 175.

[8] **crossed by fallen trees** Luigi Barzini, 'Pekin to Paris: Unbridged Rivers', *Daily Telegraph*, 29 June 1907; **[footnote]** 'Mr Edge's Great Ride: Sixty Miles an Hour for 24 Hours', *Daily Telegraph*, 29 June 1907; 'Great Motor Record: Mr Edge's Triumph', *Daily Telegraph*, 1 July 1907; Luigi Barzini, 'La Corsa Pechino–Parigi: Le Difficoltà Insuperabili della Strada Siberiana', *Corriere della Sera*, 29 June 1907; **The De Dion-Boutons and the Spyker** du Taillis, *Pékin–Paris*, pp. 182, 181; **less isolated** Cormier, *Le Raid*, p. 46; **preferred to ford | A solution** Barzini, 'Unbridged Rivers'.

[9] **lost in the** Cormier, *Le Raid*, pp. 47–48; **Just a straw mattress | dead drunk** Collignon, *Souvenirs*, p. 17; **set to ordering** Cormier, *Le Raid*, p. 48 (both Cormier and du Taillis relate this story, although du Taillis was unable

to remember exactly where it had taken place); **[footnote]** 'Peking to Paris: The Contestants Meet and Make Several Important Decisions', *Le Matin*, 22 February 1907, trans. Laura Shanahan; du Taillis, *Pékin–Paris*, p. 216; Georges Cormier, 'Pékin–Paris: Le Tour du Baikal', *L'Auto*, 4 July 1907, p. 3.

[10] **We are alive – by a miracle!** Luigi Barzini, 'Pekin to Paris: Exciting Adventure in Siberia', *Daily Telegraph*, 3 July 1907; Barzini, *Peking to Paris*, pp. 200–07.

[11] **The curtain was raised** Barzini, *Peking to Paris*, p. 210.

CHAPTER 9. SEPARATE SPHERES

[1] **Dr Bill French** Milton Lehman, 'The First Woman Driver', *Life*, 8 September 1952, pp. 83–95.

[2] **[footnote]** McShane, *Asphalt Path*, pp. 149–50; **Francis Thornton** 'Woman Teacher Talks Confidence', *New York Times*, 12 April 1914; **To think of 'The Monster'**, quoted in Scharff, *Taking the Wheel*, pp. 20–21; **Christabel Ellis** Barbara Burman, 'Racing Bodies: Dress and Pioneer Women Aviators and Racing Drivers', *Women's History Review*, 9.2 (2000), p. 299.

[3] **A female columnist** Scharff, *Taking the Wheel*, p. 16; **One involved** 'Empress's Adventure: A Motor Breakdown', *Daily Telegraph*, 27 July 1907; **The other** 'Motor-Car Fatality at Wimbledon: Lady Novelist's Mishap', *Daily Telegraph*, 27 July 1907.

[4] **Alice Roosevelt** Scharff, *Taking the Wheel*, p. 71.

[5] **[footnote]** Levitt, *Woman and the Car*, pp. 3–4; **Motoring is a pastime** Levitt, *Woman and the Car*, pp. 15–16; **in addition to pliers | [footnote]** Levitt, *Woman and the Car*, pp. 19, 30.

[6] **opportunities for racing** McShane, *Asphalt Path*, pp. 158–59; **Looking at Miss** Levitt, *Woman and the Car*, pp. 3–4; **No licence should** Scharff, *Taking the Wheel*, p. 42; **The only proper machine** Ruth Brandon, *Automobile: How the Car Changed Life* (London: Macmillan, 2002), p. 166.

[7] **electric automobiles** Gijs Mom and David Kirsch, 'Technologies in Tension: Horses, Electric Trucks, and the Motorization of American Cities, 1900–1925', *Technology and Culture*, 42.3 (2001), p. 492; **New York, Chicago and Boston** Maggie Koerth-Baker, 'Why Your Car isn't Electric', *New York Times*, 2 October 2012.

[8] **circumscribed radius | Has there ever been** Carl H. Claudy, quoted in Brandon, *Automobile*, p. 165.

[9] **One from 1906** | **Manufacturers of one** Scharff, *Taking the Wheel*, pp. 58–59, 63.

[10] **Dodge La Femme**, Benjamin Hunting, 'How the 1955 Dodge La Femme Missed the Mark on Designing Cars for Women', Hagerty, 10 February 2020.

[11] **Middletown studies** Robert Staughton Lynd and Helen Merrell Lynd, *Middletown: A Study in American Culture* (New York: Harcourt, Brace, 1929), pp. 255–56.

[12] **Edith Wharton wrote lyrically** Edith Wharton, *A Motor-Flight through France* (New York: Charles Scribner's Sons, 1908), p. 1.

CHAPTER 10. IRKUTSK – TOMSK: 1–11 JULY

[1] **much of it seemed new** Jeremiah Curtin, *A Journey in Southern Siberia: The Mongols, Their Religion and Their Myths* (London: Sampson Low, Marston, 1910), p. 19; **survivors of a fire** Beer, *House of the Dead*, p. 3; George W. DeLong, *Our Lost Explorers: The Narrative of the Jeannette Arctic Expedition* (Hartford, CN: American Publishing, 1882), pp. 158–62; **brash gold-mining millionaires** | **[footnote]** de Windt, *Pekin to Calais*, p. 398; Victor Meignan, *From Paris to Pekin over Siberian Snows*, trans. William Conn (London: Swan Sonnenschein, 1885), pp. 182–83.

[2] **sundry reports** Barzini, 'Adventure in Siberia'; **Earlier that day** Barzini, *Peking to Paris*, p. 214; **They allowed themselves** Barzini, *Peking to Paris*, pp. 215–16; **attended to the Itala** Barzini, *Peking to Paris*, p. 217.

[3] **[footnote]** Jean du Taillis, 'Celebrations in Irkutsk', *Le Matin*, 7 July 1907, trans. Laura Shanahan; **In addition to the city's** du Taillis, *Pékin–Paris*, pp. 193–94; Cormier, *Le Raid*, p. 49; **People jostle** du Taillis, 'Celebrations in Irkutsk'.

[4] **Progress is difficult** Jean du Taillis et al., 'Through The Siberian Forest', *Le Matin*, 10 July 1907, trans. Margaret Morrison; **wet, dark, and sad** Barzini, *Peking to Paris*, p. 236; **Just outside Cheremknovo** du Taillis, *Pékin–Paris*, pp. 196–97; Jean du Taillis, 'Pekin–Paris', *Le Matin*, 13 July 1907, trans. Margaret Morrison.

[5] **noted that the population** Samuel Turner, *Siberia: A Record of Travel, Climbing, and Exploration* (London: T. Fisher Unwin, 1905), pp. 22–23.

[6] **The rough ascents** Luigi Barzini, 'Pekin to Paris: Across Siberia', *Daily Telegraph*, 6 July 1907; for descriptions of villages, see Barzini, *Peking to Paris*, p. 223; Luigi Barzini, 'Pekin to Paris: Motoring in a Deluge', *Daily Telegraph*, 8 July 1907; Collignon, *Souvenirs*, pp. 22–25; Curtin, *Southern Siberia*, pp.

24–26; de Windt, *Pekin to Calais*, p. 471; **Odours of old skins** de Windt, *Pekin to Calais*, p. 471; **[footnote]** Collignon, *Souvenirs*, p. 22; Turner, *Siberia*, p. 54.

[7] **benumbed | The roads are good** Barzini, 'Across Siberia'; **would scarcely be called** de Windt, *Pekin to Calais*, pp. 475–76; **flowery meadow | We use second** Cormier, *Le Raid*, p. 60; **As Cormier's De Dion-Bouton** du Taillis, *Pékin–Paris*, p. 199; **Distances they expected | There has not been** Barzini, *Peking to Paris*, pp. 236, 232–33.

[8] **Never, even in the desert** Barzini, 'Motoring in a Deluge'.

[9] **During the winter months | another infamous case** Beer, *House of the Dead*, pp. 229–30, 216, 234, 237; **a family had been waylaid** de Windt, *Pekin to Calais*, pp. 424, 484; **left food outside** Turner, *Siberia*, pp. 56–57.

[10] **About seventy miles** Barzini, Motoring in a Deluge'.

[11] **On 8 July** Luigi Barzini, 'Pekin to Paris: Salving a Ferry-Boat', *Daily Telegraph*, 10 July 1907; **A few days later** Luigi Barzini, 'Pekin to Paris: Itala Car at Tomsk', *Daily Telegraph*, 12 July 1907.

[12] **Telegrams sent** Jean du Taillis et al., 'Through the Siberian Forest'; **Four days later** Georges Cormier et al., 'The Pekin–Paris Challenge: Titans in Automobiles', *Le Matin*, 12 July 1907, trans. Margaret Morrison.

[13] **the aristocracy** Comyns Beaumont, 'Automobile Topics', *The Bystander*, 17 July 1907, 143; **If there were any further** '113 Kilometres Per Hour', *Le Matin*, 3 July 1907, trans. Laura Shanahan.

[14] **Longoni**, Cormier, *Le Raid*, p. 56

[15] **As a sop | Despite the journalist's** Jean du Taillis and Georges Bourcier Saint-Chaffray, 'The Challenge of *Le Matin*: They Drive on a Road', *Le Matin*, 14 July 1907 , trans. Margaret Morrison; **[footnote]** '113 Kilometres Per Hour', *Le Matin*, 3 July 1907, trans. Laura Shanahan.

CHAPTER 11. HORSE POWER

[1] **In 1900** Joel A. Tarr, *The Search for the Ultimate Sink: Urban Pollution in Historical Perspective* (Akron, OH: University of Akron Press, 1997), pp. 323–24; **author of the 1893 book** William J. Gordon, *The Horse World of London (1893)* (Newton Abbot: David & Charles, 1971), p. 67.

[2] **Under this Urn** 'Monumental Inscriptions on an Urn', *Gentleman's Magazine*, 50 (1780), p. 242.

[3] **locals could immediately** A. D. Rubets, *History of Automobile Transport in*

Russia (Moscow: Eksmo, 2008), p. 12, trans. Dominique Hoffman; **Henry Meux & Co.** Susanna Forrest, *The Age of the Horse: An Equine Journey through Human History* (London: Atlantic Books, 2016), pp. 165–66.

[4] **Quite the opposite** Clay McShane and Joel A. Tarr, *The Horse in the City: Living Machines in the Nineteenth Century* (Baltimore: Johns Hopkins University Press, 2007), p. 1; **84 million tons** Forrest, *Age of the Horse*, p. 159; **specialised horse-drawn vehicles** see for example Don H. Berkebile, ed., *Horse-Drawn Commercial Vehicles: 255 Illustrations of Nineteenth-Century Stagecoaches, Delivery Wagons, Fire Engines, Etc.* (New York: Dover, 1989). For a fuller – and beautifully written – illustration of the sheer number of ways horses were put to use in nineteenth-century cities, see Forrest, *Age of the Horse*, pp. 157–63.

[5] **London in the 1890s** Gordon, *Horse World*, p. 29; **idiosyncratic names** Gordon, *Horse World*, p. 36; **in 1854** McShane and Tarr, *Horse in the City*, p. 58; **By 1890 each of the horses** Gordon, *Horse World*, pp. 10, 12; **the *Los Angeles Record*,** quoted in Scharff, *Taking the Wheel*, p. 7; **Society for the Protection of Passenger Rights** Scharff, *Taking the Wheel*, pp. 6–7.

[6] **Twelve million acres** McShane and Tarr, *Horses in the City*, p. 33; **368,000 Americans** McShane and Tarr, *Horses in the City*, p. 31; **London's Road Car Company** Gordon, *Horse World*, p. 19; **whence it comes** Gordon, *Horse World*, p. 53.

[7] **During North America's** McShane and Tarr, *Horses in the City*, p. 27; Tarr, *Ultimate Sink*, p. 327; **the oblivion of sausage meat** Gordon, *Horse World*, pp. 88–89.

[8] **there can be no doubt** Gordon, *Horse World*, p. 9; **'The Horse vs. Health',** quoted in Tarr, *Ultimate Sink*, p. 323; **Benjamin Franklin | more accidents,** quoted in Tarr, *Ultimate Sink*, p. 326; **Feeding 10,000** Gordon, *Horse World*, pp. 15–16.

[9] **In the early part** McShane and Tarr, *Horses in the City*, pp. 25–27; **Health officials in** Tarr, *Ultimate Sink*, p. 324; **In New York itself** George E. Waring Jr, *Street-Cleaning, and the Disposal of a City's Wastes: Methods and Results and the Effect upon Public Health, Public Morals, and Municipal Prosperity* (London: Gay & Bird, 1898), p. 8; **Tverskaya fire station** A. D. Rubets, *Automobile Transport*, p. 10.

[10] **In London in 1873** 'Fever in St. Giles's', *The Times*, 9 October 1873; **a reign of the cholera plague** 'Inviting Cholera', *New York Times*, 23 August 1871; **A decade later** 'Dr Peters on Filth Diseases', *New York Times*, 3 June 1881.

[11] **Paris was employing | By 1895** Waring, *Street-Cleaning*, pp. 136–37; **And We Will Keep Right On** 'Metropolitan News', *Illustrated London News*, 16 December 1848; **It is utterly hopeless** Waring, *Street-Cleaning*, p. 9.

[12] **The London Power Omnibus** 'London Traffic Problem: Motor-Bus Finance', *Daily Telegraph*, 24 July 1907; **Motorised public transport** 'Berlin's Motor-Cabs: Threatened Crisis', *Daily Telegraph*, 31 August 1907; *Le Matin* **saw nothing contradictory** 'The Festival of the Horse', *Le Matin*, 17 June 1907, trans. Laura Shanahan; **confidently predicted** 'The Triumph of the Horse', *The Economist*, 7 September 1907, pp. 1495–96.

CHAPTER 12. TOMSK – TYUMEN: 11–19 JULY

[1] **Tomsk was an impressive** For descriptions of Tomsk, see du Taillis, *Pékin–Paris*, pp. 217–18; Cormier, *Le Raid*, p. 63; Barzini, 'Itala Car at Tomsk'; Turner, *Siberia*, p. 99; **When the Itala** Luigi Barzini, 'Pekin to Paris: Startling Adventure', *Daily Telegraph*, 16 July 1907.

[2] **As they continued** Jean du Taillis, 'The Pekin–Paris Challenge: Siberian Chaos', *Le Matin*, 21 July 1907, trans. Margaret Morrison; **They have the same** | **It was a desperate** Barzini, 'Startling Adventure'.

[3] **an extinct traffic** Barzini, 'Startling Adventure'; **nothing but a deep rut** | **It is the Tract** Cormier, *Le Raid*, p. 57.

[4] **Sunday, 14 July** Barzini, 'Startling Adventure'; Barzini, *Peking to Paris*, pp. 262–63.

[5] **[footnote]** Turner, *Siberia*, pp. 64–66; **in which the largest building** Kennan, *Siberia*, I, pp. 140–41; **Along the wooden** Barzini, *Peking to Paris*, p. 265; **The town buzzed with the news** 'Reported Death of Count Tolstoy', *Daily Telegraph*, 16 July 1907; Our Own Correspondent [St Petersburg], 'Count Tolstoy', *Daily Telegraph*, 17 July 1907; 'Tolstoy's Death Rumored', *New York Times*, 16 July 1907; 'Tolstoy Building Burned', *New York Times*, 17 July 1907; **The motorists spent two days** Luigi Barzini, 'Pekin to Paris: Needful Rest at Omsk', *Daily Telegraph*, 17 July 1907; **[footnote]** du Taillis, *Pékin–Paris*, p. 228; Moretti, 'Il Terzo Uomo', p. 16, translation mine.

[6] **The Russian committee advocated** Barzini, 'Needful Rest at Omsk'; **a steady seepage** see for example 'Godard en de Spijker', *Het Nieuws van den Dag*; du Taillis and Bourcier Saint-Chaffray, 'They Drive on a Road'; Jean du Taillis and Georges Bourcier Saint-Chaffray, 'Pekin–Paris', *Le Matin*, 17 July 1907, trans. Margaret Morrison; **at three o'clock** Luigi Barzini, 'Pekin to Paris: Speedy Progress', *Daily Telegraph*, 20 July 1907; Barzini, *Peking to Paris*, p. 269.

[7] **Newspapers that** 'Peking to Paris: Plucky Motor Car Racers', *South China Morning Post*, 19 July 1907; 'Difficulties of the Paris–Peking Motor Race', *Illustrated London News*, 20 July 1907; 'Peking–Paris', *Sydney Morning Herald*, 25 July 1907; **faithful mount** 'Photographs of the

Road: Our First Snapshots of Pekin–Paris', *Le Matin*, 18 July 1907, trans. Margaret Morrison.

[8] **There is no news** Jean du Taillis, Georges Cormier et al., 'The Peking–Paris Challenge: Their Adventures', *Le Matin*, 19 July 1907, trans. Margaret Morrison.

[9] **Once past Omsk** Barzini, 'Speedy Progress'; **Borghese refused | power of resistance** Luigi Barzini, 'Conclusion of the Great Motor Race: Itala Car's Arrival in Paris', *Daily Telegraph*, 12 August 1907; **punishing regimen** Barzini, *Peking to Paris*, p. 267.

[10] **Outside every house** Barzini, 'Speedy Progress'.

[11] **The following day** Barzini, *Peking to Paris*, pp. 274–75.

[12] **On Thursday, 18 July** du Taillis, 'Siberian Chaos'; Cormier, *Le Raid*, p. 63.

[13] **The Frenchmen spent** du Taillis, 'Siberian Chaos'; Cormier, *Le Raid*, pp. 62–64; **[footnote]** du Taillis, *Pékin–Paris*, pp. 217–18; du Taillis, 'Siberian Chaos'(my thanks to Margaret Morrison for spotting the inconsistency).

CHAPTER 13. TAKING UP SPACE

[1] **pandemonium in Philadelphia** Clarence S. Brigham, 'Edgar Allan Poe's Contributions to *Alexander's Weekly Messenger*', *Proceedings of the American Antiquarian Society*, 52.1 (1942), pp. 45–125.

[2] **[footnote]** John Frost, *The Life of William Penn, the Founder of Pennsylvania* (Philadelphia: Lindsay & Blakiston, 1849), p. 113; **1896 petition** McShane, *Asphalt Path*, p. 80.

[3] **Chicago as late as 1926** Peter D. Norton, *Fighting Traffic: The Dawn of the Motor Age in the American City* (Cambridge, MA: MIT Press, 2008), p. 66; **Kensington and Chelsea, Harrow School** Martin Daunton, ed., *The Cambride Urban History of Britain, 1840–1950* (Cambridge: Cambridge University Press, 2000), III, p. 239.

[4] **Nor were cyclists** 'Outrageous Attack upon a Cyclist', *Daily Telegraph*, 31 May 1898; 'Brutal Attack on Cyclist', *Sheffield Daily Telegraph*, 17 June 1896; **In 1894 | In London** 'The Cyclist Scorcher Again', *Sheffield Weekly Telegraph*, 13 January 1894; **Fear of attack** 'Guns, Wheels, and Steel: Cyclists and Small Arms in the Late Nineteenth Century', Cycling History website, 2015; **racing with a trolley car** 'Gossip of the Cyclers: Severe Punishment Necessary to Minimise "Scorching"', *New York Times*, 3 May 1896.

[5] **a train of cars** McShane, *Asphalt Path*, pp. 12–13.

[6] **a matter of both** 'Auto-Killings Now at the Rate of Two an Hour', *Literary Digest*, 6 November 1920, p. 82; *St Louis Star* | **A judge in Philadelphia** Norton, *Fighting Traffic*, pp. 68, 69; **while another** Peter D. Norton, 'Street Rivals: Jaywalking and the Invention of the Motor Age Street', *Technology and Culture*, 48.2 (2007), p. 337.

[7] **Such efforts** 'Jaywalking: How the Car Industry Outlawed Crossing the Road', BBC News, 12 February 2014; **in Detroit in 1922** Norton, *Fighting Traffic*, pp. 72–73, 77.

[8] **The old common law** | **In Washington DC** Norton, *Fighting Traffic*, pp. 164, 142.

[9] **In America in 1906** Norton, 'Street Rivals', p. 338; **anybody thought the abolition** 'Hansard: Motor Vehicles and Speedometers' (House of Lords, 1932), pp. 169–96.

[10] **When surveyed** Norton, 'Street Rivals', p. 339; **even more determined** 'Jaywalking:', BBC News; Norton, *Fighting Traffic*, pp. 226–29.

[11] **45,000 miles** National Museum of American History, 'A Streetcar City', 2016; **General Motors and allied highway** *Reform of the Federal Criminal Laws: Hearings Before the Subcommittee on Criminal Laws and Procedures of the Committee on the Judiciary. United States Senate; Ninety-Fourth Congress; First Session on S.1 and S.1400. June 13, 17, July 19, 22, 1974 Part XI* (Washington DC: United States Senate, 1974), p. 7900; **it certainly was true that** Martha J. Bianco, 'Kennedy, 60 Minutes, and Roger Rabbit: Understanding Conspiracy-Theory Explanations of the Decline of Urban Mass Transit', Center for Urban Studies Publications and Reports, 17 (1998).

[12] **if you're a car** Quoted in 'A Chance to Transform Urban Planning', *The Economist*, 1 March 2018; **over two and a half hours** Gilles Duranton and Matthew Turner, 'The Fundamental Law of Road Congestion: Evidence from US Cities', *American Economic Review*, 101.6 (2011), p. 2616; **hefty economic impact** 'The Hidden Cost of Congestion', *The Economist*, 28 February 2018; **never necessary** J. H. Crawford, 'The Car Century Was a Mistake. It's Time to Move On', *Washington Post*, 29 February 2016; Economist Intelligence Unit, *The Global Liveability Index 2022: Recovery and Hardship* (2022).

[13] **Bristol, Amsterdam** Somini Sengupta and Nadja Popovich, 'Cities Worldwide Are Reimagining Their Relationship with Cars', *New York Times*, 14 November 2019; **Bologna** Kimberly Nicholas and Paula Kuss, 'What Are the Most Effective Ways to Get Cars out of Cities?', *The Guardian*, 16 April 2022; **Paris** Natalie Marchant, 'Paris Halves Street Parking and Asks

Residents What They Want to Do with the Space', World Economic Forum, 7 December 2020; **Heidelberg** Jack Ewing, 'The City Where Cars Are Not Welcome', *New York Times*, 28 February 2021.

[14] **5,500 square miles** Oliver Milman, 'Shifting Gears: Why US Cities Are Falling out of Love with the Parking Lot', *The Guardian*, 26 December 2022; **Washington DC** David Zipper, 'A City Fights Back against Heavyweight Cars', Bloomberg, 26 May 2022; **A 2022 California bill** Andrew Khouri, 'California Bans Mandated Parking Near Transit to Fight High Housing Prices, Climate Change', *Los Angeles Times*, 23 September 2022.

[15] **prototypes have been underwhelming** 'Elon Musk Unveils Prototype High-Speed LA Transport Tunnel', BBC News, 19 December 2018; Jack Morse, 'Oh Look, It's a Tesla Traffic Jam in Las Vegas' Boring Company Tunnel', *Mashable*, 6 January 2022; **article for the *Wall Street Journal*** Ted Mann and Julie Bykowicz, 'Elon Musk's Boring Company Ghosts Cities Across America', *Wall Street Journal*, 28 November 2022; **'ghosting' cities** Kate Linebaugh and Ryan Knutson, 'Elon Musk's Boring Company is Ghosting Cities', *The Journal*, 8 December 2022.

CHAPTER 14. TYUMEN – KAZAN: 19–24 JULY

[1] **The steppes began** Luigi Barzini, 'Pekin to Paris: Speedy Progress'.

[2] **[footnote]** Jean du Taillis, 'Peking–Matin: How the De Dions Crossed the Urals', *Le Matin*, 1 August 1907, trans. Margaret Morrison; **As they drove** Luigi Barzini, 'Pekin to Paris: Entry into Europe', *Daily Telegraph*, 22 July 1907; **The earth that squelched** Collignon, *Souvenirs*, p. 32; **bare and unfinished** de Windt, *Pekin to Calais*, p. 613.

[3] **strange signs of exorcism** Luigi Barzini, 'Pekin to Paris: Speedy Progress'; **the Italians approached** | **Our car made and left** Barzini, *Peking to Paris*, pp. 279–80; **When we reach them** Barzini, 'Speedy Progress'.

[4] **There are certain mountains** *The Russian Primary Chronicle: Laurentian Text*, trans. Samuel Hazzard Cross and Olgerd P. Sherbowitz-Wetzor (Cambridge, MA: Mediaeval Academy of America, 1953), p. 184; **highest peak does not** de Windt, *Pekin to Calais*, p. 618; **The Itala slips** Barzini, 'Entry into Europe'.

[5] **At 5.17 a.m.** | **the practical** Barzini, 'Entry into Europe'; **imposing letters carved** de Windt, *Pekin to Calais*, p. 618; **[footnote]** Knox, *Overland through Asia*, pp. 222–23.

[6] **One unfortunate** 'Police Censors of Russian Press Now: Under Gen.

Drachevsky's Order 26 Papers Are Fined and 22 Suppressed', *New York Times*, 4 August 2019; **The roads in Siberia** Editorial Board, 'Peking–Paris', *Avtomobil'*, 1 August 1907, p. 1804, trans. Dominique Hoffman; heavy-handed censorship was often mentioned by British diplomats in dispatches to the Foreign Office. See for example Charles H. Bentinck, 'Summary of Events in Russia during the Fortnight Ending June 19, 1907' (St Petersburg: British Foreign Office, 20 June 1907), pp. 392–394v; and Nevile Henderson, 'Report by Mr. Henderson on Events in Russia during the Fortnight Ended August 1, 1907' (St Petersburg: British Foreign Office, 1 August 1907), pp. 435–37; **Russia's oldest newspaper** reprinted in Lakhvari, 'Significance'.

[7] **myth of the automobile** Khilkov, quoted in Our Own Correspondent [Moscow], 'How Peking–Matin is going to Shake Up Russia', *Le Matin*, 16 August 1907, trans. Margaret Morrison.

[8] **becoming sharp** see for example du Taillis, *Pékin–Paris*, p. 239; *mangeur de poussière* Collignon, *Souvenirs*, p. 32; **admirable evenness | odd little anecdote** Jean du Taillis, 'Will we See the End of the Accursed Forest?', *Le Matin*, 22 July 1907, trans. Margaret Morrison; **The journalist griped** du Taillis, *Pékin–Paris*, p. 207.

[9] **[footnote] | dear friends** Collignon, *Souvenirs*, pp. 3, 5; **Pivo | penniless | lice** Cormier, *Le Raid*, pp. 66, 71, 87; **the epitome of sobriety** Jean du Taillis, 'From Peking to Le Matin: Dust Drinkers', *Le Matin*, 26 July 1907, trans. Margaret Morrison; **In a longer article** Georges Cormier, 'Pékin–Paris', p. 5.

[10] **responsible for the decision** Although Cormier doesn't exactly claim responsibility for this decision, he makes clear elsewhere that it is him selecting the route. See for example Cormier, *Le Raid*, p. 70; **followed a different itinerary** du Taillis, *Pékin–Paris*, p. 228; **the same explanation** Cormier, *Le Raid*, p. 68; **An exchange in the pages** Georges Cormier and Georges Bourcier Saint-Chaffray, 'Pékin–Paris', *L'Auto*, 27 July 1907, p. 5.

[11] **An hour out of Perm** Luigi Barzini, 'Pekin to Paris: Itala Car Breaks a Wheel', *Daily Telegraph*, 25 July 1907.

CHAPTER 15. POWER PLAY

[1] **As the summer** Frederick A. Talbot, *The Oil Conquest of the World* (London: William Heinemann, 1914), pp. 12–13; 'When First Oil Flowed', *New York Times*, 22 July 1934; **remedy for scurvy** Pierre L'Espagnol de la Tramerye, *The World-Struggle for Oil*, trans. C. Leonard Leese (London: Allen & Unwin, 1924), p. 25.

[2] **Work had commenced** Talbot, *Oil Conquest*, pp. 12–14; 'When First Oil Flowed', *New York Times*.

[3] **Even as these two titans** L'Espagnol de la Tramerye, *World-Struggle*, p. 53.

[4] **San Francisco earthquake** Daniel Yergin, *The Prize: The Epic Quest for Oil, Money, and Power* (New York: Simon & Schuster, 1991), p. 80.

[5] **methods of oil extraction** Brita Åsbrink, 'Nobel Spies on His Competitor, Rockefeller', Branobel History website; **By 1900** Brita Åsbrink, 'Harvest Times for the Nobel Brothers' Oil Adventure', Branobel History website; **Five years later** 'The Baku Oilfields: Great Fires Will Cripple Russia's Oil Trade', *New York Times*, 10 September 1905; **Russian trains** Albert Lidgett, *Petroleum* (London: Sir Isaac Pitman & Sons, 1919), p. 82; E. M. Movsumzade, 'The First Attempts to Use Oil as Navy Fuel', *Icon*, 6 (2000), pp. 142–48.

[6] **In 2020** N. Sonnichsen, *OECD: Oil Demand Share by Sector*, Statista, 18 January 2022; **The ready availability** Michael T. Klare, *Blood and Oil: The Dangers and Consequences of America's Growing Petroleum Dependency* (London: Hamish Hamilton, 2004), p. 8; **free and unfettered** General J. H. Binford Peay III, quoted in Klare, *Blood and Oil*, p. 3; **Protecting the Strait of Hormuz** Statista Research Department, *Oil Flows: Strait of Hormuz 2014–2020*, Statista, 6 April 2022.

[7] **At the turn of the twentieth century** 'Automobile: Early Electric Automobiles', *Britannica*,; **France, Germany and Great Britain** L'Espagnol de la Tramerye, *World-Struggle*, pp. 11, 13, 77–79.

[8] **littered with decaying** | **[footnote]** Quoted in Simon Sebag Montefiore, *Young Stalin* (London: Weidenfeld & Nicolson, 2007), p. 162; **in Batum** Yergin, *The Prize*, p. 129; **Sectarian violence** Associated Press, 'Oil Fields of Baku in Flames', *Los Angeles Herald*, 7 September 1905; Kellogg Durland, *The Red Reign: The True Story of an Adventurous Year in Russia* (New York: The Century, 1908), pp. 77–78; **By early September** 'Revolution and Fire are Devastating Baku', *New York Times*, 7 September 1905; **oilmen gloomily predicted** Raffalovich, 'Financial Progress'; **[footnote]** Durland, *Red Reign*, pp. 78–79; **[footnote]** 'Baku Oilfields', *New York Times*.

[9] **In an article published** 'What Thomas A. Edison Thinks Will Be Science's Next Most Vital Discovery', *New York Times*, 7 January 1906; **[footnote]** see for example 'What Edison Thinks', *New York Times*, 'Edison's New Battery', *New York Times*, 12 June 1906; 'Edison About to Give to the World His Greatest Wonder', *New York Times*, 21 October 1906; 'Edison, at Sixty, Outlines Wonders of the Future', *New York Times*, 19 May 1907.

[10] **An interview the following year** 'Edison, at Sixty', *New York Times Scientific American* | **General Electric** quoted in Brendan Cormier and Lizzie Bisley, eds, *Cars: Accelerating the Modern World* (London: V&A, 2019), p. 128; **a third of all vehicles** 'The History of the Electric Car', US Department

of Energy, 15 September 2014; **Edison was working on** 'Edison's New Battery', *New York Times* **greatest wonder** 'Edison About to Give', *New York Times*.

[11] **no broad understanding** See for example Lidgett, *Petroleum*, pp. 1–3; **With the constant withdrawal** Lidgett, *Petroleum*, p. 6.

[12] **steam still had** 'Steam v. Petrol', *The Autocar*, 20 April 1907, pp. 580–81; *The Bystander* | **It can be distilled** Comyns Beaumont, 'Automobile Topics', *The Bystander*, 31 July 1907, pp. 253–54; **A series of tests** 'Automobile Topics of Interest', *New York Times*, 24 August 1902; 'The Alcohol Propaganda', *The Autocar*, 17 August 1907, pp. 276–77; 'Wheels and Ways: Alcohol as a Motor Fuel', *The Field*, 14 April 1906, p. 600; **As might be expected** 'Wheels and Ways', *The Field*; 'Motor Notes', *The Field*, 7 June 1902, p. 891; **Several races** Tibballs, *Strangest Races*, pp. 28–29; 'Motor Notes', *The Field*; 'Notes: What Next?', *The Autocar*, 17 August 1907, p. 259; 'On Motoring', *Morning Post*, 25 March 1904; 'Alcohol Motor Fuel', *Scotsman*, 19 March 1919; 'Alcohol Now Urged as a Motor Fuel', *New York Times*, 31 May 1914; **It is useless** 'Notes: What Next?', *The Autocar*.

[13] **Allied armies were consuming** | **Clemenceau** Quoted in L'Espagnol de la Tramerye, *World-Struggle*, p. 82.

[14] **Sales of electric vehicles** International Energy Agency, *Global EV Outlook 2022*, May 2022.

CHAPTER 16. KAZAN – MOSCOW: 24–27 JULY

[1] **The roads had dried** | **Night and hunger** Barzini, 'Itala Car Breaks a Wheel'; **As they passed** Barzini, *Peking to Paris*, p. 294; for sundry descriptions of the reactions of Russian villagers to the sight of the car, see du Taillis, 'Dust Drinkers'; Cormier, *Le Raid*, p. 74; Collignon, *Souvenirs*, p. 30; du Taillis, *Pékin–Paris*, p. 240; Barzini, *Peking to Paris*, p. 222.

[2] **Minarets and delicate towers** Barzini, 'Itala Car Breaks a Wheel'; **garments of red** | **We crossed little tiny** Barzini, *Peking to Paris*, pp. 300, 297.

[3] **At three o'clock** Barzini, *Peking to Paris*, p. 301; **Moscow in miniature** Munro-Butler-Johnstone, *Trip up the Volga*, pp. 55–56; **the true boundary** | **the long journey from China** de Windt, *Pekin to Calais*, pp. 628–29; **Electric tramways** Cormier, *Le Raid*, p. 85.

[4] **famine belt** Durland, *Red Reign*, p. 328; **no hope** Prince Ukhtomsky, quoted in Durland, *Red Reign*, pp. 328–29.

[5] **Barzini's newspaper reports** Luigi Barzini, 'Pekin to Paris: Itala Car

Stoned by Peasants', *Daily Telegraph*, 27 July 1907; **The greatest river of Europe** Barzini, *Peking to Paris*, p. 302.

[6] **up the precipitous slope | descriptions of the fair** Munro-Butler-Johnstone, *Trip up the Volga*, pp. 69, 71; **Persians, who descend** Luigi Barzini, 'Pekin to Paris: Prince Borghese's Arrival at Moscow', *Daily Telegraph*, 29 July 1907.

[7] **The rain that had dogged** Jean du Taillis, 'From Peking to Le Matin: By Water and by Fire', *Le Matin*, 27 July 1907, trans. Margaret Morrison; **Never before have we** Our Own Correspondent [Moscow] and Jean du Taillis, 'From Peking to Le Matin: Moscow Cheers Borghese', *Le Matin*, 28 July 1907, trans. Margaret Morrison.

[8] **Nizhny's elite** Luigi Barzini, 'Pekin to Paris: Prince Borghese's Arrival at Moscow', *Daily Telegraph*, 29 July 1907.

[9] **The road is straight | Twenty miles outside** Barzini, 'Borghese's Arrival at Moscow'; **[footnote] | several foreign consuls** 'From Peking to Le Matin: Borghese', *Le Matin*, 25 July 1907, trans. Margaret Morrison.

[10] **all begrimed with** Barzini, 'Borghese's Arrival at Moscow'; **He wore** 'From Peking to Le Matin: Borghese', *Le Matin*, 25 July 1907, trans. Margaret Morrison.

[11] **The Spyker has covered** Our Own Correspondent and du Taillis, 'Moscow Cheers Borghese'.

CHAPTER 17. TRANSFORMERS

[1] **gleaming New York premises** 'The Automobile Club's Sumptuous New Home', *New York Times*, 14 April 1907; **They are expensive** *The Motor*, quoted in Brandon, *Automobile*, p. 42; **Itala catalogues** Itala brochures from David Ayre Private Collection.

[2] **Popular culture** See for example 'The Moustached Chauffeur', *The Motor*, 9 April 1907, p. 9; **Every motor garage** 'Motoring', *Daily Telegraph*, 12 January 1907; **[footnote]** 'Topics of the Times: "Cheapness" in a Multi-Millionaire', *New York Times*, 18 March 1905.

[3] **A light-hearted** Count Tishkevich, 'Guidance If You Are Invited for an Automobile Outing', *Avtomobil'*, 1 September 1907, p. 1852, trans. Dominique Hoffman; **simply smothered** 'An Innovation at Shanghai: Motor Parade', *South China Morning Post*, 13 June 1905.

[4] **regular motoring jaunts** See for example 'Henry Ford Goes to See the Start', *New York Times*, 12 January 1914; 'Burroughs Joins Edison Campers',

New York Times, 30 August 1916; Corey Ford, 'Two Weeks' Vagabonds', *New York Times*, 30 July 1922; for a detailed look at these trips see Jeff Guinn, *The Vagabonds: The Story of Henry Ford and Thomas Edison's Ten-Year Road Trip* (New York: Simon & Schuster, 2019).

[5] **under Alfred Sloan | laws of the Paris** E. Hawes, 'Making the Modern Consumer', in Cormier and Bisley, *Cars*, p. 100; **By 1957** Ralph Nader, *Unsafe at Any Speed: The Designed-In Dangers of the American Automobile* (New York: Grossman, 1965), p. 212.

[6] **so closely** Brandon, *Automobile*, p. 3; Temple University, 'People Who Really Identify with Their Car Drive More Aggressively, Study Finds', *ScienceDaily,* 17 October 2011.

[7] **On 26 March 1923** Massimo Moraglio, *Driving Modernity: Technology, Experts, Politics, and Fascist Motorways, 1922–1943*, trans. Erin O'Loughlin (New York: Berghahn, 2017), pp. 1–2.

[8] **It was not bad at all** E. Kuzmin, 'Unfamiliar Road', *Avtomobil'*, 15 May 1907, p. 1731, trans. Dominique Hoffman.

[9] **an article about** I. Yakovlev, 'On the Automobile Exhibition', *Avtomobil'*, 1 July 1907, p. 1778, trans. Dominique Hoffman.

[10] **The people need** Quoted in Edward Crankshaw, 'When Lenin Returned', *The Atlantic*, October 1954; **With this victory** Vladimir Ilyich Lenin, 'Speech Delivered at the First All-Russia Congress of Working Cossacks', trans. George Hanna, 1920.

[11] **Economic chaos** Sonia Melnikova-Raich, 'The Soviet Problem with Two "Unknowns": How an American Architect and a Soviet Negotiator Jump-started the Industrialization of Russia, Part II: Saul Bron', *IA. The Journal of the Society for Industrial Archeology*, 37.1/2 (2011), p. 9; **famine threatened annually** O'Beirne, 'Agricultural Distress'; **there were fewer** Sonia Melnikova-Raich, 'The Soviet Problem with Two "Unknowns": How an American Architect and a Soviet Negotiator Jump-started the Industrialization of Russia, Part I: Albert Kahn', *IA. The Journal of the Society for Industrial Archeology*, 36.2 (2010), p. 58.

[12] **early models were** Brandon, *Automobile*, pp. 79; 81; **[footnote]** Scharff, *Taking the Wheel*, p. 11; **It was basic** White, 'Farewell, My Lovely!'; **come through the factory** Henry Ford to John W. Anderson, investor, quoted in Brandon, *Automobile*, p. 79; **it terrified him** Brandon, *Automobile*, p. 102; **Prices for a Model T | 28 factories in America** Cormier and Bisley, *Cars*, pp. 64, 65; Walter Duranty, 'Talk of Ford Favor Thrills Moscow', *New York Times*, 17 February 1928; **In the 1920s** Melnikova-Raich, 'Soviet Problem I', p. 58; **[footnote]** Cormier and Bisley, *Cars*, p. 76.

[13] **By June 1942** 'Industry's Architect', *Time*, 29 June 1942; **In 1930, a deal** '$1,900,000,000 Building by the Soviet in 1930', *New York Times*, 11 January 1930; Melnikova-Raich, 'Soviet Problem I', pp. 59–60; **[footnote]** Melnikova-Raich, 'Soviet Problem I', p. 68.

[14] **When we place the USSR** Quoted in Lewis Siegelbaum, *Cars for Comrades: The Life of the Soviet Automobile* (Ithaca, NY: Cornell University Press, 2008), p. 20.

[15] **Each day we are shipping** Quoted in Melnikova-Raich, 'Soviet Problem', p. 74.

[16] **Almost the exact** Roland Barthes, *Mythologies*, trans. Annette Lavers (London: Vintage, 2000), p. 88.

CHAPTER 18. MOSCOW - KAUNAS: 27 JULY-4 AUGUST

[1] **Barzini's impression** Barzini, *Peking to Paris*, p. 317; Barzini, 'Borghese's Arrival at Moscow'; **Moscow was famous** de Windt, *Pekin to Calais*, p. 640; **[footnote]** Durland, *Red Reign*, p. 20.

[2] **[footnote]** Alberta Cavazza, *The Enchanted Island Isola Del Garda*, trans. Charlotte Chetwynd Talbot and Anja Pattenden (Cierre Grafica, 2016); **My travel impressions?** Our Own Correspondent and du Taillis, 'Moscow Cheers Borghese'; **trotting races | dying sun tinged** Luigi Barzini, 'Pekin to Paris: Surfeit of Hospitality', *Daily Telegraph*, 31 July 1907; **[footnote]** de Windt, *Pekin to Calais*, pp. 642–43.

[3] **Slavianski Bazaar** De Windt, *Pekin to Calais*, pp. 638–39; **a most agreeable captivity** Barzini, 'Borghese's Arrival at Moscow'.

[4] **On their final day** Barzini, 'Surfeit of Hospitality'; Editorial Board, 'Peking–Paris'; Moretti, 'Il Terzo Uomo', p. 16, translation mine; **In a report published** Special Correspondent, 'Peking–Matin: Prince Borghese Has Left Moscow', *Le Matin*, 1 August 1907, trans. Margaret Morrison; **We have only one regret** Barzini, *Peking to Paris*, p. 323; **the Russian editors** Editorial Board, 'Peking–Paris'.

[5] **Even at four o'clock** Editorial Board, 'Peking–Paris'.

[6] **a habitual pessimist | They came out of the stalls** Jean du Taillis and Scipione Borghese, 'From Peking to Le Matin: Siberian Parasites', *Le Matin*, 29 July 1907, trans. Margaret Morrison; **Ink-black storm clouds** Jean du Taillis, 'Peking–Matin: The Fury of the Urals', *Le Matin*, 3 August 1907, trans. Margaret Morrison; **Between Birsk and Yelabuga** Collignon, *Souvenirs*, pp. 32–33; **Just the day before | With the exception** Cormier, *Le Raid*, pp. 71,

76, 78; **struggled to secure fuel** Collignon, *Souvenirs*, pp. 32, 33; Cormier, *Le Raid*, p. 82; **One lunch** Collignon, *Souvenirs*, p. 33; **tins of sardines** Cormier, *Le Raid*, p. 80.

[7] **they were thinner | small, uninteresting town** Cormier, *Le Raid*, p. 78; **Police met them** Jean du Taillis, 'Peking–Matin: Dizzying Drops and Torrential Rain', *Le Matin*, 4 August 1907, trans. Margaret Morrison; **morosely commented** Jean du Taillis, 'Peking–Matin: Their Beards', trans. Margaret Morrison, *Le Matin*, 5 August 1907.

[8] **sentimental address** Du Taillis, 'Their Beards'; **The former, back in Paris** Special Correspondent and Pons, 'Borghese Arrives in St Petersburg'.

[9] **professional driver | Wall of Death** Allen Andrews, *The Mad Motorists: The Great Peking–Paris Race of '07* (London: Harrap, 1964), pp. 21–22; **The clearest account** 'Registre Matricule: Charles Godard', 1877, Departmental Archives of Marne, trans. Tui McLean.

[10] **Mr Spijker would pay him back** 'Godard en de Spijker', *Het Nieuws van den Dag*; **His most significant moment | On the third day** du Taillis, 'Lost in the Desert'; **[footnote]** Jean du Taillis, 'The Tremendous Pekin–Paris Challenge: Hunger, Thirst, Burning Sun, Cruel Cold, They Have Encountered It All in the Solitude of the Gobi Desert', *Le Matin*, 23 June 1907, trans. Laura Shanahan; **completely repaired in Tomsk** du Taillis, 'Siberian Chaos'.

[11] **Cheremkhovo to Kansk** Our Own Correspondent and du Taillis, 'Moscow Cheers Borghese'; **On 1 August** Special Correspondent and Pons, 'Borghese Arrives in St Petersburg'; **Mr Spijker himself** 'Godard en de Spijker', *Het Nieuws van den Dag*; **Impossible to take** 'Peking–Matin: Prince Borghese Continuing on His Way to Berlin Covered Another Stage Yesterday', *Le Matin*, 4 August 1907, trans. Margaret Morrison.

[12] **When Consuelo Spencer-Churchill** Consuelo Vanderbilt Balsan, *The Glitter and the Gold* (London: Hodder, 2012), pp. 131–33; **Both he and his father** Figes, *People's Tragedy*, pp. 8–9; **rouge and enamel** Durland, *Red Reign*, p. 5.

[13] **The Itala had approached** Luigi Barzini and Our Own Correspondent [St Petersburg], 'Pekin to Paris: Brilliant Reception at St. Petersburg', *Daily Telegraph*, 2 August 1907; **Their Russian hosts** Editorial Board, 'Peking–Paris'; **Covered in filth** Special Correspondent and Pons, 'Borghese Arrives in St Petersburg'; **throughout my travels** W. Windham, 'From Pekin to Paris', *The Autocar*, 17 August 1907, pp. 290–92.

CHAPTER 19. THE TOLL

[1] **William Kissam Vanderbilt II** 'W. K. Vanderbilt Jr., Mobbed and Locked Up', *New York Times*, 25 February 1906; 'Mob Attacks W. K. Vanderbilt Jr. and Wife in Italian Village', *San Francisco Call*, 25 February 1906; **[footnote]** Luigi Barzini, 'Pekin to Paris: Itala Car Stoned by Peasants'.

[2] **Bridget Driscoll** Brendan Cormier, 'The Design of Speed', in Cormier and Bisley, *Cars*, p. 32; **The Economist conducted** 'Motors and Mortality', *The Economist*, 11 November 1907; **Today, 1.53** 'Historical Car Crash Deaths and Rates' National Safety Council (2022); **Armistice Day** D. Norton, *Fighting Traffic*, p. 25; **In 1951** Nader, *Unsafe*, pp. 86–87.

[3] **In Baltimore | in 1925** Norton, *Fighting Traffic*, pp. 41, 24; **three times as many** 'Auto-Killings', *Literary Digest*.

[4] **Twenty onlookers** 'Injured List Grows at the Auto Race', *New York Times*, 11 August 1907; **Mille Miglia | Le Mans** Cormier, 'Design of Speed', pp. 50–52.

[5] **around 1.3 million | Elsewhere, the toll** World Health Organization, 'Road Traffic Injuries', 21 June 2021; **In America** Dustin Jones, 'Firearms Overtook Auto Accidents as the Leading Cause of Death in Children', *NPR*, 22 April 2022.

[6] **could have been prevented** Nader, *Unsafe*, pp. 89–90, 96–97, 100.

[7] **21 per cent | Richard Cross** quoted in Nader, *Unsafe*, pp. 324, 325.

[8] **risk of death** WHO, 'Road Traffic Injuries'; **median state-designated limit** Norton, *Fighting Traffic*, p. 53; **1903 Motor Car Act** Anthony Bird, *The Motor Car: 1765–1914* (London: Batsford, 1960), p. 150.

[9] **mandatory belt laws** Sue Anne Pressley, 'Public Caught in the Middle of National Seat Belt Debate', *Washington Post*, 11 March 1985; Leo C. Wolinsky, 'Big Lobbies Clash in Fight on Seat Belts : Hearings Open Today as California Joins Auto Safety Debate', *Los Angeles Times*, 19 February 1985; **a tussle** 'High Court Backs Airbags Mandate', *New York Times*, 25 June 1983; **In the United Kingdom** Stephen Bates, 'National Archives: Police Opposed Seat Belts Law as Waste of Their Time', *The Guardian*, 1 August 2008.

[10] **In Bangladesh** 'Why Bangladeshi Students Held up Traffic', *The Economist*, 9 August 2018; **In many countries** 'Throughout the Rich World, the Young Are Falling out of Love with Cars, *The Economist*, 16 February 2023

[11] **Juries often acquit** Martin Porter, 'Dangerous Drivers Should Not Be Allowed to Choose Trial by Jury', *The Guardian*, 8 April 2006; **A New York Times article** Daniel Duane, 'Is It OK to Kill Cyclists?', *New York Times*, 9 November 2013; **In one incident** Paul Vitello, 'Alcohol, a Car and a Fatality. Is It Murder?', *New York Times*, 22 October 2006.

[12] **extreme levels of safety** Elon Musk, quoted in '"Elon Musk's Crash Course": 3 Key Arguments from the Tesla Documentary', *Washington Post*, 20 May 2022; **rested with drivers** 'Uber's Self-Driving Operator Charged over Fatal Crash', BBC News, 16 September 2020.

[13] **As early as 1950** Nader, *Unsafe*, p. 147; **It must have been impossible for our elders** Harry A. Williams, 'Accomplishments in Air Pollution Control by the Automobile Industry', in *Proceedings. National Conference on Air Pollution. Washington, DC November 18th–20th, 1958. US Department of Health, Education, and Welfare* (Washington DC: US Government Printing Office, 1959), pp. 57–61.

[14] **numerous pollutants** For information on air pollution and health risks, see European Environment Agency, 'Transport and Public Health', *Signals*, 2016; Damian Carrington, 'Billions of Air Pollution Particles Found in Hearts of City Dwellers', *The Guardian*, 12 July 2019; Damian Carrington, 'Air Pollution Particles Found on Foetal Side of Placentas – Study', *The Guardian*, 17 September 2019; David M. Stieb et al., 'Ambient Air Pollution, Birth Weight and Preterm Birth: A Systematic Review and Meta-Analysis', *Environmental Research*, 117 (2012), pp. 100–11; B. A. Maher et al., 'Iron-Rich Air Pollution Nanoparticles: An Unrecognised Environmental Risk Factor for Myocardial Mitochondrial Dysfunction and Cardiac Oxidative Stress', *Environmental Research*, 188 (2020), art. 109816; Dean E. Schraufnagel et al., 'Air Pollution and Noncommunicable Diseases: A Review by the Forum of International Respiratory Societies' Environmental Committee, Part 1: The Damaging Effects of Air Pollution', *CHEST*, 155.2 (2018), pp. 409–16.

[15] **Dirty air** Beth Gardiner, *Choked: The Age of Air Pollution and the Fight for a Cleaner Future* (London: Granta, 2019), p. 4.

[16] **5 per cent of us** Gardiner, *Choked*, p. 18; **Of the world's** World's Most Polluted Cities in 2021 – PM2.5 Ranking, IQAir, 2021; **India's five-year** Ministry of Environment, Forest and Climate Change *NCAP: National Clean Air Program* (Government of India, 2019), pp. 18, 26–27; **acute in China** Jin Wang et al., 'Vehicle Emission and Atmospheric Pollution in China: Problems, Progress, and Prospects', *PeerJ*, 7 (2019); **The following year** 'China to Scrap Millions of Cars to Improve Air Quality', BBC News, 27 May 2014; **Five years later** 'How Smog Affects Spending in China', *The Economist*, 28 June 2018.

[17] **international climate talks** Brad Plumer and Hiroko Tabuchi, '6 Automakers and 30 Countries Say They'll Phase out Gasoline Car Sales', *New York Times*, 9 November 2021; **Sourcing, refining and processing** Hiroko Tabuchi and Brad Plumer, 'How Green Are Electric Vehicles?', *New York Times*, 2 March 2021.

CHAPTER 20. KAUNAS – LIÈGE: 4–8 AUGUST

[1] **always inclined | This wretched rain** Cormier, *Le Raid*, pp. 81, 83–84; **Cormier dutifully** Cormier, *Le Raid*, p. 85.

[2] **Before sunrise** Cormier, *Le Raid*, p. 85; du Taillis, *Pékin–Paris*, p. 261; **in a pitiful state** du Taillis, *Pékin–Paris*, pp. 261–62; **I'm not used** Cormier, *Le Raid*, p. 85.

[3] **their movements** A. S., 'Peking–Matin: Borghese Is Arriving This Evening', *Le Matin*, 10 August 1907, trans. Margaret Morrison; **Such a performance** du Taillis, *Pékin–Paris*, p. 263.

[4] **utterly exhausted** T. R. Nicholson, *Adventurer's Road* (London: Cassell, 1957), p. 114; **Here we are | blood-raw** Andrews, *Mad Motorists*, pp. 215–17.

[5] **Godard and I were interrogated** Bruno Stephan, 'Strictly Confidential: Letter to The Honourable Mr. C. Poel Jr.', 31 October 1963, Stadsarchief Amsterdam, Arch. 733/inv. 14B; **Rumours swirled** 'Automobilisme: Peking–Parijs', *De Courant* (Amsterdam), 10 August 1907, trans. Joris Canoy; **By the time he wrote his book** Cormier, *Le Raid*, p. 85.

[6] **The Russian Automobile Club** Barzini and Our Own Correspondent, 'Brilliant Reception'; **the next morning** Special Correspondent and Pons, 'Borghese Arrives in St Petersburg'; Barzini and Our Own Correspondent, 'Brilliant Reception'; **They passed underneath** Barzini, *Peking to Paris*, p. 326; **Back near Dvinsk | At Kovno** Barzini, *Peking to Paris*, pp. 328, 330.

[7] **At another town** Barzini, *Peking to Paris*, p. 330.

[8] **an intelligent fury** Luigi Barzini, 'Pekin to Paris: Prince Borghese Arrives in Germany', *Daily Telegraph*, 5 August 1907; **The journey from the border** Peking–Matin: On the Eve of Their Triumph', *Le Matin*, 6 August 1907, trans. Margaret Morrison.

[9] **almost incognito | They will rest** 'Eve of Their Triumph', *Le Matin*.

[10] **Count Adalbert von Francken-Sierstorpff** Our Own Correspondent [Paris] and Our Own Correspondent [Berlin], 'Pekin to Paris: Great Preparations in the French Capital', *Daily Telegraph*, 7 August 1907.

CHAPTER 21. ON THE OFFENSIVE

[1] **I sometimes despair** 'Hansard: House of Lords Debate' (Hansard, 1906), p. 662.

[2] **Warning tremors** Our Own Correspondent [Paris], 'Franco-German

Entente', *Daily News* (London), 2 August 1907, p. 7; Peter Frankopan, *The Silk Roads: A New History of the World* (London: Bloomsbury, 2015), p. 307; **exhaustive rounds of diplomacy** 'Royal Interviews: Kaiser and Tsar', *Daily Telegraph*, 1 August 1907, p. 9; Our Own Correspondent [Vienna], 'Royal Interviews: Approaching Meetings', *Daily Telegraph*, 31 July 1907, p. 9; 'Chorus of Praise for King Edward: Relations between England and Germany Again on Amicable Footing', *New York Times*, 18 August 1907, p. 14; 'The German Emperor: A Visit to Windsor', *Daily Telegraph*, 25 June 1907, p. 11; 'King and Kaiser: The Wilhelmshöhe Meeting', *Daily Telegraph*, 27 July 1907, p. 11; 'Royal Interviews: Tsar and the Kaiser', *Daily Telegraph*, 2 August 1907, p. 9; **unusually large** 'Six German Warships Guard Czar's Yacht', *New York Times*, 6 August 1907, p. 1.

[3] **Characterised by mistrust** 'Chorus of Praise', *New York Times*; 'Bystander Comments', *The Bystander*, 26 June 1907, p. 643; Frankopan, *Silk Roads*, p. 309; **Russia's role** Henderson, 'Events in Russia', pp. 435–437; **[Footnote]** 'Obituary: Cdr. Sir Walter Windham', *The Times*, 7 July 1942, p. 6.

[4] **The final approach** Luigi Barzini, 'Pekin to Paris: Prince Borghese Arrives in Germany', *Daily Telegraph*, 5 August 1907, p. 10.

[5] **In Germany**, Barzini, *Peking to Paris*, p. 340; **The French drivers** Jean du Taillis, 'Pekin-Matin: They Pass Muster', *Le Matin*, 27 August 1907, p. 2, trans. Margaret Morrison.

[6] **More fanciful** 'Motor Vehicles in War', *New York Times*, 28 September 1901, p. 8; **by 1903** 'Automobile Topics of Interest: New Military Motor Car to Be Tested in Washington', *New York Times*, 13 September 1903, p. 13; **Inescapable aspect** V. N., 'From the Military Life of Foreign Nations', *Avtomobil'*, 1 September 1907, pp. 1844–48, trans. Dominique Hoffman; **Both Germany** 'Mimitary Notes [*sic*]', *South China Morning Post*, 17 June 1905, p. 4; Michael Young, *Army Service Corps, 1902–1918* (Barnsley: Leo Cooper, 2000), p. 19; 'Automobiles as Army Auxiliaries', *New York Times*, 8 November 1903, p. 12; 'Motor Vehicles in War', *New York Times*, p. 8; **Earl Roberts and his subordinate** 'Automobiles as Army Auxiliaries', *New York Times*, p. 12.

[7] **for the purposes of war** 'The Permanence of the Horse: Letter to the Editor', *The Economist*, 14 September 1907, p. 1548; **The number of fatal** 'Military Automobiles', *The Volunteer Record & Shooting News*, 28 August 1901, quoted in Young, *Army Service Corps*, pp. 8, 11.

[8] **the practical value** Our Own Correspondent, 'Peking–Paris Race: Prince Borghese Arrives at St Petersburg', *Daily News* (London), 2 August 1907; **An editorial** 'The Motor Car Race', *South China Morning Post*, 5 August 1907;

A Russian magazine A.V., 'Peking–Paris', *Avtomobil'*, 1 August 1907, pp. 1815–16, trans. Dominique Hoffman.

[9] **[footnote]** 'Fifteen Animals that Went to War', Imperial War Museums website; for a thorough insight into the uses of automobiles and animals by the British during the First World War, I am greatly indebted to Michael Young; **The French | the British** Young, *Army Service Corps*, pp. 67, 38; **[footnote]** Young, *Army Service Corps*, pp. 132–33; **Itala began producing** Itala catalogues from David Ayre Private Collection.

[10] **[footnote]** '"Little Willie" Tank Marks Centenary', BBC News, 9 September 2015.

[11] **$360 million worth** 'Auto and Animal Exports', *New York Times*, 27 April 1916; **rolling eastwards** John Burroughs, *Under the Maples* (Boston: Houghton Mifflin, 1921), pp. 110–11.

[12] **American motor-car production | Renault vs Ford** James Flink, *The Automobile Age* (Cambridge, MA: MIT Press, 1990), pp. 25, 39; **Six million** 'Automobile Progress', *New York Times*, 2 February 1919.

[13] **Marie Curie** Timothy J. Jorgensen, 'How Marie Curie Brought X-Ray Machines to the Battlefield', *Smithsonian Magazine*, 11 October 2017.

[14] **feature of the recent war | global output doubled** Imperial Institute, *Petroleum: Prepared Jointly with H.M. Petroleum Department, with the Co-Operation of H.B. Cronshaw* (London: John Murray, 1921), pp. 1–2; **Wave of oil** 'Floated to Victory on a Wave of Oil', *New York Times*, 23 November 1918.

[15] **A Chinese diplomat | Mr Hoelping** 'The Pekin–Matin Racers at Berlin: Flowers for the Heroes', *Le Matin*, 25 August 1907, trans. Margaret Morrison.

CHAPTER 22. LIÈGE – PARIS: 8–10 AUGUST

[1] **There is endless** A. S., 'Borghese is Arriving'; **He had reached** Luigi Barzini, 'Pekin to Paris: The Final Stage', *Daily Telegraph*, 10 August 1907; 'Peking–Matin Triumph: All Paris Acclaims Prince Borghese', *Le Matin*, 11 August 1907, trans. Margaret Morrison.

[2] **as though I** Barzini, 'Arrival in Paris'.

[3] **the road already travelled** Barzini, 'Arrival in Paris'.

[4] **[footnote]** 'The Red Trunk: Who Struck Emma? Was It the Dark-Haired Man?', *Le Matin*, 8 August 1907, trans. Margaret Morrison; 'Porter Discovered Monte Carlo Crime', *New York Times*, 11 August 1907; **Pied Piper-like** Jean du Taillis et al., 'Peking–Matin: Borghese is Arriving', *Le Matin*, 9 August

1907, trans. Margaret Morrison; **They had left Berlin** Barzini, *Peking to Paris*, pp. 340–41; Barzini, 'The Final Stage'.

[5] **The next day** Luigi Barzini, 'Pekin to Paris: Departure from Berlin', *Daily Telegraph*, 5 August 1907; **descriptions of Borghese** 'Eve of Their Triumph', *Le Matin*; 'La Corsa Pechino–Parigi: Le Grandi Feste che si Preparano', *Corriere della Sera*, 7 August 1907, trans. Laura Shanahan; du Taillis et al., 'Borghese Is Arriving'; 'From Pekin to Paris: Prince Borghese Reaches Paris', *Daily News* (Perth), 12 August 1907; 'Pekin to Paris: Great Contest Finished', *The Herald* (Melbourne), 12 August 1907; 'Motoring: From Pekin to Paris', *Manchester Guardian*, 2 August 1907; **descriptions of Guizzardi** 'All Paris Acclaims', *Le Matin*; 'La Corsa Pechino–Parigi: L'Ultima Tappa', *Corriere della Sera*, 10 August 1907; **witty face** du Taillis et al., 'Borghese Is Arriving'; **Adamantine fibre** Eugenio Checchi, 'Barzini Tratteggiato da Eugenio Checchi', *Corriere della Sera*, 10 August 1907.

[6] **Just after | You – a prince!** Barzini, 'Final Stage'.

[7] **Alongside the Itala** Georges Bourcier Saint-Chaffray, 'On the Road to Victory', *Le Matin*, 17 August 1907, trans. Margaret Morrison; **still chewing over | Godard told me** 'Peking–Matin: Mud Torture!', *Le Matin*, 13 August 1907, trans. Margaret Morrison; **A few hours of wrangling** Stephan, 'Strictly Confidential'.

[8] **Leaving Kazan** 'Mud Torture!', *Le Matin*; **A correspondent in Moscow** Our Own Correspondent [Moscow], 'Pekin–Matin: Delighted to Rediscover the Joys of Civilisation', *Le Matin*, 15 August 1907, trans. Margaret Morrison; **cursed the rain** 'Peking–Matin: While Borghese Receives Ovations, our Other Heroes Fight the Floodwaters', *Le Matin*, 12 August 1907, trans. Margaret Morrison.

[9] **Unknowingly** Stephan, 'Strictly Confidential'.

[10] **Sixty-five years . . .** This section quotes extensively from the letter Bruno Stephan wrote to Poel: Stephan, 'Strictly Confidential'.

[11] **One would need . . .** Stephan, 'Strictly Confidential'.

[12] **festivities** 'Eve of Their Triumph', *Le Matin*; 'The Borghese Festivities: A Fairy Tale of Light, 10 August, at the Tuileries', *Le Matin*, 8 August 1907, trans. Margaret Morrison; **Tactfully soothing** du Taillis et al., 'Borghese Is Arriving'; **He was perhaps** 'Borghese Reaches Paris from Peking: Winner of the Great Automobile Race Makes a Triumphal Entry', *New York Times*, 11 August 1907.

[13] **Horns sounded** Barzini, 'Arrival in Paris'; **a good-natured lecture** 'All Paris Acclaims', *Le Matin*; **Henri Fournier** Doug Nye, 'Paris to Vienna, 1902', *The Telegraph*, 4 April 2017.

[14] **Descriptions of the entry into Paris** Barzini, 'Arrival in Paris'; 'All Paris Acclaims', *Le Matin*.

[15] **We launched the** 'La Festività dell'Ambiente: Il Discorso Telefoni di Caponi', *Corriere della Sera*, 12 August 1907.

CHAPTER 23. AFTER THE RACE

[1] **My machine** 'Pons to Drive a Car', *New York Times*, 19 January 1908; **Two days later** 'Sizaire-Naudin on the Way', *New York Times*, 17 February 1908; **By 18 February** 'Sizaire-Naudin Withdrawn', *New York Times*, 19 February 1908.

[2] **The race announced** Guido, 'Ancora il Raid Automobilistico Pechino–Parigi: Verso la Fine', *L'Illustrazione Italiana*, 11 August 1907, pp. 128–30; **Rather than basking** Luigi Barzini and Our Own Correspondent [Milan], 'Itala Car's Return', *Daily Telegraph*, 17 August 1907.

[3] **Is it not most admirable** Bourcier Saint-Chaffray, 'On the Road to Victory'; **By the time** Gaston Leroux, 'Pekin–Matin: At Warsaw all Races Jostle to see the Passage of Progress', *Le Matin*, 21 August 1907, trans. Margaret Morrison; Gaston Leroux and Jean du Taillis, 'Pekin–Matin: An Epic of Endurance', *Le Matin*, 23 August 1907, trans. Margaret Morrison; **Belated but jubilant** From Our Own Correspondent, 'Pekin to Paris', *Daily Telegraph*, 31 August 1907.

[4] For the biographical information on Georges Cormier and Jean du Taillis I am indebted to the work of genealogist Murièle Ochoa Gadaut, who combed through the French official archives finding birth, marriage and death certificates, military records and other information to help flesh out the lives of the Peking–Paris participants on my behalf. The information here is taken from a selection of these records.

[5] **health reasons** Jean du Taillis, 'Pekin–Matin: They Are Approaching', *Le Matin*, 26 August 1907, trans. Margaret Morrison; **sporadically adopting** see for example 'Motobloc Loses Its Way: French Crew Is Caught in a Storm Also', *New York Times*, 16 March 1908; **one grand stiff** 'Motobloc at Buffalo', *New York Times*, 19 February 1908; **falling behind** 'Motobloc Loses Its Way', *New York Times*; 'Sandstorms Meet the Leading Racer', *New York Times*, 18 March 1908; Julie M. Fenster, *Race of the Century: The Heroic True Story of the 1908 New York to Paris Auto Race* (New York: Crown, 2005), pp. 173–74.

EPILOGUE

[1] Information in this chapter is taken from telephone interviews and email exchanges with Anton Gonnissen, his co-driver, and the Historic Endurance Rallying Association, organisers of the modern Peking to Paris Motor Challenges.

BIBLIOGRAPHY

Newspapers referred to in endnotes

Corriere della Sera (Milan)
De Courant (Amsterdam)
Daily Telegraph (London)
Daily News (London)
Daily News (Perth)
Evening Bulletin (Honolulu)
Evening Standard (London)
The Guardian (London)
Hawaiian Gazette (Honolulu)
The Herald (Melbourne)
Hong Kong Telegraph
Lancashire Daily Post
Los Angeles Herald
Los Angeles Times
Manchester Guardian
New York Times
Het Nieuws van den Dag (Amsterdam)
North-China Herald (Shanghai)
Le Matin (Paris)
Morning Post (London)
The Scotsman (Edinburgh)
Sheffield Daily Telegraph
Sheffield Weekly Telegraph
South China Morning Post (Hong Kong)
Sunday Times (London)
Sydney Morning Herald
The Times (London)
Wall Street Journal (New York)
Washington Post

Andrews, Allen, *The Mad Motorists: The Great Peking–Paris Race of '07*
   (London: Harrap, 1964)
Åsbrink, Brita, 'Harvest Times for the Nobel Brothers' Oil Adventure',
   Branobel History, http://www.branobelhistory.com/themes/oil-production/
   harvest-times-for-the-nobel-brothers-oil-adventure/

Åsbrink, Brita, 'Nobel Spies on His Competitor, Rockefeller', Branobel History, http://www.branobelhistory.com/themes/oil-distribution/nobel-spies-on-his-competitor-rockefeller/

Atkins, William, 'On the Island of the Black River', *Granta*, 21 November 2019, https://granta.com/on-the-island-of-the-black-river/

The Autocar, 'The Alcohol Propaganda', 17 August 1907, pp. 276–77

The Autocar, 'From Paris to Pekin', 9 February 1907, p. 198

The Autocar, 'Notes: What Next?', 17 August 1907, p. 259

The Autocar, 'The Pekin to Paris Drive', 2 March 1907, p. 297

The Autocar, 'Steam v. Petrol', 20 April 1907, pp. 580–81

The Autocar, 'The World Girdlers', 29 March 1902, p. 305

Automotor Journal, 'Paris–Madrid', 30 May 1903, pp. 546–57

A.V., 'Peking–Paris', *Avtomobil'*, 1 August 1907, pp. 1815–16

Baatz, Simon, *The Girl on the Velvet Swing: Sex, Murder, and Madness at the Dawn of the Twentieth Century* (New York: Mulholland, 2018)

Baekeland, Leo, Diary Volume 1, 1907–1908, Leo Baekeland Papers, Archives Center, National Museum of American History, Smithsonian Institution, Washington DC

Barthes, Roland, *Mythologies*, trans. Annette Lavers (London: Vintage, 2000)

Barty-King, Hugh, *Stupendous Engine: The First Fifty Years of Telegraphy* (Penzance: Cable and Wireless Porthcurno and Collections Trust, 2005)

Barzini, Luigi, *Peking to Paris: Across Two Continents in an Itala*, trans. L.P. de Castelvecchio (Harmondsworth: Penguin, 1986)

Barzini, Luigi, *Vita Vagabonda: Ricordi di un Giornalista* (Milan: Rizzoli, 1948)

BBC News, 'China to Scrap Millions of Cars to Improve Air Quality', 27 May 2014, https://www.bbc.com/news/business-27583404

BBC News, 'Elon Musk Unveils Prototype High-Speed LA Transport Tunnel', 19 December 2018, https://www.bbc.com/news/world-us-canada-46616902

BBC News, 'Jaywalking: How the Car Industry Outlawed Crossing the Road', 12 February 2014, https://www.bbc.com/news/magazine-26073797

BBC News, 'Uber's Self-Driving Operator Charged over Fatal Crash', 16 September 2020, https://www.bbc.com/news/technology-54175359

Beaumont, Comyns, 'Automobile Topics', *The Bystander*, 31 July 1907, pp. 253–54

Beer, Daniel, *The House of the Dead: Siberian Exile under the Tsars*, 2nd edn (London: Penguin, 2017)

Bentinck, Charles H., 'Summary of Events in Russia during the Fortnight Ending June 19, 1907' (St Petersburg: British Foreign Office, 20 June 1907), pp. 392–394v, British Foreign Office Russia: Correspondence 1907–1917, FO 371, vol. 318

Berkebile, Don H., ed., *Horse-Drawn Commercial Vehicles: 255 Illustrations of Nineteenth-Century Stagecoaches, Delivery Wagons, Fire Engines, Etc.* (New York: Dover, 1989)

Bianco, Martha J., 'Kennedy, 60 Minutes, and Roger Rabbit: Understanding Conspiracy-Theory Explanations of the Decline of Urban Mass Transit', Center for Urban Studies Publications and Reports, 17 (1998)

Birch, John Grant, *Travels in North and Central China* (London: Hurst & Blackett, 1902)

Bird, Anthony, *The Motor Car: 1765–1914* (London: Batsford, 1960)

Bird, Isabella L., *Chinese Pictures: Notes on Photographs Made in China* (London: Cassell, 1900)

Borgeson, Griffith, 'The Automotive World of Albert de Dion', *Automobile Quarterly*, 15.3 (1977), pp. 266–85

Bowers Blog, 'Car Fit for an Empress Dowager', 2017, https://www.bowers. org/index.php/collection/collection-blog/car-fit-for-an-empress-dowager

Bowlt, John E., *Moscow and St. Petersburg in Russia's Silver Age: 1900–1920* (London: Thames & Hudson, 2008)

Brandon, Ruth, *Automobile: How the Car Changed Life* (London: Macmillan, 2002)

Brigham, Clarence S., 'Edgar Allan Poe's Contributions to *Alexander's Weekly Messenger*', *Proceedings of the American Antiquarian Society*, 52.1 (1942), pp. 45–125

Burman, Barbara, 'Racing Bodies: Dress and Pioneer Women Aviators and Racing Drivers', *Women's History Review*, 9.2 (2000), pp. 299–326, https:// doi.org/10.1080/09612020000200243

Burroughs, John, *Under the Maples* (Boston: Houghton Mifflin, 1921)

Cavazza, Alberta, *The Enchanted Island: Isola Del Garda*, trans. Charlotte Chetwynd Talbot and Anja Pattenden (Cierre Grafica, 2016)

Chang, Jung, *Empress Dowager Cixi: The Concubine Who Launched Modern China* (London: Vintage Books, 2014)

Collignon, Victor, *Mes Souvenirs de Route: 16,000 Kilometres sur de Dion-Bouton (Pneus Dunlop)*, 1908

Conger, Sarah Pike, *Letters from China, with Particular Reference to the Empress Dowager and the Women of China* (Chicago: A.C. McClurg, 1909)

Cormier, Brendan, 'The Design of Speed', in Brendan Cormier and Lizzie Bisley, eds, *Cars: Accelerating the Modern World* (London: V&A, 2019), pp. 30–55

Cormier, Brendan, and Lizzie Bisley, eds, *Cars: Accelerating the Modern World* (London: V&A, 2019)

Cormier, Georges, *Le Raid Pékin–Paris en 1907*, 3rd edn (Paris: Édition 'Publi-Inter', 1954)

335

Cormier, Georges, 'Mon Tour d'Europe', *La Vie au Grand Air*, 13 December 1902, pp. 842–43

Cormier, Georges, 'Pékin–Paris', *L'Auto*, 16 July 1907, pp. 3, 5

Cormier, Georges, 'Pékin–Paris: Le Tour du Baikal', *L'Auto*, 4 July 1907, p. 3

Cormier, Georges, and Georges Bourcier Saint-Chaffray, 'Pékin–Paris', *L'Auto*, 27 July 1907, p. 5

Crankshaw, Edward, 'When Lenin Returned', *The Atlantic*, October 1954, https://www.theatlantic.com/magazine/archive/1954/10/when-lenin-returned/303867/

Crossley, Pamela, 'In the Hornets' Nest', *London Review of Books*, 17 April 2014, https://www.lrb.co.uk/the-paper/v36/n08/pamela-crossley/in-the-hornets-nest

Curtin, Jeremiah, *A Journey in Southern Siberia: The Mongols, Their Religion and Their Myths* (London: Sampson Low, Marston, 1910)

Cycling History, 'Guns, Wheels, and Steel: Cyclists and Small Arms in the Late Nineteenth Century', 2015, https://cyclehistory.wordpress.com/2015/10/03/guns-wheels-and-steel-cyclists-and-small-arms-in-the-late-19th-century/

Dauncey, Hugh, *French Cycling: A Social and Cultural History* (Liverpool: Liverpool University Press, 2012)

Daunton, Martin, ed., *The Cambridge Urban History of Britain, 1840–1950*, III (Cambridge: Cambridge University Press, 2000)

DeLong, George W., *Our Lost Explorers: The Narrative of the Jeannette Arctic Expedition* (Hartford, CN: American Publishing, 1882)

De Windt, Harry, *From Pekin to Calais by Land* (London: Chapman & Hall, 1889)

Drew, Edward B., 'Sir Robert Hart and His Life Work in China', *Journal of Race Development*, 4.1 (1913), pp. 1–33, https://doi.org/10.2307/29737977

Duranton, Gilles, and Matthew Turner, 'The Fundamental Law of Road Congestion: Evidence from US Cities', *American Economic Review*, 101.6 (2011), pp. 2616–52

Durland, Kellogg, *The Red Reign: The True Story of an Adventurous Year in Russia* (New York: The Century, 1908)

The Economist, 'A Chance to Transform Urban Planning', 1 March 2018, https://www.economist.com/special-report/2018/03/01/a-chance-to-transform-urban-planning

The Economist, 'The Hidden Cost of Congestion', 28 February 2018, https://www.economist.com/graphic-detail/2018/02/28/the-hidden-cost-of-congestion

The Economist, 'How Smog Affects Spending in China', 28 June 2018, https://www.economist.com/graphic-detail/2018/06/28/how-smog-affects-spending-in-china

The Economist, 'Motors and Mortality', 11 November 1907

The Economist, 'The Triumph of the Horse', 7 September 1907, pp. 1495–96

Economist Intelligence Unit, *The Global Liveability Index 2022: Recovery and Hardship* (2022)

Editorial Board, 'Peking–Paris', *Avtomobil'*, 1 August 1907, pp. 1803–14

European Environment Agency, 'Transport and Public Health', *Signals,* 2016, https://www.eea.europa.eu/signals/signals-2016/articles/transport-and-public-health

Fédération Internationale de l'Automobile, *A Report on the Global Contribution of Motor Sport to Economy and Community Development* (EY-Parthenon/FIA, July 2021), https://www.fia.com/multimedia/publication/report-global-contribution-motor-sport-economy-and-community-development

Fenster, Julie M., *Race of the Century: The Heroic True Story of the 1908 New York to Paris Auto Race* (New York: Crown, 2005)

The Field, 'Motor Notes', 7 June 1902, p. 891

The Field, 'Wheels and Ways: Alcohol as a Motor Fuel', 14 April 1906, p. 600

Figes, Orlando, *Natasha's Dance: A Cultural History of Russia* (London: Allen Lane, 2002)

Figes, Orlando, *A People's Tragedy: The Russian Revolution 1891–1924,* 2nd edn (London: Bodley Head, 2014)

Firley-Kanarsky, N., 'Letter from the Provinces', *Avtomobil'*, 15 May 1907, pp. 1728–30

Flink, James, *The Automobile Age* (Cambridge, MA: MIT Press, 1990)

Forrest, Susanna, *The Age of the Horse: An Equine Journey through Human History* (London: Atlantic, 2016)

Frankopan, Peter, *The Silk Roads: A New History of the World* (London: Bloomsbury, 2015)

Freeth, F. A., 'James Swinburne. 1858–1958', *Biographical Memoirs of Fellows of the Royal Society,* 5 (February, 1960), pp. 253–68

Frost, John, *The Life of William Penn, the Founder of Pennsylvania* (Philadelphia: Lindsay & Blakiston, 1849)

Gardiner, Beth, *Choked: The Age of Air Pollution and the Fight for a Cleaner Future* (London: Granta, 2019)

Gauguin, Paul, *Noa Noa: The Tahitian Journal,* trans. O. F. Theis (New York: Dover, 1985)

Gentleman's Magazine, 'Monumental Inscriptions on an Urn', 50 (1780), p. 242

Gordon, William J., *The Horse World of London (1893)* (Newton Abbot: David & Charles, 1971)

Gorky, Maxim, 'England and the Russian Revolution', *Social Democrat,* 11.7 (1907), pp. 420–24

Guido, 'Ancora il Raid Automobilistico Pechino–Parigi: Verso la Fine',
   L'Illustrazione Italiana, 11 August 1907, pp. 128–30
Guinn, Jeff, *The Vagabonds: The Story of Henry Ford and Thomas Edison's Ten-
   Year Road Trip* (New York: Simon & Schuster, 2019)
'Hansard: House of Lords Debate' (Hansard, 1906), p. 662, https://api.par-
   liament.uk/historic-hansard/lords/1906/jul/10/imperial-defence
'Hansard: Motor Vehicles and Speedometers' (House of Lords, 1932), pp.
   169–96, https://api.parliament.uk/historic-hansard/lords/1932/dec/01/
   motor-vehicles-and-speedometers
Harp, Stephen L., *Marketing Michelin: Advertising and Cultural Identity in
   Twentieth-Century France* (Baltimore: Johns Hopkins University Press, 2001)
Hart, Robert, *Entering China's Service: Robert Hart's Journals, 1854–1863*,
   ed. Katherine Frost Bruner, John King Fairbank and Richard J. Smith
   (Cambridge, MA: Harvard University Press, 1986)
Hart, Robert, *The I.G. in Peking: Letters of Robert Hart, Chinese Maritime
   Customs 1868–1907*, I, ed. John King Fairbank, Katherine Frost Bruner
   and Elizabeth MacLeod Matheson (Cambridge, MA: Belknap Press,
   1975)
Hart, Robert, *Robert Hart and China's Early Modernization: His Journals,
   1863–1866*, ed. Richard J. Smith, John King Fairbank and Katherine Frost
   Bruner (Cambridge, MA: Harvard University Press, 1991)
Heaton, J. Henniker, 'An Imperial Telegraph System', *Nineteenth Century*,
   45 (1899), pp. 906–14
Henderson, Nevile, 'Report by Mr. Henderson on Events in Russia dur-
   ing the Fortnight Ended August 1, 1907' (St Petersburg: British Foreign
   Office, 1 August 1907), pp. 435–37, British Foreign Office Russia:
   Correspondence 1907–1917, FO 371, vol. 318
Hodgson Liddell, T., *China: Its Marvel and Mystery* (London: George Allen
   & Sons, 1909)
Horseless Age, 'End of the Pekin–Paris Run', 28 August 1907, p. 283
Huang, Xuelei, 'Deodorizing China: Odour, Ordure, and Colonial
   (Dis)Order in Shanghai, 1840s–1940s', *Modern Asian Studies*, 50.3
   (2016), pp. 1092–122
Hunting, Benjamin, 'How the 1955 Dodge La Femme Missed the Mark on
   Designing Cars for Women', Hagerty, 10 February 2020, https://www.hag-
   erty.com/media/car-profiles/1955-dodge-la-femme-missed-the-mark/
Huntington, Rania, 'Chaos, Memory, and Genre: Anecdotal Recollections
   of the Taiping Rebellion', *Chinese Literature: Essays, Articles, Reviews*, 27
   (2005), pp. 59–91
Illustrated London News, 'Difficulties of the Paris–Peking Motor Race',
   20 July 1907
Illustrated London News, 'Taking Their Road with Them: The Ingenious
   Equipment for the Peking to Paris Motor Race', 27 April 1907

Imperial Institute, *Petroleum: Prepared Jointly with H.M. Petroleum Department, with the Co-Operation of H.B. Cronshaw* (London: John Murray, 1921)

Imperial War Museums, 'Fifteen Animals That Went to War', https://www.iwm.org.uk/history/15-animals-that-went-to-war

International Energy Agency, *Global EV Outlook 2022*, May 2022, https://www.iea.org/data-and-statistics/data-product/global-ev-outlook-2022

Isted, G. A., 'Guglielmo Marconi and the History of Radio – Part I', *GEC Review*, 7.1 (1991), pp. 45–56

Isted, G. A., 'Guglielmo Marconi and the History of Radio – Part II', *GEC Review*, 7.2 (1991), pp. 110–22

Jones, Dustin, 'Firearms Overtook Auto Accidents as the Leading Cause of Death in Children', *NPR*, 22 April 2022, https://www.npr.org/2022/04/22/1094364930/firearms-leading-cause-of-death-in-children

Jorgensen, Timothy J., 'How Marie Curie Brought X-Ray Machines to the Battlefield', *Smithsonian Magazine*, 11 October 2017, https://www.smithsonianmag.com/history/how-marie-curie-brought-x-ray-machines-to-battlefield-180965240/

Kennan, George, *Siberia and the Exile System*, I (London: James R. Osgood, McIlvaine, 1891)

King, Frank H., 'The Boxer Indemnity: "Nothing but Bad"', *Modern Asian Studies*, 40.3 (2006), pp. 663–89

Klare, Michael T., *Blood and Oil: The Dangers and Consequences of America's Growing Petroleum Dependency* (London: Hamish Hamilton, 2004)

Knox, Thomas W., *Overland through Asia: Pictures of Siberian, Chinese, and Tartar Life* (London: Trübner & Co., 1871)

Kuhn, Philip A., 'Origins of the Taiping Vision: Cross-Cultural Dimensions of a Chinese Rebellion', *Comparative Studies in Society and History*, 19.3 (1977), pp. 350–66

Kuzmin, E., 'Letter to the Editor: Paris–Peking', *Avtomobil'*, 1 April 1907, pp. 1675–77

Kuzmin, E., 'Unfamiliar Road', *Avtomobil'*, 15 May 1907, pp. 1730–32

Lakhvari, 'Significance of the Peking–Paris Automobile Contest', *Avtomobil'*, 1 August 1907, pp. 1814–15

La Motte, Ellen N., *Peking Dust* (New York: The Century, 1919)

Lehman, Milton, 'The First Woman Driver', *Life*, 8 September 1952, pp. 83–95

Lenin, Vladimir Ilyich, 'Speech Delivered at the First All-Russia Congress of Working Cossacks', trans. George Hanna, Marxists.Org, 1920, https://www.marxists.org/archive/lenin/works/1920/mar/01.htm

L'Espagnol de la Tramerye, Pierre, *The World-Struggle for Oil*, trans. C. Leonard Leese (London: Allen & Unwin, 1924)

Levitt, Dorothy, *The Woman and the Car: A Chatty Little Handbook for All*

Women Who Motor or Who Want to Motor (London: John Lane, Bodley Head, 1909)

Lidgett, Albert, *Petroleum* (London: Sir Isaac Pitman & Sons, 1919)

Literary Digest, 'Auto-Killings Now at the Rate of Two an Hour', 6 November 1920, pp. 80–88

Lynd, Robert Staughton, and Helen Merrell Lynd, *Middletown: A Study in American Culture* (New York: Harcourt, Brace, 1929)

Maher, B. A., A. González-Maciel, R. Reynoso-Robles, R. Torres-Jardón and L. Calderón-Garcidueñas, 'Iron-Rich Air Pollution Nanoparticles: An Unrecognised Environmental Risk Factor for Myocardial Mitochondrial Dysfunction and Cardiac Oxidative Stress', *Environmental Research*, 188 (2020), art. 109816, https://doi.org/10.1016/j.envres.2020.109816

Marchant, Natalie, 'Paris Halves Street Parking and Asks Residents What They Want to Do with the Space', World Economic Forum, 7 December 2020, https://www.weforum.org/agenda/2020/12/paris-parking-spaces-greenery-cities/

McShane, Clay, *Down the Asphalt Path: The Automobile and the American City* (New York: Columbia University Press, 1994)

McShane, Clay, and Joel A. Tarr, *The Horse in the City: Living Machines in the Nineteenth Century* (Baltimore: Johns Hopkins University Press, 2007)

Meignan, Victor, *From Paris to Pekin over Siberian Snows: A Narrative of a Journey by Sledge over the Snows of European Russia and Siberia, by Caravan through Mongolia, across the Gobi Desert and the Great Wall, and by Mule Palanquin through China to Pekin*, trans. William Conn (London: William Swan Sonnenschein, 1885)

Melnikova-Raich, Sonia, 'The Soviet Problem with Two "Unknowns": How an American Architect and a Soviet Negotiator Jump-started the Industrialization of Russia, Part I: Albert Kahn', *IA. The Journal of the Society for Industrial Archeology*, 36.2 (2010), pp. 57–80

Melnikova-Raich, Sonia, 'The Soviet Problem with Two "Unknowns": How an American Architect and a Soviet Negotiator Jump-started the Industrialization of Russia, Part II: Saul Bron', *IA. The Journal of the Society for Industrial Archeology*, 37.1/2 (2011), pp. 5–28

Miéville, China, *October: The Story of the Russian Revolution* (London: Verso)

Ministry of Environment, Forest and Climate Change, *NCAP: National Clean Air Program* (Government of India, 2019), https://moef.gov.in/wp-content/uploads /2019/05/NCAP_Report.pdf

Mironenko, S.V., and Andrei Maylunas, *A Lifelong Passion: Nicholas and Alexandra, Their Own Story* (London: Doubleday, 1997)

Mom, Gijs, and David Kirsch, 'Technologies in Tension: Horses, Electric Trucks, and the Motorization of American Cities, 1900–1925', *Technology and Culture*, 42.3 (2001), pp. 489–518

Moraglio, Massimo, *Driving Modernity: Technology, Experts, Politics, and Fascist Motorways, 1922–1943*, trans. Erin O'Loughlin (New York: Berghahn, 2017)

Moretti, Valerio, 'Il Terzo Uomo della Pechino Parigi', *Auto Italiana*, 18 July 1963, trans. Kassia St Clair, David Ayre Private Collection

Morse, Jack, 'Oh Look, It's a Tesla Traffic Jam in Las Vegas' Boring Company Tunnel', *Mashable*, 6 January 2022, https://mashable.com/article/ces-las-vegas-boring-tunnel-tesla-traffic

Morus, Iwan Rhys, 'The Electric Ariel: Telegraphy and Commercial Culture in Early Victorian England', *Victorian Studies*, 39.3 (1996), pp. 339–78

Morus, Iwan Rhys, ' "The Nervous System of Britain": Space, Time and the Electric Telegraph in the Victorian Age', *British Journal for the History of Science*, 33.4 (2000), pp. 455–75

The Motor, 'The Moustached Chauffeur', 9 April 1907, p. 9

The Motor, 'News', 30 July 1907, p. 795

Movsumzade, E. M., 'The First Attempts to Use Oil as Navy Fuel', *Icon*, 6 (2000), pp. 142–48

Moyle, John, *Timelines of Telegraphy* (Penzance: Cable and Wireless Porthcurno and Collections Trust, 2008)

Munro-Butler-Johnstone, H. A., *A Trip up the Volga to the Fair of Nijni-Novgorod* (Oxford: James Parker, 1875)

Nader, Ralph, *Unsafe at Any Speed: The Designed-in Dangers of the American Automobile* (New York: Grossman, 1965)

National Museum of American History, 'A Streetcar City', 2016, https://americanhistory.si.edu/america-on-the-move/streetcar-city

Nicholson, T. R., *Adventurer's Road: The Story of Pekin–Paris, 1907 and New York–Paris, 1908* (London: Cassell, 1957)

Norton, Peter D., *Fighting Traffic: The Dawn of the Motor Age in the American City* (Cambridge, MA: MIT Press, 2008)

Norton, Peter D. 'Street Rivals: Jaywalking and the Invention of the Motor Age Street', *Technology and Culture*, 48.2 (2007), pp. 331–59

O'Beirne, Hugh, 'Report by Mr. O'Beirne Respecting Agricultural Distress in Russia' (St Petersburg: British Foreign Office, 30 January 1907), pp. 109–110v, British Foreign Office Russia: Correspondence 1907–1917, FO 371, vol. 321

Otter, Christopher, 'Cleansing and Clarifying: Technology and Perception in Nineteenth-Century London', *Journal of British Studies*, 43.1 (2004), pp. 40–64

Pons, Auguste, 'Lache en Plein Desert!', *La Vie au Grand Air*, 24 August 1907, pp. 129–32

Prade, Georges, 'La Course Bordeaux–Périgueux–Bordeaux: Toujours Plus Oultre', *La Vie au Grand Air*, 17 June 1900, p. 522

Putnam Weale, B. L., Photographs Belonging to Sir Robert Hart, Sent to

Him by B. Lenox Simpson (pen name: Putnam Weale), 1907, Sir Robert Hart Collection, Queen's University Belfast, MS 15.1.86.69a (ii–xi)

Raboy, Marc, *Marconi: The Man Who Networked the World* (Oxford: Oxford University Press, 2016)

Reform of the Federal Criminal Laws: Hearings before the Subcommittee on Criminal Laws and Procedures of the Committee on the Judiciary, United States Senate, 94th Congress, 1st Session on S.1 and S.1400, 13, 17 June and 19, 22 July 1974, Part XI (Washington DC: US Senate, 1974)

'Registre Matricule: Charles Godard', 1877, Departmental Archives of Marne, https://archives.marne.fr/ark:/86869/41d73qwkz2j0/ fcbf2007-4a51-4ce9-87b4-5b6a36111126

Reiss, Tom, 'Imagining the Worst', *New Yorker*, 28 November 2005, pp. 106, 108, 110

Rogers, John, 'Rare Art from China's 19th Century Woman Ruler Come to US', AP News, 11 November 2017, https://apnews.com/article/ f9297a2ddbfe4a908fe139cccd24510d

Rubets, A. D., *History of Automobile Transport in Russia*, (Moscow: Eksmo, 2008)

Rubinstein, David, 'Cycling in the 1890s', *Victorian Studies*, 21.1 (1977), pp. 47–71

Russell, Claud, 'Summary of Events in Russia for the Fortnight Ended April 24, 1907' (St Petersburg: British Foreign Office, 25 April 1907), pp. 366–366v, British Foreign Office Russia: Correspondence 1907–1917, FO 371, vol. 318

Russell, Claud, 'Summary of Events in Russia for the Fortnight Ended June 6, 1907' (St Petersburg: British Foreign Office, 5 June 1907), pp. 386–386v, British Foreign Office Russia: Correspondence 1907–1917, FO 371, vol. 318

Russian Primary Chronicle: Laurentian Text, trans. Samuel Hazzard Cross and Olgerd P. Sherbowitz-Wetzor (Cambridge, MA: Mediaeval Academy of America, 1953)

Sarkar, Tapan K., Robert J. Mailloux, Arthur A. Oliner, Magdalena Salazar-Palma and Dipak L. Sengupta, *History of Wireless* (Hoboken, NJ: John Wiley, 2006)

Savorgnan di Brazza, F., 'Da Pechino a Parigi in Triciclo', *L'Illustrazione Italiana*, 16 June 1907, p. 583

Scharff, Virginia, *Taking the Wheel: Women and the Coming of the Motor Age* (Albuquerque: University of New Mexico Press, 1992)

Schraufnagel, Dean E., John R. Balmes, Clayton T. Cowl, Sara De Matteis, Soon-Hee Jung, Kevin Mortimer et al., 'Air Pollution and Noncommunicable Diseases: A Review by the Forum of International Respiratory Societies' Environmental Committee, Part 1: The Damaging Effects of Air Pollution', *CHEST*, 155.2 (2018), pp. 409–16

Scott, Ernest, 'Report by Mr. Scott on Events in Russia for the Fortnight Ended January 2, 1907' (St Petersburg: British Foreign Office, 2 January 1907), pp. 285–286v, British Foreign Office Russia: Correspondence 1907–1917, FO 371, vol. 318

Scott, Ernest, 'Report on the Events in Russia for the Fortnight Ended February 13, 1907' (St Petersburg: British Foreign Office, 13 February 1907), pp. 323–25, British Foreign Office Russia: Correspondence 1907–1917, FO 371, vol. 318

Sebag Montefiore, Simon, *The Romanovs: 1613–1918* (London: Weidenfeld & Nicolson, 2016)

Sebag Montefiore, Simon, *Young Stalin* (London: Weidenfeld & Nicolson, 2007)

Service, Robert, *The Last of the Tsars: Nicholas II and the Russian Revolution* (London: Macmillan, 2017)

Siegelbaum, Lewis, *Cars for Comrades: The Life of the Soviet Automobile* (Ithaca, NY: Cornell University Press, 2008)

Smith, Douglas, *Former People: The Last Days of the Russian Aristocracy* (London: Pan, 2012)

Sonnichsen, N., 'OECD: Oil Demand Share by Sector', Statista, 18 January 2022, https://www.statista.com/statistics/307194/top-oil-consuming-sectors-worldwide/

State Council Information Office: The People's Republic of China, 'What Are Six Economic Corridors Under Belt and Road Initiative?', 2020, http://english.scio.gov.cn/beltandroad/2020-08/04/content_76345602.htm

Statista Research Department, 'Oil Flows: Strait of Hormuz 2014–2020' Statista, 6 April 2022, https://www.statista.com/statistics/277157/key-figures-for-the-strait-of-hormuz/

Stephan, Bruno, 'Strictly Confidential: Letter to The Honourable Mr. C. Poel Jr.', 31 October 1963, Stadsarchief Amsterdam, Arch. 733/inv. 14B

Stieb, David M., Li Chen, Maysoon Eshoul and Stan Judek, 'Ambient Air Pollution, Birth Weight and Preterm Birth: A Systematic Review and Meta-Analysis', *Environmental Research*, 117 (2012), pp. 100–11

Taillis, Jean du, *Pékin–Paris: Automobile en Quatre-Vingts Jours*, (Paris: Félix Juven, 1907)

Talbot, Frederick A., *The Oil Conquest of the World* (London: William Heinemann, 1914)

Tarr, Joel A., *The Search for the Ultimate Sink: Urban Pollution in Historical Perspective* (Akron, OH: University of Akron Press, 1997)

The Tatler, 'The Motor World – Week by Week', 8 October 1902, p. 55

Taylor, Michael, 'The Bicycle Boom and the Bicycle Bloc: Cycling and Politics in the 1890s', *Indiana Magazine of History*, 104.3 (2008), pp. 213–40

Temple University, 'People Who Really Identify with Their Car Drive More

Aggressively, Study Finds', *ScienceDaily*, 17 October 2011, https://www.sciencedaily.com/releases/2011/10/111017124346.htm

Tibballs, Geoff, *Motor Racing's Strangest Races* (London: Portico, 2016)

Time, 'Industry's Architect', 29 June 1942, http://content.time.com/time/subscriber/article/0,33009,795936,00.html

Tishkevich, Count, 'Guidance If You Are Invited for an Automobile Outing', *Avtomobil'*, 1 September 1907, p. 1852

Traill, H. D., *Number Twenty: Fables and Fantasies* (London: Henry Holt, 1892)

Trevithick, Francis, *Life of Richard Trevithick, With an Account of His Inventions* (London: E. & F. N. Spon, 1872)

Turner, Samuel, *Siberia: A Record of Travel, Climbing, and Exploration* (London: T. Fisher Unwin, 1905)

Uruburu, Paula, *American Eve: Evelyn Nesbit, Stanford White, the Birth of the 'It' Girl, and the Crime of the Century* (New York: Riverhead, 2008)

Vanderbilt Balsan, Consuelo, *The Glitter and the Gold* (London: Hodder, 2012)

La Vie au Grand Air, 'Le Raid Pékin–Paris', 7 September 1907, pp. 166–74

La Vie au Grand Air, 'Résultats Techniques de la Semaine', 8 December 1905, p. 1026

V. N., 'From the Military Life of Foreign Nations', *Avtomobil'*, 1 September 1907, pp. 1844–48

The Volunteer Record & Shooting News, 'Military Automobiles', 28 August 1901, pp. 280–81

Vynckier, Henk, and Chihyun Chang, ' "Imperium in Imperio": Robert Hart, the Chinese Maritime Customs Service, and Its (Self-)Representations', *Biography*, 37.1 (2014), pp. 69–92

Walsh, George Ethelbert, 'Cable Cutting in War', *North American Review*, 167.503 (1898), pp. 498–502

Wang, Jin, Qiuxia Wu, Juan Liu, Hong Yang, Meiling Yin, Shili Chen et al., 'Vehicle Emission and Atmospheric Pollution in China: Problems, Progress, and Prospects', *PeerJ*, 7 (2019), https://www.ncbi.nlm.nih.gov/pmc/articles/PMC6526014/

Waring Jr, George E., *Street-Cleaning, and the Disposal of a City's Wastes: Methods and Results and the Effect upon Public Health, Public Morals, and Municipal Prosperity* (London: Gay & Bird, 1898)

Wharton, Edith, *A Motor-Flight through France* (New York: Charles Scribner's Sons, 1908)

White, E. B., 'Farewell, My Lovely!', *New Yorker*, 16 May 1936, https://www.newyorker.com/magazine/1936/05/16/farewell-my-lovely

Williams, Harry A., 'Accomplishments in Air Pollution Control by the Automobile Industry', in *Proceedings. National Conference on Air Pollution. Washington, DC November 18th–20th, 1958. US Department of Health,*

Education, and Welfare (Washington DC: United States Government Printing Office, 1959), pp. 57–61

Windham, W., 'From Pekin to Paris', *The Autocar*, 17 August 1907, pp. 290–92

World Health Organization, 'Road Traffic Injuries', 21 June 2021, https://www.who.int/news-room/fact-sheets/detail/road-traffic-injuries

'World's Most Polluted Cities in 2021 – PM2.5 Ranking', IQAir, 2021, https://www.iqair.com/in-en/world-most-polluted-cities

Ximenes, Eduardo, 'Corriere di Parigi', *L'Illustrazione Italiana*, 25 August 1907, pp. 200–03

Yakovlev, I., 'On the Automobile Exhibition', *Avtomobil'*, 1 July 1907, pp. 1778–81

Yergin, Daniel, *The Prize: The Epic Quest for Oil, Money, and Power* (New York: Simon & Schuster, 1991)

Young, Michael, *Army Service Corps, 1902–1918* (Barnsley: Leo Cooper, 2000)

Zedong, Mao, 'To the Glory of the Hans: Toward a New Golden Age', *Hsiang-Chiang p'inq-Lun*, July 1919, https://www.marxists.org/reference/archive/mao/selected-works/volume-6/mswv6_03.htm

Zipper, David, 'A City Fights Back Against Heavyweight Cars', Bloomberg, 26 May 2022, https://www.bloomberg.com/news/articles/2022-05-26/a-new-way-to-curb-the-rise-of-oversized-pickups-and-suvs

INDEX

wireless communication *see*
    telegraphs
Witte, Sergei 97, 103, 104, 185
women 119–23, 124–6, 142–3,
    190–1, 262
World Trade Organization (WTO)
    48

Xi Jinping 48–9

Yekaterinburg 173,
    174

Zanetti, Gian Luca 26
Zhangjiakou *see* Kalgan